International Series in Operations Research & Management Science

Volume 372

The book series **International Series in Operations Research and Management Science** encompasses the various areas of operations research and management science. Both theoretical and applied books are included. It describes current advances anywhere in the world that are at the cutting edge of the field. The series is aimed especially at researchers, advanced graduate students, and sophisticated practitioners.

The series features three types of books:

- Advanced expository books that extend and unify our understanding of particular areas.
- Research monographs that make substantial contributions to knowledge.
- Handbooks that define the new state of the art in particular areas. Each handbook will be edited by a leading authority in the area who will organize a team of experts on various aspects of the topic to write individual chapters. A handbook may emphasize expository surveys or completely new advances (either research or applications) or a combination of both.

The series emphasizes the following four areas:

Mathematical Programming: Including linear programming, integer programming, nonlinear programming, interior point methods, game theory, network optimization models, combinatorics, equilibrium programming, complementarity theory, multiobjective optimization, dynamic programming, stochastic programming, complexity theory, etc.

Applied Probability: Including queuing theory, simulation, renewal theory, Brownian motion and diffusion processes, decision analysis, Markov decision processes, reliability theory, forecasting, other stochastic processes motivated by applications, etc.

Production and Operations Management: Including inventory theory, production scheduling, capacity planning, facility location, supply chain management, distribution systems, materials requirements planning, just-in-time systems, flexible manufacturing systems, design of production lines, logistical planning, strategic issues, etc.

Applications of Operations Research and Management Science: Including telecommunications, health care, capital budgeting and finance, economics, marketing, public policy, military operations research, humanitarian relief and disaster mitigation, service operations, transportation systems, etc.

This book series is indexed in Scopus.

Yacob Khojasteh • S.M. Mousavi Jahan Abadi
Editors

Digital Transformation of Supply Chains

Business Objectives and Applications

Springer

Editors
Yacob Khojasteh
Sophia University
Tokyo, Japan

S.M. Mousavi Jahan Abadi
Sophia University
Tokyo, Japan

ISSN 0884-8289 ISSN 2214-7934 (electronic)
International Series in Operations Research & Management Science
ISBN 978-3-032-01205-0 ISBN 978-3-032-01206-7 (eBook)
https://doi.org/10.1007/978-3-032-01206-7

This Springer imprint is published by the registered company Springer Nature Switzerland AG
The registered company address is: Gewerbestrasse 11, 6330 Cham, Switzerland

Preface

In the contemporary global economy, supply chains serve as the backbone of commerce and industry. The smooth flow of goods, services, and information across borders and sectors is a key determinant of organizational competitiveness and resilience. Amid rising customer expectations, increasing market volatility, and rapid globalization, supply chain management has evolved from a supporting function into a strategic enabler of business success. Its role in ensuring efficiency, cost-effectiveness, and customer satisfaction underscores the critical need for organizations to continuously refine and optimize their supply chain practices.

Concurrently, digital transformation has emerged as a powerful force reshaping industries. Across sectors, from finance and healthcare to manufacturing and retail, organizations are adopting digital transformation to enhance agility, foster innovation, and generate new sources of value. This sweeping shift goes beyond the adoption of new technologies; it signifies a fundamental reconfiguration of organizational processes, culture, and business models. The impact is extensive, enabling organizations to respond more effectively to market demands, streamline operations, and gain practical insights through data-driven decision-making.

Within the domain of supply chain management, digital transformation holds particular significance. The complex, interconnected nature of modern supply chains makes them especially responsive to the efficiencies and capabilities offered by digital advancements. Digital transformation in supply chains enables end-to-end visibility, real-time responsiveness, and enhanced collaboration among stakeholders. As supply chains grow more dynamic and global, the integration of digital solutions becomes essential for mitigating risks, optimizing performance, and sustaining competitive advantage.

The business objectives that drive digital transformation in supply chains are diverse and strategic. Companies seek to improve operational efficiency, reduce costs, and enhance service levels while simultaneously building resilience against disruptions. Moreover, digital transformation supports sustainability goals by optimizing resource utilization and reducing environmental impact. It also empowers organizations to deliver superior customer experiences through faster fulfillment, greater customization, and improved transparency. Aligning digital transformation initiatives with these objectives is crucial for organizations aiming to succeed in today's demanding business environment.

Applications of digital transformation in supply chains are both broad and impactful. From demand forecasting and inventory optimization to transportation management and supplier collaboration, digital solutions are applied across every link in the chain. These applications facilitate greater automation, predictive analytics, and intelligent decision-making, transforming traditional supply chains into agile, responsive networks. The ability to leverage data and digital tools effectively has become a key competitive advantage, enabling organizations to anticipate market shifts, adapt swiftly, and sustain long-term growth.

This book, Digital Transformation of Supply Chains: Business Objectives and Applications, aims to explore these themes in depth, offering insights into how organizations can harness digital transformation to achieve their supply chain goals. By exploring the key drivers, strategic objectives, and real-world applications, this book offers a comprehensive understanding of the evolving supply chain landscape and serves as a practical guide for practitioners, scholars, and decision-makers navigating the digital future.

The book is structured in three parts: (1) Business Objectives and Motivations for Digital Transformation in Supply Chain, (2) Applications of Digital Transformation in Retails and Logistics, and (3) Control Towers in Supply Chains.

The first part of the book explores why and how digital transformation is crucial in today's supply chain landscape. It explores the key business drivers, strategic objectives, and performance indicators that are pushing companies to embrace digital technologies. The second part shifts the focus on the practical applications of digital transformation within specific sectors of the supply chain. It provides real-world case studies, demonstrating how companies are using digital tools to optimize their supply chains, improve customer service, and gain a competitive advantage. The last part of the book focuses on the crucial role of control towers in managing global supply chains. It explores the functionalities, benefits, and implementation challenges of control towers, showcasing how these centralized hubs can enhance visibility, improve decision-making, and increase supply chain agility.

In the **first chapter** entitled "Leveraging digital transformation in supply chains: strategies for agility, performance, and resilience", we examine strategic approaches to digital transformation for building agile, high-performing, and resilient supply chains. Tracing the evolution toward Supply Chain 4.0, this chapter highlights the transformative role of AI, IoT, and big data across areas like transportation, warehousing, logistics, and procurement. The chapter also outlines KPIs for measuring success and discusses the difference between basic digitalization and a comprehensive digital state, providing insights for creating efficient and adaptive supply chains. In **Chap.** 2, Mucahit Ozden critically highlights the necessity of aligning digital transformation efforts with clear strategic business goals. He cautions against technology-driven initiatives lacking a broader organizational vision, emphasizing that digital transformation should serve business objectives, not be an end in itself. By discussing the strategic application of technologies like AI and blockchain and tools like balanced scorecards, the chapter highlights the importance of leadership, collaboration, and continuous evaluation for creating sustainable business value. In **Chap.** 3, Maria Jesus Saenz, Christoph Lennartz, and Carl Marcus Wallenburg explore the psychological and organizational motivations behind successful digital supply chain transformation. They present a framework that examines factors such as discomfort and psychological safety during the momentum and continuity stages of change. Through a case study of international companies, they identify psychological safety as a key dimension. Their findings offer actionable insights for fostering motivation by involving employees, providing training, and adapting IT infrastructure to support effective end-to-end digital initiatives.

In **Chap.** 4, Alexander Samuels and Blessing Takawira examine the significant transformative impact of Industry 4.0 technologies, including AI, IoT, and blockchain, on global supply chains. They explore how these technologies enhance efficiency, improve transparency, and build greater resilience within complex networks. The chapter highlights how real-time data, and advanced analytics enable proactive decision-making and operational optimization. Emphasizing the lessons from recent disruptions like COVID-19, it underscores the growing importance of these technologies for creating adaptable and robust supply chains. **Chapter** 5 explores the development of a customer value-oriented digital servitization strategy, using Komatsu Ltd. as a detailed case study. In this chapter, Junichi Tomita examines how Komatsu leverages technology, smart products, and digital transformation to gain a competitive edge. The analysis highlights Komatsu's focus on customer touchpoints, extensive data collection through Komtrax, AI-driven analysis, and the provision of optimized solutions via Smart Construction. The case illustrates how focusing on customer needs can drive successful, competitive digital transformation in traditional industries. In **Chap.** 6, Aleksi Harju, Mika Immonen, and Jukka Hallikas investigate the crucial role of data management and analytics in enhancing environmental sustainability during the digital transformation of supply chains. They introduce a framework for leveraging sustainability-related data and discuss the barriers and drivers influencing its adoption. Through qualitative research and a case study, they emphasize the importance of data quality and integration for achieving supply chain transparency and sustainability. The chapter highlights how advanced analytics can inform decision-making and improve operational efficiency in alignment with environmental objectives.

The **second part** of the book focuses on the practical applications of digital transformation within specific sectors of the supply chain. It begins with a chapter that examines the challenges and opportunities facing traditional brick-and-mortar stores in the age of e-commerce. In this chapter (**Chap.** 7), Alan D. Smith explores how those retailers can leverage digital technologies to enhance the in-store experience, improve their supply chain efficiency, and compete more effectively with online giants. The study contrasts digital-first companies like Amazon with brick-and-mortar leaders like Walmart, identifying strengths and challenges in each model. In **Chap.** 8, Goutam Sutar, Krantiraditya Dhalmahapatra, and Sayan Chakraborty provide a comprehensive review of the critical factors, such as temperature monitoring, route optimization, and inventory management, influencing the efficiency of cold chain logistics, essential for perishable goods. They categorize research trends, challenges, and the role of digital transformation in optimizing cold chain operations.

In **Chap.** 9, Manish B.R. Veeramani and Marcelo Martins de Sá identify and evaluate critical success factors for digital transformation in agri-food supply chains, focusing on the Brazilian and Indian craft beer industries. By comparing the two markets, the chapter offers strategic insights and practical recommendations for improving traceability, reducing waste, and supporting SMEs in the agri-food sector through smart farming, collaboration, and AI-driven decision-making. **Chapter** 10 delves into the order-to-cash (O2C) process for logistics service providers. In this chapter, Christiaan de Goeij and Luca Gelsomino use Transaction Cost Economics to analyze the factors that contribute to inefficiencies in the O2C process, such as complex pricing structures and lack of internal coordination. Based on case studies from Dutch firms, they highlight the challenges and opportunities of digitalizing the O2C process. They offer practical recommendations for logistics service providers seeking to improve their financial performance and operational efficiency. In **Chap.** 11, Ozgu Turgut addresses the potential of digital supply chain management tools to empower smallholder farmers, especially in developing countries. She highlights the challenges they face due to market isolation and proposes a digital market model integrated with AI and optimization-based decision support. The chapter explains the analytical components of this tool and presents a case study to illustrate its potential to improve market access and reduce logistical inefficiencies for these crucial agricultural producers.

The **last part** of the book focuses on the crucial role of control towers in managing global supply chains. **Chapter** 12 provides a foundational overview of supply chain control towers (SCCTs). In this chapter, Jose M. Framinan and Paz Perez-Gonzalez discuss the concept of SCCTs, their key components and functionalities, and explore their potential benefits for improving supply chain performance. They present a comprehensive review of the literature on SCCTs, summarizing the current state of research and practice in this area. **Chapter** 13 delves into the specific role of control towers in enhancing supply chain agility. In this chapter, Natalia Szozda discusses how control towers can be used to monitor supply chain events, predict potential disruptions, and respond quickly to changing market conditions. She presents case studies of companies that have successfully implemented control towers to improve their agility. The chapter also explores the challenges of implementing control towers and offers practical advice for companies seeking to leverage this technology.

In the last chapter (**Chap.** 14), Emre İpekçi and Arda Toygar investigate the application of digital control towers in the automotive industry. They discuss the specific challenges facing automotive supply chains, such as the need for just-in-time inventory management and the complexity of global sourcing. They explore how digital control towers can be used to optimize material flows, improve supplier collaboration, and enhance overall supply chain visibility. The chapter presents a SWOT and TOWS analysis of digital control tower implementation in the automotive industry, highlighting both the opportunities and threats.

We would like to thank all the authors who contributed to this book. We appreciate their considerable efforts in preparing and revising their chapters, especially those who were asked to make multiple revisions.

We are also grateful to Camille C. Price, the former Series Editor of the *International Series in Operations Research & Management Science*, and Jialin Yan, Senior Editor, for their support throughout this project. Our thanks also go to the Springer team, including the project coordinator, Sneha Arunagiri, for their assistance and dedication.

Finally, we extend our heartfelt thanks to our families for their unwavering support and for allowing us the time needed to complete this book. We lovingly dedicate this book to them.

Tokyo, Japan
May 2025

Yacob Khojasteh
S.M. Mousavi Jahan Abadi

Contents

About the Editors

Yacob Khojasteh is a Professor of Operations Management at the Graduate School of Global Studies, Sophia University in Tokyo, Japan. He was a Visiting Professor at the Department of Mechanical and Industrial Engineering, Toronto Metropolitan University (formerly Ryerson University), in Toronto, Canada, during 2021–2022. He received his Ph.D. in Systems and Information Engineering from the University of Tsukuba, Japan. He is the author or editor of several books, including *Supply Chain Risk Mitigation: Strategies, Methods, and Applications* (2022, Springer), *Supply Chain Risk Management* (2018, Springer), *Production Management* (2017, CRC Press), and *Production Control Systems* (2016, Springer). He has several years of professional experience in industry and consulting. His research interests include production and operations management, supply chain management, logistics and warehousing, systems modeling and optimization, digital transformation, and business analytics.

S.M. Mousavi Jahan Abadi is a Senior Research Manager at FUJIFILM Software Co., Ltd. in Tokyo, Japan, where he has led research and development initiatives for over two decades. His work has focused on advancing technologies in digital transformation, cloud computing, data science, and artificial intelligence, contributing significantly to the innovation of numerous products and solutions. In addition to his corporate leadership, Dr. Mousavi Jahan Abadi possesses deep expertise in Japanese business practices and organizational culture. He holds a Ph.D. in Systems Engineering and, since 2021, he served as an Adjunct Professor at Sophia University in Tokyo, where he teaches undergraduate and graduate courses, including *Business Intelligence* and *Management Information Systems*. His combined experience in industry and academia reflects a strong commitment to bridging cutting-edge research with real-world business applications.

Part I
Business Objectives and Motivations for Digital Transformation in Supply Chain

Chapter 1
Leveraging Digital Transformation in Supply Chains: Strategies for Agility, Performance, and Resilience

Yacob Khojasteh, S.M. Mousavi Jahan Abadi, and Chelsea Ng

Abstract In this chapter, we investigate how organizations can strategically leverage digital transformation to design superior, future-ready supply chains. It examines the evolution from early digitalization to the comprehensive Supply Chain 4.0 paradigm, outlining strategic applications and considerations for enhancing critical capabilities: agility, performance, and resilience. This enhancement is driven by advanced technologies such as Artificial Intelligence, the Internet of Things, and big data analytics. The discussion illustrates impactful key applications of these technologies across essential sectors including intelligent transportation, automated warehousing, optimized logistics, and strategic procurement. Furthermore, the chapter outlines essential Key Performance Indicators for effectively measuring transformation success and explores crucial implementation considerations. It critically distinguishes between merely 'doing digital' and achieving a holistic, enterprise-wide 'being digital' state, offering practical insights for developing efficient, adaptive, and robust supply chains capable of prospering in today's complex and dynamic global landscape.

Keywords Digital transformation · Supply chain management · Supply chain agility · Supply chain performance · Supply chain resilience

Y. Khojasteh
Graduate School of Global Studies, Sophia University, Tokyo, Japan

S.M. M. Jahan Abadi (✉)
Graduate School of Global Studies, Sophia University, Tokyo, Japan

FUJIFILM Software Co., Ltd., Yokohama, Japan
e-mail: mousavi@sophia.ac.jp

C. Ng
Graduate Program in Global Studies (GPGS), Sophia University, Tokyo, Japan

Y. Khojasteh, S.M. M. Jahan Abadi (eds.), *Digital Transformation of Supply Chains*, International Series in Operations Research & Management Science 372, https://doi.org/10.1007/978-3-032-01206-7_1

1.1 Introduction

A supply chain constitutes a network of interconnected organizations, each pursuing its own objectives and operating under distinct protocols yet collaborating to facilitate the flow of goods and services. These inter-organizational relationships, governed by formal and informal agreements, become crucial managerial resources. They enable firms to develop competitive advantages while pursuing efficiency and profitability. Stemming from foundational lean and agile principles, contemporary supply chains have historically sought to achieve a balance between efficiency (leanness) and adaptability (agility), a hybrid approach termed *leagility*. This equilibrium aimed to address dynamic market demands while simultaneously minimizing waste and preserving operational flexibility (Goldsby et al., 2006).

However, organizational focus has expanded beyond solely pursuing cost efficiency as the primary operational objective. Firms are increasingly prioritizing the enhancement of supply chain resilience (Ivanov, 2022), primarily driven by the sophisticated application of digital technologies and the principles of Industry 4.0.

Advancements in digital technology, notably the integration of Industry 4.0 principles, have substantially reshaped the contemporary supply chain landscape. Emerging technologies, including the Internet of Things (IoT), big data analytics, cloud computing, and Artificial Intelligence (AI), are instrumental in realizing the smart factory paradigm central to Industry 4.0. In a recent study, Khojasteh et al. (2026) examined how these technologies are reshaping supply chains to enhance resilience, efficiency, and performance. Within a smart factory environment, machines, conveyance systems, and products are capable of communication and negotiation, thereby facilitating highly flexible production processes and enabling the dynamic reconfiguration of systems to manufacture diverse product variants.

Extending this paradigm, Supply Chain 4.0 signifies the integration of digital tools and technologies to enhance the management of supply chain operations. Technologies such as Radio-Frequency Identification (RFID), sensors, Global Positioning Systems (GPS), Electronic Data Interchange (EDI), and advanced information capture equipment enable precise tracking and comprehensive monitoring of supply chain activities. The synergistic integration of these tools, alongside AI and blockchain technology, has fundamentally reshaped supply chain operations, yielding significant improvements in efficiency, transparency, and coordination (Khojasteh et al., 2026).

Notwithstanding these technological advancements, the advent of the Covid-19 pandemic starkly illustrated the unpreparedness of many enterprises for significant disruptions, resulting in widespread factory closures, severe shipping bottlenecks, and substantial cost escalations. As reported by Harapko (2023), only 2% of survey participants indicated full preparedness for the pandemic, whereas a substantial 72% reported experiencing negative effects. Consequently, numerous organizations transitioned from a 'just-in-time' to a 'just-in-case' inventory strategy, emphasizing increased stock levels as a protective measure against potential disruptions (Pellicelli, 2022).

Findings by Dieaconescu et al. (2022) indicated that 42% of surveyed respondents perceived the pandemic as having accelerated their digitization initiatives. Industry trends further suggest that 64% of surveyed executives anticipate an accelerated adoption of digital technologies specifically aimed at enhancing supply chain resilience. Projections accompanying these trends forecast that by the year 2035, approximately 45% of supply chains may operate in a predominantly autonomous manner (Harapko, 2023). However, the realization of a fully digitized and autonomous supply chain necessitates more than mere technology adoption; it fundamentally requires the seamless integration of interconnected systems spanning planning, procurement, manufacturing, and logistics functions. This critical distinction between "doing digital" and "being digital" underscores the imperative for comprehensive, enterprise-wide digital integration (Harapko, 2023).

In this chapter, we explore how digital transformation initiatives can effectively enhance the agility, performance, and resilience of supply chains within the contemporary global business environment. Building upon this overarching objective, the subsequent sections systematically delve into the critical aspects of leveraging digital transformation within complex supply chain networks. Specifically, Sect. 1.2 is dedicated to discussing the foundational supply chain capabilities—notably agility, performance, and resilience—and strategic objectives enabled through digital technologies. This section also introduces relevant key performance indicators for evaluating the success of these digital transformation initiatives. Section 1.3 then provides illustrative examples of digital transformation applications across pivotal supply chain sectors and processes, including Transportation, Warehousing, Logistics, and Procurement, serving to demonstrate the practical implementation of these technologies. Section 1.4 outlines the essential strategic and implementation considerations deemed necessary for achieving successful digital adoption within supply chains. This section distinguishes between merely 'doing digital' and the fundamental organizational reshaping driven by digital capabilities termed 'being digital.' Finally, Sect. 1.5 concludes the chapter by summarizing the main insights and pointing out possible directions for future research and practical use in digital supply chain transformation.

1.1.1 Evolution of Digital Transformation in Supply Chains

The digital transformation of supply chains is not a recent phenomenon but rather a progressive journey that has significantly accelerated and deepened in scope over recent decades. Early digitalization efforts, while impactful for their time, primarily focused on optimizing discrete functions and internal processes. Technologies such as EDI facilitated standardized communication for transactions, while Material Requirements Planning (MRP) and later Manufacturing Resource Planning (MRP II) systems aimed to improve production scheduling and inventory control. The subsequent advent of Enterprise Resource Planning (ERP) systems marked a significant step towards integration, consolidating data and processes across different organizational functions. While these foundational technologies enhanced efficiency and provided a degree of operational control, they often operated in silos, lacking the capability for seamless cross-functional data integration, comprehensive root cause analysis for performance issues, or precise, dynamic demand forecasting (Gezgin et al., 2017). Initial technology deployments were often driven by a desire for localized, faster, and more efficient results in specific areas like production or logistics; however, this "utilized" IT approach frequently fell short of delivering lasting, sustainable, enterprise-wide change (Bauer et al., 2024).

The emergence of Industry 4.0 marked a significant shift, driving a deeper transformation toward smart manufacturing and integrated supply networks. Industry 4.0 emphasizes the integration of advanced information and communication technologies to connect the physical and virtual worlds, thereby creating cyber-physical systems that unify resources, information, and people. The core objective is to enable the intelligent transformation of traditional manufacturing and supply chain processes by leveraging technologies such as the IoT, AI, robotics, cloud computing, and big data analytics. These innovations facilitated enhanced efficiency, greater flexibility, and real-time responsiveness, fostering a shift from conventional linear supply chains to dynamic, interconnected, and data-driven networks. A framework for implementation of Industry 4.0 in supply chains developed by Ghadge et al. (2020) is shown in Fig. 1.1.

The digital transformation of supply chains involves more than just implementing new tools; it requires a clear strategic vision of how digital applications can enhance key metrics such as service quality, cost efficiency, agility, and inventory management. The consistent implementation of process improvements and organizational changes, supported by these technologies, is what ultimately drives operational excellence and resilience in the modern era (Gezgin et al., 2017). This strategic, holistic approach guides the transition from traditional methods to more agile, responsive, and intelligent supply chain models, highlighting that digital transformation is a continuous journey of evolving processes, structures, and capabilities to maintain competitiveness in a dynamic global landscape.

1.2 Supply Chain Capabilities and Strategic Digital Objectives

Supply chain capability is conceptualized as a firm's capacity for effective collaboration with its supply chain partners, thereby enhancing productivity, augmenting network visibility, and facilitating real-time feedback throughout all supply chain

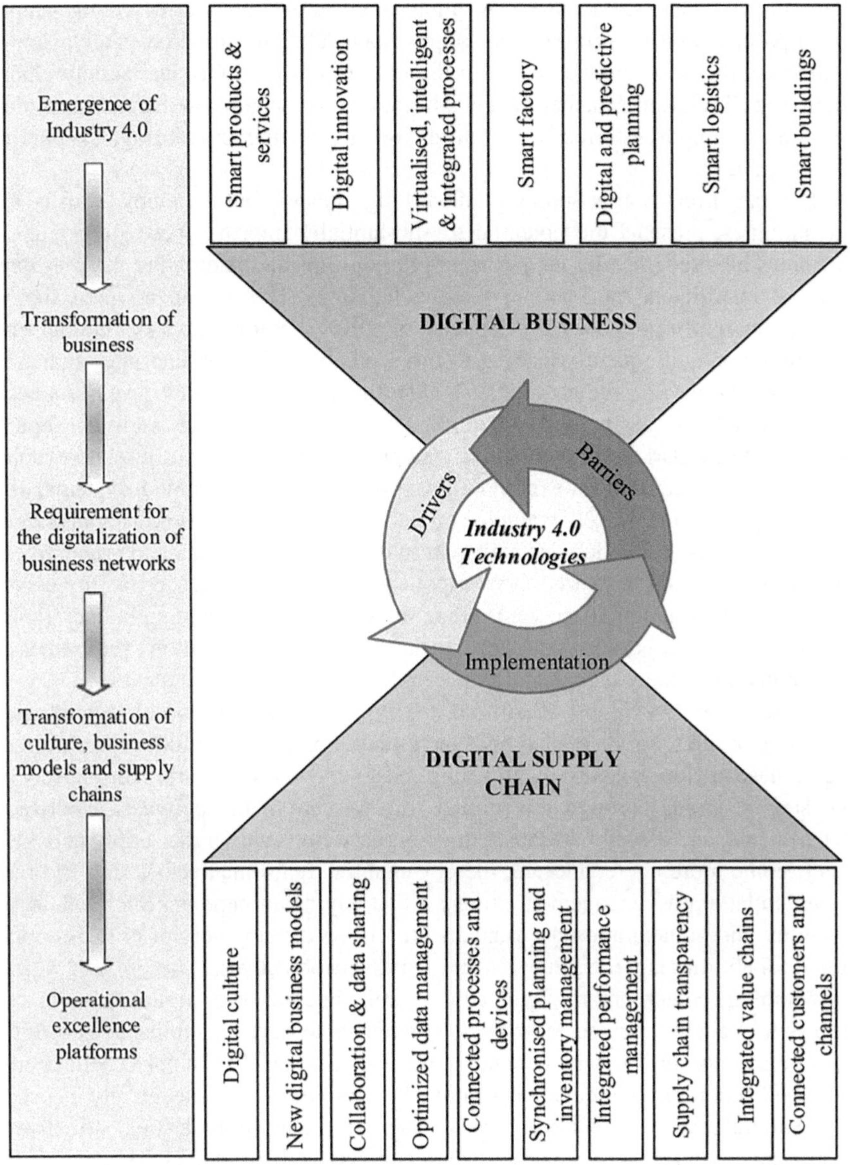

Fig. 1.1 A framework for implementation of Industry 4.0 in supply chains (Ghadge et al., 2020)

stages. Amidst the continuous evolution of digital technologies, collaboration and end-to-end visibility have emerged as critical factors in driving future advancements within both supply chains and operations management. Supply chain collaboration contributes to the achievement of competitive advantage through the strategic coordination of inter-organizational relationships, alignment of mutual interests, and equitable risk sharing, achieved by improving mutual interaction among supply chain partners and fostering robust supply chain relationships. Conversely, supply chain visibility denotes the ability to digitally represent a physical supply chain, including all relevant data that can be collected, processed, updated, and accessed in real-time to support informed decision-making in planning, monitoring, and control (Ivanov, 2021).

The transition towards Supply Chain 4.0, significantly influenced by Industry 4.0 technologies, presents the potential to substantially improve these supply chain attributes by strengthening inter-partner relationships and addressing the prevalent issue of insufficient transparency (Pellicelli, 2022). The digitalization of supply chains inherently possesses the capacity to enhance information availability and collaboration, consequently leading to more reliable, efficient, and agile management paradigms (Baziyad et al., 2022). Historically, supply chains have often been characterized by segmented operational structures, wherein various departments, such as marketing, production, and procurement, function in relative isolation, with limited real-time communication. Within these siloed systems, the marketing department, for example, might generate a sales forecast that subsequently informs production decisions; these decisions, in turn, dictate procurement orders for raw materials, parts, and components. However, forecast errors frequently lead to the propagation of problems along the supply chain, resulting in operational inefficiencies such as overproduction, stockouts, or the misallocation of resources—a phenomenon widely recognized as the 'bullwhip effect' (Lee et al., 1997).

In the context of Supply Chain 4.0, the synergistic integration of technologies including the IoT, big data analytics, and cloud computing facilitates the eliminating of these operational silos by affording end-to-end visibility across the entirety of the supply chain. Through this digital transformation, the boundaries between departments, and indeed between distinct organizations within the supply network, can become more interconnected, thereby enabling real-time data sharing and fostering collaborative engagement among all supply chain partners. Such enhanced visibility and collaboration are fundamental to the development of critical operational capabilities and the achievement of strategic objectives.

Within the contemporary digital economy, the integration of digital technologies into supply chain processes empowers companies to develop sophisticated digital operational systems, thereby enhancing the efficiency of their supply chain operations and concurrently reducing operational expenditures. By strategically incorporating digital technologies into both products and services, firms can more effectively gather and analyze data originating from suppliers and consumers. This analytical capability subsequently aids in refining product development processes, optimizing marketing strategies, addressing diverse consumer needs more effectively, and ultimately enhancing overall supply chain performance (Ivanov et al., 2022).

Furthermore, the adoption of a digital business model enables companies to leverage both online and offline delivery channels, thereby diversifying revenue streams. The digitally enabled supply chain also inherently supports open interfaces, which promote seamless connectivity with supply chain partners, foster the development of an open and collaborative system, and contribute to sustained long-term profitability. The essential strategic imperative, therefore, lies in leveraging these digital advancements to cultivate specific, measurable, and impactful capabilities.

The subsequent subsections will delve into a detailed exploration of how digital transformation supports the development of critical supply chain capabilities, specifically agility, performance, and resilience, and will concurrently discuss the Key Performance Indicators (KPIs) essential for rigorously measuring the success and tangible impact of these transformative digital initiatives.

1.2.1 Supply Chain Agility

The contemporary operational environment, characterized by rapid transformations and worsened by events such as the global pandemic and natural disasters, has underscored supply chain agility as an essential determinant for organizational survival and the sustainment of competitive advantage. Defined as a dynamic capability, supply chain agility enables firms to secure a competitive advantage through swift and effective responses to market volatilities, inherent uncertainties, and operational disruptions within the supply chain. This capability emphasizes rapidity and adaptability to shifting demand parameters, including variations in product variety, volume, and value propositions. Furthermore, it acknowledges the critical importance of addressing both demand-side and supply-side factors in the development of supply chain agility (Brusset, 2016).

Supply chain agility plays a critical role in inventory optimization and empowers organizations to enhance responsiveness to customer demand, improve adaptability to market dynamics, and foster more effective collaboration with both customers and suppliers (Mason et al., 2002). Such enhancements in collaboration and adaptability are substantially enhanced by digital transformation initiatives, which furnish the necessary tools for real-time information dissemination and synchronized actions across the extended supply network.

The degree of supply chain agility directly influences an organization's capacity to develop, produce, and deliver innovative products in a manner that is both timely and economically efficient (Swafford et al., 2006). Consequently, organizations are required to carefully monitor their operational environments to facilitate prompt responses and adjustments to fluctuating customer demands, evolving supply conditions, and emergent supply chain disruptions.

In essence, supply chain agility is formally defined as the inherent capability of a supply chain to exhibit rapid adaptation to environmental changes, existing uncertainties, and disruptive events, all while sustaining operational efficiency and fulfilling customer demands. This capability assumes a critical function in enabling businesses to maintain competitiveness within the context of contemporary

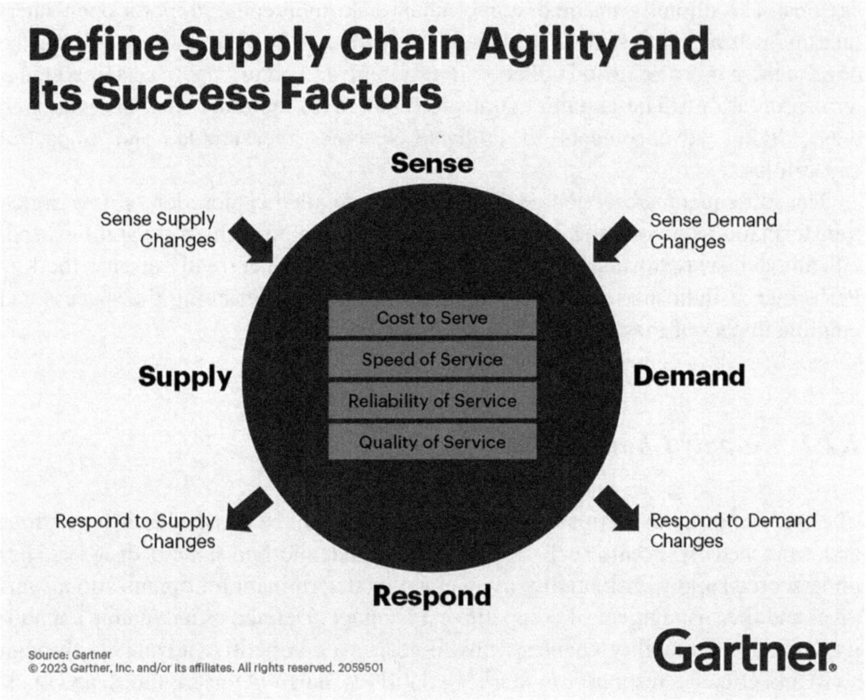

Fig. 1.2 Graphical representation illustrating the definition of supply chain agility and its constituent success factors (Gartner, 2024b)

fast-paced and unpredictable market environments. As it is shown in Fig. 1.2, the successful realization of supply chain agility is dependent on four interconnected factors: the capacity to sense, manage supply, manage demand, and respond effectively (Gartner, 2024b). The digital transformation of supply chains furnishes the essential technological infrastructure supporting these factors. Primarily, an agile supply chain must possess the capability to detect changes in supply conditions, including disruptions in procurement processes or fluctuations in material availability.

Digital technologies, including IoT sensors, RFID, and real-time data analytics, provide unprecedented visibility into supply chain events, thereby facilitating the early detection of potential issues. Analogously, it is imperative for the supply chain to sense changes in demand, identifying changes in customer requirements, preferences, or emergent market trends in real time. Big data analytics and AI-powered demand forecasting instruments are capable of analyzing extensive datasets derived from diverse sources (e.g., social media platforms, point-of-sale data) to identify emerging demand patterns with enhanced accuracy and rapidity (Dubey et al., 2019).

Ultimately, an agile supply chain must demonstrate effective responsiveness to demand fluctuations, thereby ensuring the timely delivery of appropriate products to designated locations, even amidst unforeseen surges or declines in customer demand (Gartner, 2024b). Intelligent logistics solutions, which leverage GPS and AI for dynamic routing, alongside flexible manufacturing systems integral to smart factory paradigms, contribute significantly to this responsive capability (Kagermann et al., 2013).

For organizations to achieve success within the prevailing highly competitive environment, it is imperative for both individual companies and their encompassing supply chains to exhibit agility. Research conducted by Alfalla-Luque et al. (2023) indicates a statistically significant positive correlation between supply chain agility and organizational performance, irrespective of the specific performance metrics employed, the scale used to measure supply chain agility, the industry sector, or the economic region under consideration.

Through the development of digitally enabled collaboration, the enhancement of adaptability via advanced technological integration, and the strategic leveraging of data-driven innovation, companies can strengthen resilience and sustain their competitive advantage, regardless of the encountered challenges. A comprehensive understanding and thorough implementation of both proactive and reactive enablers of supply chain agility, supported by a resilient digital infrastructure, are fundamental to the development of a robust and adaptable supply chain. By carefully balancing these two strategic approaches, businesses are empowered not only to anticipate and mitigate risks with greater efficacy through digital foresight but also to respond with proficiency to unforeseen disruptions by utilizing real-time information and integrated systems, thereby ensuring sustained success within a dynamic landscape.

1.2.2 Supply Chain Performance

The crucial requirement for digital transformation within supply chains is significantly driven by its potential to enhance overall performance. The advancement of digital technologies facilitates not only the digitalization of information flows but also, in some contexts, the "unphysicalisation" of supply chain processes, emphasizing key elements such as increased connectivity, seamless integration, and robust collaboration among participating firms. Digitalization, characterized by the transition from analogue to digital information, coupled with unphysicalisation—the strategic reduction of physical flows where possible (e.g., through digital products or direct-to-consumer models enabled by e-commerce)—collectively contributes to improvements in market performance, internal productivity, and overall supply chain efficiency. Ultimately, a company's profitability is inherently linked to the performance of its supply chain, which, in the contemporary landscape, is profoundly influenced by the strategic adoption and integration of digital technologies.

Recognizing this critical linkage, organizations are increasingly channeling investments into supply chain technologies and broader digital transformation initiatives. These investments are pivotal for accelerating development and enabling active participation in integrated supply chains, which, as discussed in Sect. 1.2, thrive on collaboration and end-to-end visibility. Such strategic financial commitments not only bolster operational efficiency but also cultivate deeper collaboration among stakeholders, empowering businesses to adapt more rapidly to dynamic market demands and unforeseen disruptions. Gartner (2024b) underscores the necessity for close collaboration between business leaders and IT departments to identify and implement advanced technologies. These technologies must support real-time, secure data sharing and seamless multi-enterprise process execution, which are fundamental for constructing transparent, resilient, and scalable supply chain networks that drive innovation and generate sustainable long-term value. Through such integration, companies can strengthen partnerships, enhance the quality and speed of decision-making, and unlock new avenues for growth within an increasingly complex global market.

The transformative impact of specific digital technologies on supply chain performance is evident across various operational aspects. For instance, blockchain technology can simplify buyer-supplier relationships by reducing transaction costs, automating order fulfillment via smart contracts, and leveraging big data analytics to identify more suitable suppliers, thereby enhancing procurement efficiency and transparency. AI/ML play a crucial role in optimizing inventory levels by delivering more accurate demand predictions, aligning stock with anticipated needs, and consequently reducing warehousing costs. Concurrently, the IoT, often utilizing technologies like RFID, ensures real-time transparency and tracking of inventory, which improves service levels and strengthens supply chain relationships through enhanced visibility.

In operational planning and execution, AI-powered algorithms can predict demand fluctuations with greater precision, optimize production schedules, and proactively identify potential risks. Complementary technologies such as Geographic Information Systems (GIS), cloud computing, and digital twins further enhance planning accuracy and operational efficiency by enabling advanced simulations and providing holistic views of the supply chain. These digital tools also contribute to significant cost reductions, improvements in productivity, and improved energy management. For example, AI and IoT sensors can monitor energy consumption in real time, detect inefficiencies, and support predictive maintenance, while ML algorithms can assist in mitigating operational and environmental risks. Within the transportation domain, AI optimizes routes and load capacities, while IoT and geospatial tools provide continuous tracking of shipments, enhancing logistics performance. Blockchain further fosters trust, transparency, and secure transactions across the supply network, thereby strengthening collaborative partnerships (Perano et al., 2023).

These examples collectively illustrate how digital transformation fundamentally enhances the performance of supply chains. Organizations can leverage technologies like Robotic Process Automation (RPA) to address inefficiencies in time-consuming, repetitive tasks; employ audit software licenses to enhance demand planning accuracy; utilize AI for spend analysis to integrate procurement data and identify savings opportunities; and invest in comprehensive SCM software. Such software empowers sourcing and procurement teams to optimize processes, manage spending more effectively, and capitalize on extended supplier networks (Gartner, 2024b). Overall, the strategic integration of these digital tools demonstrates that digital transformation not only streamlines supply chain operations but also serves as a critical driver for long-term growth, enhanced resilience (which will be further explored in Sect. 1.2.3), and sustained efficiency. By embedding digital capabilities such as AI, blockchain, and IoT into their core processes, companies can better adapt to fluctuating market conditions, optimize end-to-end operations, and reinforce their competitive advantage in a rapidly evolving global business landscape.

1.2.3 Supply Chain Resilience

Supply chain resilience is defined as the adaptive capability of a supply chain to prepare for unexpected events, respond to disruptions, and recover from them while maintaining continuity of operations at a desired level of connectedness and control over structure and function (Ponomarov & Holcomb, 2009). The digital transformation of supply chains—driven by Supply Chain 4.0 technologies such as big data analytics, AI, ML, cloud computing, IoT, and robotics—has emerged as a critical enabler of this essential resilience. These technological advancements not only accelerate the development of new products and services but also fundamentally reshape supply chain management paradigms. They promote significant improvements in operational efficiency, including cost reductions, minimized lost sales, and optimized inventory levels, thereby enabling companies to operate with greater speed, agility, precision, and overall effectiveness.

The development of supply chain resilience is often conceptualized through a framework encompassing three interconnected core capabilities: absorptive, adaptive (or response), and restorative (or recovery) capabilities (Hosseini et al., 2019; Zhao et al., 2023). Digital transformation initiatives play a crucial role in strengthening each of these dimensions.

First, **absorptive capability** refers to a supply chain's capacity to recognize, understand, and proactively prepare for potential disruptions by absorbing impacts with minimal deviation from normal operations. Key elements supporting this capability include situational awareness, redundancy, and, critically, visibility. Digital technologies are instrumental in enhancing absorptive capacity. For instance, big data analytics and AI/ML algorithms can analyze vast datasets from diverse sources (e.g., social media, news feeds, historical disruption data, supplier risk profiles) to identify early warning signals and predict potential risks with greater accuracy (Ivanov, 2021). IoT sensors deployed across the supply chain provide real-time data on asset conditions, inventory levels, and environmental factors, contributing to enhanced situational awareness. Digital twin technology allows for the creation of virtual replicas of the supply chain, enabling organizations to simulate various disruption scenarios, stress-test their systems, and proactively develop mitigation strategies, such as optimizing redundant inventory or diversifying supplier networks, without impacting physical operations. Enhanced visibility, facilitated by integrated digital platforms, ensures that supply chain partners have timely and accurate information about demand and supply, which is critical for minimizing disruption impacts by recognizing early risk signals (Kalaiarasan et al., 2022; Pettit et al., 2010).

Second, **response (or adaptive) capability** relates to the supply chain's ability to effectively and promptly react to disruptions by adjusting workflows, reallocating resources, and implementing alternative plans to maintain functionality (Hosseini & Barker, 2016; Sheffi & Rice, 2005). This capability acts as a critical line of defense when disruptions occur and allows firms to potentially capitalize on market changes, securing competitive advantages (Sheffi & Rice, 2005). Digital transformation significantly bolsters response capability. Real-time data streams from IoT devices and integrated enterprise systems, fed into advanced control towers, enable rapid assessment of an ongoing disruption's scope and impact. AI/ML algorithms can support dynamic decision-making by suggesting optimal rerouting of shipments, identifying alternative suppliers from pre-qualified digital networks, or adjusting production schedules in smart factories. Cloud-based collaborative platforms facilitate seamless communication and coordination among supply chain partners, enabling synchronized and agile responses. This rapid, data-driven responsiveness, inherently linked to the supply chain agility, allows organizations to quickly formulate strategies and implement adaptive measures to reduce disruption impacts, supported by effective risk management processes and collaborative inter-firm relationships.

Third, **recovery (or restorative) capability** focuses on a firm's ability to return to its original operational state, or ideally an improved one, efficiently and cost-effectively after a disruption has occurred. This becomes crucial when absorptive and response capabilities are insufficient to fully mitigate the disruption's effects (Hosseini et al., 2019). Digital tools support recovery by facilitating the efficient reconfiguration of resources and systems. Data analytics can be employed for post-incident analysis, identifying root causes and lessons learned, which are then fed into digital knowledge management systems to refine future contingency plans and operational strategies (Sheffi & Rice, 2005). Blockchain technology can enhance transparency in post-disruption scenarios, for example, by providing immutable records for insurance claims or verifying the origin of alternative components. Furthermore, integrated digital platforms can streamline communication and coordination during complex recovery processes, ensuring all stakeholders are aligned (Adobor & McMullen, 2018).

Central to enhancing all three resilience capabilities—absorptive, response, and recovery—is **end-to-end supply chain visibility**. As highlighted in Sect. 1.2, visibility, defined as the digital representation of the physical supply chain with real-time data access functions as a proactive warning system and a facilitator of coordinated action. The crucial requirement for end-to-end visibility has been particularly amplified by large-scale disruptions like the Covid-19 pandemic, which affected multiple echelons of supply chains and markets over extended periods (Ivanov, 2021). Digital transformation, through technologies such as IoT-enabled track and trace systems (e.g., RFID, GPS), AI-powered data analytics, digital supply chain twins, collaborative digital platforms, blockchain, and smart contracts, is fundamental to achieving this level of comprehensive insight. Such visibility allows for the timely sharing of information, enhancing the readiness of supply chain entities to manage risks, respond to events, recover swiftly, and ultimately improve overall performance.

To effectively leverage these digital advancements for resilience, companies must innovate their supply chain design and management, transitioning towards data-driven, adaptable, and sustainable networks that can address both the opportunities presented by digital technologies and the challenges posed by significant disruptions. The application of data analytics and AI, for instance, spans procurement risk assessment, manufacturing process monitoring, demand sensing in omnichannel models, logistics routing optimization, real-time traffic monitoring, and proactive safety management, all contributing to a more resilient posture (Ivanov, 2021).

Risk management systems can integrate diverse data sources (e.g., material requirements planning, purchasing databases, sales volume planning) within a quantitative risk exposure model visualized through intuitive dashboards, supported by comprehensive supply chain mapping (Ivanov, 2021). These digital tools essentially augment human cognitive capabilities, such as pattern recognition, information integration, decision-making, and forecasting, by embedding them into machine-driven processes, enabling predictive approaches that facilitate rapid analysis, superior risk mitigation, and an enhanced overall ability to respond to and recover from disruptions, thereby increasing overall supply chain resilience.

In summary, the digital transformation of supply chains is not merely an operational upgrade but a strategic imperative for building resilience. By systematically enhancing absorptive, response, and recovery capabilities through the intelligent application of technologies like AI, big data analytics, IoT, and blockchain, organizations can significantly improve their ability to anticipate, withstand, and adapt to disruptions. This digitally enabled resilience, built upon a foundation of end-to-end visibility and robust collaborative frameworks, is essential for navigating the complexities of the global supply chains, ensuring sustained performance, and maintaining competitive advantage.

1.2.4 Key Performance Indicators (KPIs) for Digital Transformation Success

The successful implementation of digital transformation initiatives within supply chains, as discussed in the preceding sections focusing on enhancing agility, performance, and resilience, necessitates a robust framework for measuring impact and progress. Defining appropriate KPIs is critical not only for quantifying these improvements but also for guiding strategic decisions and demonstrating the value derived from substantial investments in digital technologies. A summary of the overall objectives and roles of digital transformation, drawn from the earlier discussions, is presented in Table 1.1 to set the context for KPI selection.

As businesses commence on digital transformation journeys, it is crucial to recognize that traditional KPIs, while still relevant for baseline measurement, are often insufficient to capture the full range and nuanced impact of these efforts. Digital business KPIs must evolve to become leading indicators of enterprise performance, actively guiding strategic choices and clearly demonstrating the tangible benefits of transformation initiatives (Gartner, 2024a). Digital transformation is not merely about modernizing existing systems; it fundamentally involves introducing innovative solutions and new organizational structures aimed at significantly improving both financial and operational outcomes. Given that digital transformation is related to multiple functions—including operations, finance, and customer experience—relying solely on traditional KPIs would provide a limited, one-dimensional perspective on these profound changes. This necessitates a customized set of KPIs specifically designed to assess the progress and success of the digital business transformation effort. Some of these indicators will serve as "transitional" KPIs, used

Table 1.1 Core digital transformation objectives and their implications for KPI selection

Objective/role of digital transformation	Implication for KPI development
Enhancing Competitive Advantage: Leveraging IT resources (e.g., AI, IoT, cloud) to create firm-specific assets and outperform competitors (Yeniyurt et al., 2019)	KPIs should measure market positioning, innovation speed, and customer value creation attributable to digital capabilities
Strengthening Dynamic Capabilities: Building relational capabilities (collaboration) and informational capabilities (transparency, data accuracy) for superior supply chain management	KPIs should track levels of collaboration, data quality, end-to-end visibility, and the effectiveness of information sharing in decision-making
Transitioning from "Doing Digital" to "Being Digital": Moving beyond technology adoption to holistic, integrated digital operations across all supply chain functions (Kodukula, 2025)	KPIs must differentiate between mere technology deployment and genuine business model transformation, measuring integration depth, process automation, and cultural assimilation of digital practices
Driving Efficiency and Agility: Supporting end-to-end visibility and coordination to enable dynamic adaptation to disruptions, thereby bolstering resilience and operational efficiency	KPIs need to capture improvements in operational costs, cycle times, responsiveness to change, and the ability to maintain service levels during disruptions
Utilizing Advanced Systems: Employing tools like supply chain control towers that leverage AI and real-time data for enhanced decision-making and automation (IBM, 2024; Kuntze et al., 2020)	KPIs should assess the adoption rate of advanced systems, the quality of insights generated, and the impact on decision-making speed and accuracy

during the transformation phase, while others will mature into permanent measures of business performance as digital capabilities become deeply embedded in standard operations (Anderson & Proctor, 2017).

Gartner (2024a) proposes a structured framework for developing digital business KPIs tailored to an organization's specific context, emphasizing five key elements for each KPI:

- What is being measured? (The specific metric)
- Where are we today? (The baseline value)
- What is our target goal? (The desired future state)
- What is our desired business outcome/benefit? (The strategic value)
- What is our balance point? (The trade-offs or constraints to consider)

This framework underscores the need for KPIs to be actionable, outcome-oriented, and aligned with strategic objectives. The Gartner Digital Execution Scorecard™ (DES) further highlights that top-performing organizations—those excelling at generating new revenue through digital means—consistently outperform others across a set of key strategic digital KPIs. These ten KPIs, categorized by Gartner (2024a) as depicted in Table 1.2, reflect broad organizational capabilities crucial for digital success.

While these ten KPIs provide a high-level view of digital maturity, in this chapter which focuses on supply chain transformation, it is essential to translate these and supplement them with more specific metrics that directly measure enhancements in supply chain agility, performance, and resilience. The subsequent sub-sections explore such KPIs.

1.2.4.1 KPIs for Enhanced Supply Chain Agility

As detailed in Sect. 1.2.1, supply chain agility is the capability to sense and respond swiftly and effectively to changes and disruptions. Digital transformation is a key enabler of agility through enhanced visibility, real-time data analytics, and

Table 1.2 Ten digital KPIs where top performers excel

KPIs		
Digital culture and skills	Agile and iterative ways of working	Ecosystems and customer platforms
(a) Digital Culture (b) Digital Technology Skills (c) Digital Leadership Culture	(a) Adaptive Funding Allocation (b) Agile Methods and Technical Practices Adoption (c) Product-Centric Organization Adoption (d) High-Frequency Releases	(a) Cross-Channel Customer Data Access (b) Modern APIs/Streaming Adoption (c) Modern Customer UI Adoption

Adapted from Gartner (2024a)

collaborative platforms. KPIs for digitally enabled supply chain agility should reflect these improvements:

- **Time-to-Respond to Unplanned Events:** Measures the duration taken to detect an unplanned event (e.g., supplier disruption, demand surge) and implement a viable solution (e.g., reroute shipment, activate alternative supplier). Digital tools like real-time alerts and integrated planning systems directly impact this.
- **Demand Forecast Accuracy Improvement:** While traditional, the improvement rate due to AI/ML adoption for demand sensing and forecasting is a key digital KPI. Measures the percentage reduction in forecast error (e.g., MAPE, MPE) post-digital implementation.
- **Order Fulfillment Cycle Time Variability:** Lower variability indicates a more predictable and agile system. Digitalization aims to reduce unexpected delays through better coordination and visibility (Metric: Standard deviation of order fulfillment cycle time).
- **Supply Chain Responsiveness Index:** A composite measure that could include factors like speed of new product introduction, ability to adjust production volumes quickly, and time to incorporate engineering changes. This reflects the "Adaptive Funding Allocation" and "Product-Centric Organization Adoption" from Gartner's list.
- **Rate of Adoption of Real-Time Visibility Tools:** Percentage of critical supply chain nodes (e.g., key suppliers, logistics hubs, inventory points) integrated into real-time visibility platforms (e.g., control towers, IoT tracking). This relates to "Cross-Channel Customer Data Access" and "Modern APIs/Streaming Adoption."
- **Scenario Simulation and Response Effectiveness:** For organizations using digital twins, KPIs can include the number of critical risk scenarios simulated and the quantified improvement in planned responses versus non-simulated responses.

1.2.4.2 KPIs for Improved Supply Chain Performance

Section 1.2.2 highlighted how digital transformation enhances supply chain performance through efficiency gains, cost reduction, and improved service levels. KPIs in this domain must capture these diverse benefits:

- **Perfect Order Rate (POR):** The percentage of orders delivered to the right place, with the right product, at the right time, in the right condition, with the right packaging, and with the right documentation. Digitalization (e.g., RFID, automated quality checks, integrated order management) significantly improves POR.
- **Cash-to-Cash Cycle Time:** Measures the time between paying for raw materials and collecting payment for finished goods. Digital transformation can shorten this through better inventory management, faster invoicing, and improved demand-supply synchronization.

- **Total Landed Cost Reduction:** Percentage reduction in total costs associated with sourcing, manufacturing, and delivering products, achieved through digital optimization of logistics, procurement (e.g., AI for spend analysis), and inventory.
- **Inventory Turnover and Days of Supply:** Optimized through better forecasting, real-time inventory visibility (IoT), and demand-driven replenishment enabled by digital systems.
- **Automation Rate of Supply Chain Processes:** Percentage of previously manual tasks (e.g., order processing, warehouse picking, invoice matching) now automated through RPA, AI, or integrated SCM systems. This reflects the "Doing Digital" to "Being Digital" transition.
- **Overall Equipment Effectiveness (OEE) in Smart Manufacturing:** For digitally transformed manufacturing nodes, OEE (Availability × Performance × Quality) is a critical KPI.
- **Digital Revenue Contribution:** As per Anderson and Proctor (2017), measuring the percentage of revenue generated from digitally enabled products, services, or channels. This ties into "Digital Culture" and "Digital Technology Skills" fostering innovation.

Anderson and Proctor (2017) highlight that as organizations shift toward digital business models, KPIs must go beyond traditional financial metrics to include measures of digital revenue generation and market share expansion. For supply chain performance, this requires tracking how digital initiatives, such as e-commerce integration, direct-to-consumer strategies, and data-driven product customization, drive top-line growth and enhance profitability.

1.2.4.3 KPIs for Enhanced Supply Chain Resilience

Section 1.2.3 discussed supply chain resilience as the ability to prepare for, respond to, and recover from disruptions. Digital technologies are critical in building absorptive, adaptive, and restorative capacities. KPIs for resilience should measure these:

- **Time to Recover (TTR) from Disruptions:** Average time taken for the supply chain to return to pre-disruption operational levels (or a new, stable state) following a significant event. Digital tools like predictive analytics for early warning and collaborative platforms for coordinated response directly influence TTR.
- **Supply Chain Visibility Index:** A measure of the extent to which an organization has real-time visibility across its multi-tier supply network (e.g., percentage of key suppliers and their operational status visible). This links to Gartner's "Cross-Channel Customer Data Access" and "Modern APIs/Streaming Adoption."
- **Predictive Risk Identification Rate:** Percentage of potential disruptions successfully identified by AI-powered predictive analytics tools before they significantly impact operations.
- **Supplier Risk Score Improvement:** Average reduction in risk scores for critical suppliers achieved through digital monitoring, enhanced collaboration, and proactive risk mitigation strategies.

- **Effectiveness of Contingency Plan Activation:** Measured by the speed and success rate of activating digitally managed contingency plans (e.g., switching to alternative suppliers or routes identified through digital platforms). This reflects "Adaptive Funding Allocation" and "Agile Methods".
- **Percentage of Supply Chain Mapped and Assessed via Digital Twin:** For organizations leveraging digital twins, the completeness of the virtual model and the frequency/effectiveness of its use for resilience testing.
- **Ecosystem Collaboration Index:** Measures the maturity of digital collaboration with key partners (suppliers, logistics providers, customers) using metrics like data integration levels, joint planning frequency, and shared platform usage. This is supported by Gartner's "Modern APIs/Streaming Adoption" and creating "Ecosystems and Customer Platforms."

In essence, measuring the success of digital transformation in supply chains may require a balanced scorecard of KPIs. These indicators should transcend traditional operational metrics to include the strategic impact on agility, overall performance (including new value creation), and resilience. The Gartner (2024a) KPIs related to Digital Culture, Skills, Leadership, Agile Ways of Working, and Ecosystems provide an essential foundation, indicating an organization's capacity to transform. However, these must be complemented by specific supply chain KPIs that measure the outcomes of leveraging that capacity. For businesses digitalizing their supply chain processes, KPIs measuring the degree of digitalization, such as the percentage of digitally interconnected assets or transactions, become crucial. These helps evaluate progress and efficiency improvements, and linking them directly to enhancements in supply chain agility, performance, and resilience ensures that the transformation efforts are indeed building a more competitive and robust supply chain for the future.

1.3 Digital Transformation in Action: Applications Across Key Supply Chain Areas

This section explores the impact of digital transformation across four critical areas: transportation, warehousing, logistics, and procurement. Each subsection illustrates how these innovations enhance efficiency, agility, and resilience by transforming traditional processes into data-driven, interconnected systems. By leveraging these technologies, organizations can attain exceptional levels of performance and gain a strong competitive edge in their global supply chains

1.3.1 Supply Chain 4.0 and Transportation

The development of Supply Chain 4.0, driven by the widespread use of digital technologies, has influenced all areas of supply chain management. This includes major improvements in transportation, warehousing, logistics, and procurement. Among these areas, transportation logistics has experienced one of the most significant changes, moving from basic management approaches to advanced, data-driven systems that are reshaping industry practices and capabilities (Baziyad et al., 2022; Pellicelli, 2022). This change is especially important because transportation networks are essential to global trade, and their efficiency, flexibility, and reliability strongly affect the overall performance of the supply chain.

Historically, transportation logistics, particularly in areas like trucking, was primarily concerned with the basic tracking of vehicle locations against predefined schedules. The introduction of GPS represented a notable advancement, yet the fundamental objective remained largely confined to monitoring physical locations and ensuring compliance to planned routes. This approach, while offering a degree of control, lacked the dynamism and predictive capabilities necessary to navigate the complexities and uncertainties inherent in modern supply chains, often leading to inefficiencies stemming from unforeseen delays, suboptimal asset utilization, and limited responsiveness to real-time conditions.

Supply Chain 4.0 has fundamentally expanded the objectives and capabilities within transportation logistics, moving far beyond elementary tracking. The convergence of advanced digital technologies now facilitates a comprehensive, real-time analysis and optimization of multi-dimensional transportation operations. Key technologies driving this transformation include:

- **IoT and Telematics:** IoT devices, including advanced sensors and telematics systems embedded in vehicles and cargo, provide a continuous stream of granular data. This extends beyond simple GPS coordinates to include vehicle diagnostics (e.g., engine health, fuel consumption, tire pressure), driver behavior (e.g., speed, braking patterns), and cargo conditions (e.g., temperature, humidity, shock, security) (Hofmann & Rüsch, 2017; Tijan et al., 2019). Such real-time data enables proactive maintenance, improved safety, enhanced cargo integrity, and optimized fuel efficiency. Geofencing capabilities allow for automated alerts and actions when vehicles enter or exit predefined virtual boundaries, improving security and operational control.
- **AI/ML:** AI/ML algorithms are central to interpreting the vast datasets generated by IoT devices and other sources. These technologies are critical for dynamic route optimization, which considers variables such as current speed limits, real-time traffic congestion, weather forecasts, temporary road closures, port congestion, and fluctuating fuel costs to suggest optimal paths and provide accurate Estimated Times of Arrival (ETAs). Beyond routing, AI/ML applications in transportation include predictive maintenance scheduling for fleets (reducing downtime and costs), intelligent load balancing and consolidation, demand forecasting for transportation capacity, dynamic pricing strategies, automated dispatching, and the ongoing development of autonomous vehicles (trucks, delivery drones), which offer further efficiencies and safety improvements (Ghadge et al., 2020).
- **Big Data Analytics:** The vast volume, velocity, and variety of data generated within modern transportation networks necessitate sophisticated big data analytics capabilities. By analyzing historical and real-time data from diverse sources—

including telematics, traffic feeds, weather services, shipping manifests, and market trends—organizations can derive actionable insights for strategic decision-making. This includes optimizing transportation network design, improving carrier selection and performance management, benchmarking operational efficiency, identifying systemic bottlenecks, and enhancing sustainability through initiatives like carbon footprint reduction and optimized fuel consumption strategies (Dubey et al., 2019; Wang et al., 2016).

- **Cloud Computing:** Cloud platforms provide the scalable infrastructure required for storing, processing, and analyzing the massive datasets generated by transportation operations. They also facilitate the development and deployment of collaborative platforms, enabling seamless information sharing and coordination among shippers, carriers, freight forwarders, and customers in real-time. This enhanced informational capability fosters more integrated and responsive transportation systems.
- **Blockchain Technology:** Blockchain offers significant potential to enhance transparency, security, and efficiency in transportation logistics, particularly in complex international supply chains. Applications include creating immutable and transparent records for freight documentation (e.g., bills of lading, customs declarations), streamlining and securing payment settlements through smart contracts, improving traceability of goods to counteract counterfeiting and ensure compliance, and facilitating more trustworthy data exchange among multiple, often competing, stakeholders (Kshetri, 2018).
- **Digital Twins:** The concept of the digital twin is increasingly being applied to transportation networks. By creating dynamic digital models of physical transportation assets and systems (e.g., fleets, routes, logistics hubs), organizations can simulate various scenarios, such as the impact of disruptions (e.g., port strikes, extreme weather), test new operational strategies (e.g., introduction of electric vehicles, alternative routing), optimize resource allocation, and enhance alternative planning without impacting real-world operations (Fuller et al., 2020; Ivanov & Dolgui, 2021).

The collaborative application of these technologies in transportation directly contributes to the core supply chain objectives of agility, performance, and resilience, as outlined in Sect. 1.2. Enhanced performance is achieved through optimized routes leading to reduced fuel consumption and emissions, lower maintenance costs due to predictive analytics, improved asset utilization (e.g., reduced empty miles), faster transit times, and more accurate delivery windows, all contributing to cost savings and enhanced customer satisfaction (Baryannis et al., 2019; Goldsby et al., 2006). Increased agility is fostered by the ability to dynamically reroute shipments in response to real-time conditions, rapidly adjust capacity to fluctuating demand, and quickly adapt to unforeseen disruptions, ensuring operational continuity (Gligor & Holcomb, 2012).

Real-time visibility and AI-driven decision support enable quicker, more informed responses. Strengthened resilience is a direct outcome of enhanced visibility across the transportation network, the predictive capabilities of AI for risk identification, and the adaptive capacity provided by digital tools to navigate disruptions. For example, digital platforms can facilitate rapid identification and engagement of alternative carriers or modes of transport during crises. Concurrently, the real-time, transparent sharing of critical transportation data (e.g., shipment status, ETAs, potential delays) improves informational capabilities, reducing uncertainty and enabling more synchronized planning and execution across the entire supply network.

In conclusion, Supply Chain 4.0 has initiated a paradigm shift in transportation logistics, transforming it from a traditionally reactive function into a proactive, intelligent, and highly integrated component of the broader supply chain. By leveraging the power of IoT, AI/ML, big data analytics, cloud computing, blockchain, and digital twins, organizations can achieve unprecedented levels of efficiency, responsiveness, and resilience in their transportation operations, thereby securing a critical competitive advantage in the dynamic global marketplace.

1.3.2 Supply Chain 4.0 and Warehousing

The transformative influence of Supply Chain 4.0 extends deeply into warehousing operations, a critical node in the modern supply chain. Warehousing has evolved significantly from its traditional role as a static storage space to become a dynamic, highly responsive, and technologically advanced component essential for meeting contemporary market demands. The rise in e-commerce and shifting retail demands, including smaller batch sizes, shorter product life cycles, and faster delivery expectations, have increased uncertainty and complexity in inventory and warehouse resource management (Hofmann & Rüsch, 2017; Kagermann et al., 2013). These pressures necessitate a paradigm shift towards intelligent warehousing solutions. Consequently, recent decades have seen a transition from traditional, manual warehouses to smart, efficient facilities that leverage digital technologies and operate paperlessly, aiming for heightened levels of operational excellence.

According to Winkelhaus and Grosse (2022), a smart warehouse is "a highly integrated warehouse that uses advanced digital technologies and automation for efficient and effective operations to adapt to the dynamic business environment of today's economy." The strategic importance of such technologies is underscored by industry perceptions; a global survey by DHL on Warehousing 4.0 found that 39% of participants view robotics as the most critical mechanical technology for managing their supply chain in the coming years (Harrington, 2022).

The integration of Supply Chain 4.0 technologies into warehousing is multidimensional, aiming to enhance visibility, speed, accuracy, and flexibility. Key technologies driving this transformation include:

- **Robotics and Automation:** Advances in computing power have significantly enhanced the operational capabilities of robots, allowing them to analyse their surroundings and process information at incredible speeds. In the 1970s, one of the first autonomous robots at Stanford University required over an hour to travel a single meter. Today, with far more sophisticated computational processes, these tasks are completed in milliseconds, enabling robots to handle complex operations such as picking and packing with greater efficiency. This includes Automated Guided Vehicles (AGVs) for predictable, repetitive movements, more versatile Autonomous Mobile Robots (AMRs) that can navigate dynamic environments, collaborative robots (cobots) working alongside human staff, and sophisticated Automated Storage and Retrieval Systems (AS/RS) that dramatically improve space utilization and retrieval speeds (Khojasteh & Son, 2016).

- **IoT:** IoT devices, including sensors, RFID tags, and beacons, are critical for creating a connected warehouse environment. They enable real-time tracking of inventory, assets (like forklifts and picking carts), and personnel, significantly improving inventory accuracy and reducing search times (Ben-Daya et al., 2019). Smart shelves can monitor stock levels automatically, triggering replenishment orders, while environmental sensors monitor conditions like temperature and humidity, crucial for perishable or sensitive goods.
- **AI/ML:** AI/ML algorithms analyze data from IoT devices and Warehouse Management Systems (WMS) to optimize various processes. Applications include intelligent slotting (dynamically assigning storage locations based on demand patterns and item characteristics), predictive maintenance for automated equipment, optimizing picking routes for human or robotic pickers, demand forecasting for better inventory positioning within the warehouse, and automated quality inspection using computer vision.
- **Warehouse Management Systems (WMS) and Warehouse Control Systems (WCS):** Modern WMS are the central nervous system of the smart warehouse, integrating data from various technologies and coordinating activities. Advanced WMS incorporate real-time analytics, AI-driven decision support, and seamless integration with ERP systems and Transportation Management Systems (TMS). WCS, often working in conjunction with WMS, directly manage and synchronize automated material handling equipment.
- **Big Data Analytics and Digital Twins:** The vast amounts of data generated by smart warehouses require robust analytics capabilities to derive actionable insights for continuous improvement in layout design, process flows, and resource allocation (Wang et al., 2016). Digital twin technology allows warehouses to create virtual replicas of their physical operations, enabling simulation of different scenarios (e.g., impact of new automation, changes in order profiles), testing optimizations, and training staff in a risk-free environment before implementing changes in the live facility (Fuller et al., 2020; Ivanov & Dolgui, 2021).
- **Wearable Technology and Augmented Reality (AR):** Wearable devices, such as smart glasses and voice-directed picking systems, provide hands-free guidance to warehouse staff, improving picking accuracy and efficiency. AR can

overlay digital information onto the worker's field of view, assisting with tasks like item identification, navigation, and order verification.

Supply Chain 4.0 will further shift operational capacity to warehouses, necessitating these technological solutions to meet the growing demand, particularly during peak periods. Robotics, already a key element in logistics management, will play an even greater role as warehouses strive to enhance productivity and cope with the rising demands of e-commerce and other industries (Pellicelli, 2022). The impact of these technologies is evident in the enhanced operational capabilities and process efficiencies:

- **Receiving and Delivery:** Automated identification (e.g., RFID portals), dimensioning, weighing, and sorting (DWS) systems streamline inbound processes. AI can optimize delivery routes and storage location assignments based on real-time conditions and historical data.
- **Inventory Management:** Real-time visibility through IoT and advanced WMS drastically reduces discrepancies, enables automated cycle counting, and minimizes stockouts or overstocking. This aligns with the objective of improving overall supply chain performance as discussed in Sect. 1.2.2.
- **Order Picking and Packing:** This is an area of significant automation focus. Goods-to-person systems (e.g., AS/RS feeding picking stations, AMRs bringing shelves to pickers) and robotic picking arms enhance speed and accuracy. AI-optimized batching and routing further improve efficiency. Automated packing solutions can select appropriate box sizes and materials, reducing waste and shipping costs.
- **Sortation and Shipping:** High-speed automated sortation systems accurately route packages to the correct shipping lanes. Integration with TMS ensures seamless handover to transportation partners, improving the efficiency discussed in Sect. 1.3.1.

In conclusion, Supply Chain 4.0 is revolutionizing warehousing by transforming traditional facilities into intelligent, interconnected, and highly automated hubs. The strategic deployment of robotics, IoT, AI, advanced WMS, and analytics empowers warehouses to meet the escalating demands of modern commerce, significantly enhancing operational performance, agility, and resilience, thereby playing an essential role in the overall efficiency and competitiveness of the end-to-end supply chain.

1.3.3 Supply Chain 4.0 and Logistics

Logistics, in its broadest sense, encompasses the strategic planning, implementation, and control of the efficient, effective forward and reverse flow and storage of goods, services, and related information from point of origin to point of consumption to meet customer requirements. While transportation and warehousing represent critical operational components within this framework, the overarching domain of logistics orchestrates these elements, ensuring that products reach customers at the right place, at the right time, and in the desired condition (Inbound Logistics, 2024). Key strategic objectives traditionally include the selection of optimal transport modes, the design and management of efficient warehousing and distribution networks, strategic inventory control across the supply chain, and the establishment of a cohesive and responsive overall logistics network architecture. This involves managing the complex movement of products via land, air, and sea, alongside robust fleet management, continuous shipment monitoring, and transparent information sharing among all network partners to sustain both speed and operational efficiency.

Historically, many organizations perceived logistics primarily as an operational, often cost-centric, function, frequently reporting to sales or manufacturing divisions. The primary focus was on the physical execution of tasks: delivering raw materials to production facilities and subsequently distributing finished products to customers. Consequently, significant portions of these operational logistics activities were often outsourced to third-party logistics providers (3PLs). However, the emergence of Supply Chain 4.0 and the widespread integration of smart digital technologies have accelerated a fundamental transformation, elevating logistics from a purely tactical execution arm to a strategic, data-driven, and highly integrated core competency. This strategic evolution incorporates advanced planning capabilities, sophisticated analytics, and increasingly autonomous systems that can optimize and operate with minimal human intervention, thereby enhancing overall supply chain performance, agility, and resilience.

The digital transformation of logistics within the Supply Chain 4.0 paradigm is driven by a confluence of technologies that enable unprecedented levels of visibility, coordination, and intelligence:

- **Advanced Planning and Scheduling (APS) Systems:** Modern APS systems, enhanced by AI/ML go beyond basic planning. They enable dynamic demand forecasting across the entire supply network, optimize inventory placement at different levels (multi-echelon inventory optimization, or MEIO), and support advanced logistics network design and scenario analysis. These systems work smoothly with Sales and Operations Planning (S&OP) and Integrated Business Planning (IBP) processes, ensuring logistics activities align with broader business goals.
- **Logistics Control Towers:** Representing a significant advancement in operational oversight, digital control towers provide end-to-end, real-time visibility across the entire logistics network. They integrate data from disparate sources (TMS, WMS, IoT sensors, carrier updates, weather services, traffic data) into a centralized dashboard. AI/ML algorithms within control towers enable predictive analytics for early detection of potential disruptions (e.g., port congestion,

carrier delays, weather impacts) and prescriptive analytics to recommend optimal responses, such as rerouting shipments or reallocating inventory (Kuntze et al., 2020). This capability is fundamental to managing complex, global logistics operations proactively.

- **Big Data Analytics:** The logistics system generates vast quantities of data. Big data analytics platforms are essential for processing and analyzing this information to uncover actionable insights. Applications include optimizing logistics network configurations, benchmarking carrier performance, identifying systemic bottlenecks and inefficiencies, conducting detailed cost-to-serve analyses, assessing and mitigating logistics risks, and driving sustainability initiatives through optimized routing and load consolidation (Dubey et al., 2019; Wang et al., 2016).
- **IoT and Real-Time Tracking:** While central to transportation and warehousing, IoT-generated data (e.g., GPS, RFID, environmental sensors on shipments and assets) is a critical input for overarching logistics management. It provides real-time updates on shipment status, location, and condition, feeding into control towers and planning systems to enable dynamic adjustments and improve the accuracy of Estimated Times of Arrival (ETAs) across the entire journey.
- **Cloud Computing:** Cloud platforms provide the scalable, flexible, and cost-effective infrastructure necessary for hosting advanced logistics applications, storing large datasets, and facilitating real-time collaboration among diverse logistics partners (shippers, carriers, 3PLs, customs brokers, customers). Cloud-based TMS, WMS, and collaborative logistics platforms enable seamless information sharing and process synchronization across organizational boundaries.
- **Blockchain Technology:** In multi-party logistics networks, blockchain offers the potential to enhance transparency, security, and trust. Applications include creating immutable records for trade documentation (e.g., bills of lading, certificates of origin), streamlining cross-border customs clearance, facilitating secure and automated payment settlements via smart contracts, and providing auditable end-to-end traceability for high-value or sensitive goods (Kshetri, 2018; Saberi et al., 2019).
- **Digital Twins of Logistics Networks:** Extending the concept from individual assets, organizations are beginning to create digital twins of their entire logistics networks. These dynamic digital models allow for the simulation of network flows, the impact assessment of various disruptions (e.g., supplier outages, transport lane closures), the testing of alternative network designs or policies, and the optimization of inventory placement and flow paths in a risk-free environment (Fuller et al., 2020; Ivanov & Dolgui, 2021).
- **Robotic Process Automation (RPA):** RPA can automate repetitive, rule-based administrative tasks within logistics, such as data entry, freight auditing, invoice processing, and status update communications, freeing human resources for more strategic activities and reducing errors (Shamsuzzoha & Pelkonen, 2025).

The integration of these technologies is profoundly reshaping core logistics processes and capabilities:

- **Integrated Logistics Planning:** Moving from isolated planning to synchronized, end-to-end planning that considers all constraints and capacities across transportation, warehousing, and inventory.
- **Dynamic Network Orchestration:** The ability to intelligently coordinate and manage the flow of goods through complex, multi-modal, and multi-partner networks in real-time.
- **Optimized Order Fulfillment:** Enhanced ability to fulfill customer orders accurately and on time by optimizing sourcing decisions, inventory allocation, and delivery routing dynamically.
- **Proactive Risk Management and Event Response:** Utilizing predictive analytics and real-time visibility to anticipate potential disruptions and orchestrate mitigation strategies effectively.
- **Enhanced Reverse Logistics:** Digital tools improve the efficiency, visibility, and cost-effectiveness of managing product returns, repairs, and recycling, an increasingly important aspect of sustainable supply chains.

In conclusion, Supply Chain 4.0 is revolutionizing logistics by transforming it into an intelligent, integrated, and strategically vital function. The sophisticated application of digital technologies empowers organizations to orchestrate complex global flows with unprecedented efficiency, responsiveness, and foresight. This digitally enabled logistics capability is no longer just about moving and storing goods; it is about strategically managing information and material flows to create sustainable competitive advantage, enhance customer satisfaction, and build robust, resilient supply chains capable of prospering in a volatile global environment.

1.3.4 Supply Chain 4.0 and Procurement

Procurement, traditionally viewed as a transactional function focused on acquiring goods and services at the lowest possible cost, has evolved into a strategic discipline critical to overall organizational success and supply chain performance. In the contemporary integrated global economy, procurement acts as a vital interface, managing the complex network of relationships and information flows between a firm's internal operations and its external supplier network (Lysons & Farrington, 2020). The digital toolkit for Procurement 4.0 includes a broader array of interconnected technologies as follows:

- **AI/ML:** Beyond pattern recognition, AI/ML applications in procurement are extensive. Predictive analytics, powered by ML, enhance demand forecasting, predict commodity price movements, identify potential supplier disruptions (e.g., financial distress, geopolitical instability), and automate risk scoring. AI-driven supplier discovery platforms can identify and evaluate potential new suppliers globally, while NLP facilitates the automated review and analysis of complex contract clauses, identifying risks and opportunities. AI-powered chatbots are increasingly used for handling routine supplier inquiries and providing internal procurement support, freeing up human resources for more strategic tasks (Herold et al., 2025).

- **Big Data Analytics:** The procurement function generates and consumes vast amounts of data from internal systems (ERP, P2P platforms) and external sources (market intelligence providers, supplier databases). Big data analytics capabilities are crucial for harnessing this data to achieve comprehensive spend visibility, identify cost-saving opportunities through advanced modeling, benchmark supplier performance thoroughly, optimize sourcing strategies, and conduct sophisticated risk assessments (Schoenherr & Speier-Pero, 2015).
- **Cloud-Based E-Procurement Platforms:** Modern procurement heavily relies on integrated, cloud-based source-to-pay (S2P) suites. These platforms offer modules for e-sourcing (e.g., RFx management, e-auctions), contract lifecycle management (CLM), supplier relationship management (SRM), procure-to-pay (P2P) automation, and spend analytics (Gartner, 2022). Cloud deployment ensures scalability, accessibility, and facilitates easier collaboration with suppliers who can access shared portals for information exchange and transaction management.
- **Robotic Process Automation (RPA):** RPA is widely applied to automate high-volume, repetitive, rule-based tasks within procurement, such as purchase order creation from approved requisitions, invoice data entry, supplier master data updates, and status query responses. This improves efficiency, reduces errors, and allows procurement staff to focus on more value-added activities.
- **Blockchain Technology:** Blockchain offers significant potential to enhance transparency, security, and efficiency in procurement processes. Applications include creating immutable records for supplier credentials and certifications, enhancing traceability for ethically sourced materials, facilitating secure and transparent contract management through smart contracts (which can automate payments upon fulfillment of predefined conditions), and reducing risks of fraud in P2P processes (Kshetri, 2018; Saberi et al., 2019). While adoption is still emerging, its utility in complex, multi-tier supply chains is increasingly recognized.
- **Supplier Relationship Management (SRM) Systems:** Digitally enhanced SRM systems move beyond basic supplier databases. They incorporate tools for collaborative planning, performance tracking dashboards, risk monitoring alerts, and platforms for joint innovation initiatives. AI can augment SRM by providing insights into supplier sentiment or predicting relationship health.
- **Digital Twins in Sourcing:** Although a more advanced application, the concept of creating digital twins for sourcing scenarios is emerging. This would allow procurement teams to simulate the impact of different sourcing strategies, sup-

plier choices, or potential disruptions on cost, resilience, and lead times before making real-world commitments.

The transformative impact of these technologies aligns with critical strategic shifts in the procurement domain. These shifts, amplified and enabled by Procurement 4.0, are essential for building agile and resilient supply chains:

- **Building and Maintaining Strong Supplier Relationships:** Procurement 4.0 moves beyond a purely transactional, cost-focused approach towards fostering strategic partnerships. Digital SRM platforms, collaborative cloud-based tools, and transparent data sharing mechanisms enable closer collaboration, improved communication, and mutual trust-building with key suppliers. This enhanced relational capability is crucial for navigating disruptions, managing supply risks proactively, and co-creating value and innovation.
- **Enhancing Visibility and Streamlining Operations through Automation:** Digital technologies provide unprecedented visibility into procurement processes and the wider supply market. Integrated S2P platforms and advanced analytics offer decision-makers comprehensive insights into spend, supplier performance, contract compliance, and market dynamics. Automation of transactional tasks through RPA and AI-powered workflows streamlines operations, reduces cycle times, and improves accuracy. This heightened informational capability allows for more informed and faster decision-making.
- **Boosting Agility and Resilience:** Procurement plays a pivotal role in enhancing overall supply chain agility and resilience. Digital tools accelerate the sourcing process, enabling rapid identification and onboarding of alternative suppliers during disruptions. Predictive analytics provide early warnings of potential supply risks, allowing for preemptive action. Enhanced visibility into multi-tier supply networks, potentially supported by blockchain for traceability, helps to understand and mitigate vulnerabilities. The ability to quickly analyze market changes and adjust sourcing strategies accordingly is a characteristic of agile procurement.
- **Improving Supplier Compliance, Resilience, and Value Addition:** Digital platforms can facilitate the collection, verification, and monitoring of supplier compliance data related to ethical standards, sustainability, and regulatory requirements. Blockchain can provide auditable proof of compliance. Analytics can help identify and develop a diverse supplier base, including smaller, local, or minority-owned vendors, which can enhance resilience and contribute to corporate social responsibility goals. By moving beyond cost reduction to focus on total value, including risk mitigation, innovation, and sustainability, procurement becomes a significant contributor to the company's overall strategic objectives.

In essence, Procurement 4.0 indicates a fundamental transformation of the procurement function from a primarily administrative and cost-control center to a strategic, data-driven enabler of competitive advantage. By leveraging the full spectrum of digital technologies, organizations can optimize procurement performance (e.g., reduced costs, improved efficiency), enhance agility in responding to market dynamics, and build greater resilience against supply chain disruptions. This digital evolution not only streamlines existing processes but also unlocks new sources of value, fosters innovation through deeper supplier collaboration, and positions procurement as a critical orchestrator of the external supply network, contributing significantly to the objectives of Supply Chain 4.0.

1.4 Key Considerations for Successful Digital Transformation in Supply Chains

The preceding sections have explained the profound potential of digital transformation to enhance supply chain agility, performance, and resilience, alongside outlining specific applications across key sectors. However, the journey towards realizing these benefits is complex and challenging. Merely investing in new technologies—"doing digital"—is insufficient to achieve the transformative outcomes associated with Supply Chain 4.0. Instead, organizations must aim to "being digital," a state characterized by the fundamental integration of digital capabilities into the core strategy, operations, and culture of the enterprise (Kodukula, 2025). This distinction highlights that successful digital transformation is not solely a technological endeavor but a multi-dimensional strategic imperative requiring careful consideration of several critical factors. This section outlines these key considerations, providing a framework for organizations navigating the path towards a truly digitized and intelligent supply chain.

1.4.1 Strategic Alignment and Holistic Vision

A critical consideration for successful digital transformation in supply chains is the establishment of a clear, comprehensive strategic vision that is clearly aligned with business objectives. The adoption of digital technologies should not be an ad-hoc pursuit of the latest innovations but a deliberate strategy aimed at solving specific business problems, creating new value, or building sustainable competitive advantages (Westerman et al., 2014). This requires moving beyond isolated, departmental initiatives towards an enterprise-wide approach that integrates digital efforts across all supply chain functions and, indeed, the entire organization (Ross et al., 2019). A holistic vision ensures that digital investments are purposeful, contributing to defined outcomes such as enhanced customer value, improved operational efficiency, or increased market share, directly linking to the KPIs discussed in Sect. 1.2.4. Furthermore, this strategic alignment must balance short-term performance improvements with the long-term goal of building a resilient and adaptive digital supply network. This long-term perspective is crucial for achieving the state of "being digital," where digital capabilities are woven into the organizational fabric, driving continuous innovation and adaptation rather than being limited to discrete projects.

1.4.2 Leadership Commitment and Effective Governance

The driving force and consistent support of top leadership are essential for navigating the complexities of digital transformation. Leaders must not only advocate the vision but also be actively involved in driving the change, allocating necessary resources, and promoting a culture that adopts digital innovation (Kane et al., 2015). Effective governance structures are equally critical. This includes establishing clear roles, responsibilities, and decision-making processes for digital initiatives, often involving cross-functional steering committees to ensure enterprise-wide alignment and support. Such governance frameworks should guide priorities, manage the digital transformation portfolio, monitor progress against strategic objectives, and ensure accountability. A digital-first mindset, disseminated from the leadership team, can permeate the organization, encouraging proactive engagement with digital opportunities and challenges. Without strong leadership and robust governance, digital transformation efforts risk becoming fragmented, under-resourced, and ultimately falling short of their transformative potential.

1.4.3 Managing Change and Cultivating a Digital-Ready Culture

Digital transformation inherently involves significant organizational change, impacting processes, roles, and ways of working. Effectively managing this change is a critical determinant of success. Organizations must anticipate and address potential resistance by implementing comprehensive change management strategies that include clear communication, active stakeholder engagement, and robust training programs. A core component of this is cultivating a digital-ready culture. This involves developing digital literacy and upskilling or reskilling the workforce to equip employees with the competencies required to leverage new technologies and operate in a data-driven environment (Bughin et al., 2018). Furthermore, it entails promoting a culture that values continuous learning, experimentation, agility, and data-informed decision-making, aligning with the agile ways of working identified as crucial by Gartner (2024a). The human element is often the most challenging aspect of digital transformation, and neglecting it can undermine even the most technologically advanced initiatives (Nadkarni & Prügl, 2021). Therefore, investing in people and cultural adaptation is as vital as investing in technology itself.

1.4.4 The Role of Data: Strategy, Management, and Advanced Analytics

In the digital era, data is a foundational strategic asset. A robust data strategy, encompassing effective data management and advanced analytics capabilities, is crucial for unlocking the value potential of digital transformation in supply chains.

This begins with ensuring data quality, accuracy, consistency, and accessibility across the organization (Redman, 2008). Significant challenges often arise from integrating data from diverse and disparate sources, including legacy enterprise systems, IoT devices, external partner data, and unstructured data streams. Establishing strong data governance frameworks, adhering to principles such as those outlined in the DAMA Data Management Body of Knowledge (DAMA International, 2017), is essential for managing data as an enterprise asset. Beyond management, organizations must invest in advanced analytics capabilities—spanning descriptive, diagnostic, predictive, and prescriptive analytics—to transform raw data into actionable insights that drive improved decision-making and process optimization. Furthermore, ensuring data security, integrity, and compliance with privacy regulations (e.g., GDPR) is non-negotiable in an environment of increased data sharing and connectivity.

1.4.5 Technology Selection, Integration, and Scalability

While strategy and people are essential, the appropriate selection and implementation of technology remain a core consideration. Organizations must avoid the "shiny object syndrome," where technology is adopted without a clear understanding of its strategic fit or value proposition. Instead, technology choices should be driven by specific business needs and their potential to deliver on the objectives of agility, performance, and resilience. A significant challenge lies in integrating new digital technologies (e.g., AI, IoT, blockchain) with existing legacy systems, such as ERP and older SCM platforms, which often lack modern APIs and interoperability standards (Brown & Bessant, 2003). Ensuring seamless data flow and process integration across the technology landscape is critical. Solutions must also be scalable to accommodate future business growth, evolving operational requirements, and increasing data volumes. The total cost of ownership (TCO), implementation timelines, and the complexity of managing a diverse technology stack are all important factors that require careful evaluation during the selection process.

1.4.6 Cybersecurity and Risk Mitigation in Integrated Supply Chains

The increased connectivity and data sharing inherent in digital supply chains significantly expand the cybersecurity attack surface, exposing organizations to a wider range of threats. As supply chains become more reliant on integrated digital systems, including IoT devices, cloud platforms, and shared data networks, the potential impact of a cyber incident, such as data breaches, ransomware attacks, or operational disruptions, escalates. Therefore, a proactive and comprehensive cybersecurity strategy is not merely an IT concern but a fundamental business imperative. This strategy should encompass robust measures for prevention, detection, rapid response, and effective recovery, often guided by frameworks like the NIST

Cybersecurity Framework (NIST, 2018). Protecting sensitive supply chain data, customer information, and intellectual property is crucial. Moreover, risk mitigation must extend to the broader digital system, including managing vulnerabilities associated with third-party vendors, suppliers, and logistics partners, as a breach in one part of the network can have cascading effects. Building cyber resilience in conjunction with operational resilience is essential for safeguarding the integrity and continuity of digitally transformed supply chains.

1.4.7 Fostering Collaboration and Supply Network Engagement

Digital transformation in supply chains fundamentally extends beyond the boundaries of a single organization. The full potential of technologies like blockchain, collaborative platforms, and real-time data sharing can only be realized through effective collaboration and engagement with the entire supply network, including suppliers, customers, logistics providers, and technology partners. This necessitates a shift towards greater transparency, trust-building, and the development of shared data standards and protocols.

Developing robust relational capabilities is vital for fostering these inter-organizational synergies. Collaborative platforms and shared data environments can improve collective situational awareness, enable synchronized planning and execution, and facilitate joint problem-solving during disruptions. Furthermore, co-innovation with supply chain partners can lead to the development of novel solutions and value propositions that benefit all stakeholders. The role of industry consortia and standardization bodies can also be significant in promoting interoperability and common approaches to digital integration across supply networks.

1.4.8 Measuring ROI and Driving Continuous Improvement

Demonstrating the value of digital transformation initiatives is crucial for sustaining momentum and securing ongoing investment. This requires defining clear metrics and KPIs (as discussed in Sect. 1.2.4) to track both the financial return (e.g., cost savings, revenue growth) and the operational benefits (e.g., improved efficiency, reduced cycle times, enhanced customer satisfaction). However, quantifying the intangible benefits, such as increased agility, enhanced resilience, or improved employee morale, can be challenging yet is vital for a holistic assessment of impact.

Organizations should establish robust feedback loops and performance monitoring systems to continuously evaluate the effectiveness of digital solutions and identify areas for improvement. Adopting an agile methodology for implementation—iterating, testing, learning, and scaling successful initiatives—can help manage risk and accelerate value delivery (Gartner, 2024a). Critically, digital transformation should be viewed not as a finite project with a defined endpoint, but as an ongoing journey of continuous improvement, adaptation, and innovation in response to evolving technologies, market dynamics, and business priorities.

In conclusion, the journey to a digitally transformed supply chain is a strategic undertaking that extends far beyond technology deployment. By carefully addressing these key considerations—from strategic alignment and leadership to change management, data governance, technology integration, cybersecurity, supply network collaboration, and performance measurement—organizations can navigate the complexities effectively. A holistic and considered approach is fundamental to transitioning from merely "doing digital" to truly "being digital," thereby unlocking the full potential of digital transformation to create supply chains that are not only efficient and agile but also resilient, sustainable, and future-ready.

1.5 Conclusions

This chapter has systematically explored how digital transformation can be strategically leveraged to enhance the agility, performance, and resilience of contemporary supply chains, alongside examining its key applications. We began by tracing the evolution of digital transformation, highlighting its progression from foundational digitalization efforts like EDI and ERP towards the integrated, intelligent supply network of Supply Chain 4.0. This evolution reflects a critical shift in organizational priorities, moving beyond traditional lean and agile paradigms to an increasing emphasis on resilience, accelerated by Industry 4.0 technologies and the lessons learned from global disruptions such as the Covid-19 pandemic.

A central theme throughout this chapter has been the significant impact of digital transformation on core supply chain capabilities and objectives. We detailed how digital technologies—including IoT, AI, big data analytics, and blockchain—are instrumental in fostering enhanced supply chain agility, enabling firms to sense and respond rapidly to market volatility and disruptions. Furthermore, the chapter underscored how these digital tools drive significant improvements in overall supply chain performance through greater efficiency, cost reduction, improved service levels, and enhanced collaboration. Crucially, the role of digital transformation in building robust supply chain resilience was emphasized, focusing on developing absorptive, adaptive, and restorative capacities. The necessity of defining and tracking appropriate KPIs to measure the tangible success and guide these digital transformation initiatives was also established.

The practical demonstrations of these capabilities were illustrated through an examination of digital transformation applications across key supply chain sectors and processes. We explored how digital transformation is reshaping transportation with intelligent routing and real-time tracking; revolutionizing warehousing through automation and smart inventory management; optimizing broader logistics functions via integrated planning and control towers; and transforming procurement with data-driven sourcing and supplier relationship management. These examples collectively demonstrate the tangible benefits and operational enhancements achievable through the strategic deployment of digital tools.

However, the journey to a fully digitized supply chain is complex. In Sect. 1.4 we outlined important considerations for successful digital transformation, stressing that technology adoption alone is insufficient. True transformation requires strategic alignment, leadership commitment, effective change management, a robust data strategy, seamless technology integration, observative cybersecurity, strong supply network collaboration, thorough ROI measurement, and a commitment to ethical and sustainable practices. The distinction between merely "doing digital" and the more significant state of "being digital"—where digital capabilities are fundamentally embedded within the organization's strategy, operations, and culture—was highlighted as a critical success factor.

In essence, this chapter proposes that digital transformation is no longer an optional upgrade but a strategic requirement for modern supply chains. By thoughtfully integrating advanced digital technologies and adopting a holistic approach to change, organizations can unlock significant improvements in agility, performance, and resilience. This enables them to not only navigate the complexities and uncertainties of the contemporary global business environment but also to build more competitive, efficient, and future-ready supply networks.

References

Adobor, H., & McMullen, R. S. (2018). Supply chain resilience: A dynamic and multidimensional construct. *The International Journal of Logistics Management, 29*(4), 1451–1481.

Alfalla-Luque, R., Luján García, D. E., & Marin-Garcia, J. A. (2023). Supply chain agility and performance: Evidence from a meta-analysis. *International Journal of Operations & Production Management, 43*(10), 1587–1633.

Anderson, J., & Proctor, P. (2017, September 28). *Digital business KPIs: Defining and measuring success*. Gartner. Retrieved April 25, 2025, from https://www.gartner.com/en/doc/3803509-digital-business-kpis-defining-and-measuring-success.

Baryannis, G., Validi, S., Dani, S., & Antoniou, G. (2019). Supply chain risk management and artificial intelligence: state of the art and future research directions. *International Journal of Production Research, 57*(7), 2179–2202.

Bauer, F., Dilda, V., Gong, L., Gröger, C., Jensen, B., Spüntrup, F. S., & Vedpathak, K. (2024, January 29). *Tech-enabled transformations: Three supply chain success stories*. McKinsey. Retrieved April 25, 2025, from https://www.mckinsey.com/capabilities/operations/our-insights/tech-enabled-transformations-three-supply-chain-success-stories.

Baziyad, H., Kayvanfar, V., & Kinra, A. (2022). The internet of things—An emerging paradigm to support the digitalization of future supply chains. In B. L. MacCarthy & D. Ivanov (Eds.), *The digital supply chain* (pp. 61–76). Elsevier.

Ben-Daya, M., Hassini, E., & Bahroun, Z. (2019). Internet of things and supply chain management: a literature review. *International Journal of Production Research, 57*(15–16), 4719–4742.

Brown, S. L., & Bessant, J. R. (2003). The manufacturing strategy-capabilities link in mass customisation and Agile manufacturing—An exploratory study. *International Journal of Operations & Production Management, 23*(7), 707–730.

Brusset, X. (2016). Does supply chain visibility enhance agility?. *International Journal of Production Economics*, 171, 46–59.

Bughin, J., Hazan, E., Lund, S., Dahlström, P., Wiesinger, A., & Subramaniam, A. (2018). *Skill shift: Automation and the future of the workforce*. McKinsey Global Institute. Retrieved April 25, 2025, from https://www.mckinsey.com/featured-insights/future-of-work/skill-shift-automation-and-the-future-of-the-workforce.

DAMA International. (2017). *DAMA-DMBOK: Data Management Body of Knowledge* (2nd ed.). Technics.

Dieaconescu, R. I., Belu, M. G., Paraschiv, D. M., & Joldes, C. S. (2022). Supply chain digital transformation. In *BASIQ 2022 international conference on new trends in sustainable business and consumption*. https://doi.org/10.24818/basiq/2022/08/079.

Dubey, R., Gunasekaran, A., Childe, S. J., Papadopoulos, T., Blome, C., & Luo, Z. (2019). Antecedents of resilient supply chains: An empirical study. *IEEE Transactions on Engineering Management, 66*(1), 8–19.

Fuller, A., Fan, Z., Day, C., & Barlow, C. (2020). Digital twin: Enabling technologies, challenges and open research. *IEEE Access, 8*, 108952–108971.

Gartner. (2022). Magic quadrant for procure-to-pay suites. Retrieved April 25, 2025, from https://www.gartner.com/en/documents/4021524.

Gartner. (2024a). CIOs: Use the right KPIs to measure digital performance. Gartner. Retrieved April 25, 2025, from https://www.gartner.com/en/chief-information-officer/insights/digital-performance-kpis.

Gartner. (2024b). Supply chain management: How it works & why it's important. Retrieved April 25, 2025, from https://www.gartner.com/en/supply-chain/topics/supply-chain-management.

Gezgin, E., Huang, X., Samal, P., & Silva, I. (2017). *Digital transformation: Raising supply-chain performance to new levels*. McKinsey. Retrieved April 25, 2025, from https://www.mckinsey.com/capabilities/operations/our-insights/digital-transformation-raising-supply-chain-performance-to-new-levels.

Ghadge, A., Er Kara, M., Moradlou, H., & Goswami, M. (2020). The impact of industry 4.0 implementation on supply chains. *Journal of Manufacturing Technology Management, 31*(4), 669–686.

Gligor, D. M., & Holcomb, M. C. (2012). Antecedents and consequences of supply chain agility: Establishing the link to firm performance. *Journal of Business Logistics, 33*(4), 295–308.

Goldsby, T. J., Griffis, S. E., & Roath, A. S. (2006). Modeling lean, agile, and leagile supply chain strategies. *Journal of Business Logistics, 27*(1), 57–80.

Harapko, S. (2023). *How Covid-19 impacted supply chains and what comes next*. EY. Retrieved April 25, 2025, from https://www.ey.com/en_pl/insights/supply-chain/how-covid-19-impacted-supply-chains-and-what-comes-next.

Harrington, L. (2022). *Digitalization business brief—Warehousing 4.0: The age of the smart DC, DHL supply chain*. DHL. Retrieved April 25, 2025, from https://www.dhl.com/content/dam/dhl/global/dhl-supply-chain/documents/pdf/SCI_Warehousing-4-Brief.pdf.

Herold, S., Heller, J., Rozemeijer, F., & Mahr, D. (2025). Brave new procurement deals: An experimental study of how generative artificial intelligence reshapes buyer–supplier negotiations. *Journal of Purchasing and Supply Management, 31*(4), Article 101012.

Hofmann, E., & Rüsch, M. (2017). Industry 4.0 and the current status as well as future prospects on logistics. *Computers in Industry, 89*, 23–34.

Hosseini, S., & Barker, K. (2016). A Bayesian network model for resilience-based supplier selection. *International Journal of Production Economics, 180*, 68–87.

Hosseini, S., Ivanov, D., & Dolgui, A. (2019). Review of quantitative methods for supply chain resilience analysis. *Transportation Research Part E: Logistics and Transportation Review, 125*, 285–307.

IBM. (2024). *What is a supply chain control tower?* IBM. Retrieved April 25, 2025, from https://www.ibm.com/topics/control-towers

Inbound Logistics. (2024). Logistics management: Definition, functions, and benefits. Retrieved April 25, 2025, from https://www.inboundlogistics.com/articles/logistics-management/.

Ivanov, D. (2021). Digital supply chain management and technology to enhance resilience by building and using end-to-end visibility during the COVID-19 pandemic. *IEEE Transactions on Engineering Management, 71*, 10485–10495.

Ivanov, D. (2022). Viable supply chain model: Integrating agility, resilience and Sustainability Perspectives—lessons from and thinking beyond the COVID-19 pandemic. *Annals of Operations Research, 319*(1), 1411–1431.

Ivanov, D., & Dolgui, A. (2021). A digital supply chain twin for managing the disruption risks and resilience in the era of Industry 4.0. *Production Planning & Control, 32*(9), 775–788.

Ivanov, D., Dolgui, A., & Sokolov, B. (2022). Cloud supply chain: Integrating Industry 4.0 and digital platforms in the "supply chain-as-a-service". *Transportation Research Part E: Logistics and Transportation Review, 160*, 102676.

Kagermann, H., Wahlster, W., & Helbig, J. (Eds.). (2013). *Recommendations for implementing the strategic initiative Industrie 4.0: Final report of the Industrie 4.0 Working Group*. Forschungsunion.

Kalaiarasan, R., Olhager, J., Agrawal, T. K., & Wiktorsson, M. (2022). The ABCDE of supply chain visibility: A systematic literature review and framework. *International Journal of Production Economics, 248*, 108464.

Kane, G. C., Palmer, D., Phillips, A. N., Kiron, D., & Buckley, N. (2015). Strategy, not technology, drives digital transformation. *MIT Sloan Management Review and Deloitte University Press, 14*(1), 1–25.

Khojasteh, Y. & Son, J.D. (2016) A travel time model for order picking systems in automated warehouses, *International Journal of Advanced Manufacturing Technology, 86*(5), 2219–2229.

Khojasteh, Y., Mousavi Jahan Abadi, S.M., & Ng, C. (2026). Enabling supply chain transformation with cutting-edge digital tools. In Y. Khojasteh & S.M. Mousavi Jahan Abadi (Eds.), *Emerging technologies in supply chains: Towards digital transformation*. Springer. https://doi.org/10.1007/978-3-032-01218-0_1

Kodukula, R. (2025). *What is digital transformation and its importance?* Antino. Retrieved April 25, 2025, from https://www.antino.com/blog/digital-transformation.

Kshetri, N. (2018). 1 Blockchain's roles in meeting key supply chain management objectives. *International Journal of Information Management, 39*, 80–89.

Kuntze, C., Lal, S., & Seibert, K. (2020). *Launching the journey to autonomous supply chain planning*. McKinsey. Retrieved April 25, 2025, from https://www.mckinsey.com/capabilities/operations/our-insights/launching-the-journey-to-autonomous-supply-chain-planning.

Lee, H. L., Padmanabhan, V., & Whang, S. (1997). Information distortion in a supply chain: The Bullwhip effect. *Management Science, 43*(4), 546–558.

Lysons, K., & Farrington, B. (2020). *Procurement and supply chain management* (10th ed.). Pearson.

Mason, S. J., Cole, M. H., Ulrey, B. T., & Yan, L. (2002). Improving electronics manufacturing supply chain agility through outsourcing. *International Journal of Physical Distribution & Logistics Management, 32*(7), 610–620.

Nadkarni, S., & Prügl, R. (2021). Digital transformation: a review, synthesis, and opportunities for future research. *Management Review Quarterly, 71*(2), 233–341.

NIST. (2018). *Framework for improving critical infrastructure cybersecurity, version 1.1*. National Institute of Standards and Technology.

Pellicelli, M. (2022). *The digital transformation of supply chain management*. Elsevier.

Perano, M., Cammarano, A., Varriale, V., Del Regno, C., Michelino, F., & Caputo, M. (2023). Embracing supply chain digitalization and unphysicalization to enhance supply chain performance: A conceptual framework. *International Journal of Physical Distribution & Logistics Management, 53*(5/6), 628–659.

Pettit, T. J., Fiksel, J., & Croxton, K. L. (2010). Ensuring supply chain resilience: development of a conceptual framework. *Journal of Business Logistics, 31*(1), 1–21.

Ponomarov, S. Y., & Holcomb, M. C. (2009). Understanding the concept of supply chain resilience. *International Journal of Logistics Management, 20*(1), 124–143.

Redman, T. C. (2008). *Data driven: Profiting from your most important business asset*. Harvard Business Press.

Ross, J. W., Beath, C. M., & Mocker, M. (2019). *Designed for digital: How to architect your business for sustained success*. MIT Press.

Saberi, S., Kouhizadeh, M., Sarkis, J., & Shen, L. (2019). Blockchain technology and its relationships to sustainable supply chain management. *International Journal of Production Research, 57*(7), 2117–2135.

Schoenherr, T., & Speier-Pero, C. (2015). Data science, predictive analytics, and big data in supply chain management: Current state and future potential. *Journal of Business Logistics, 36*(1), 120–132.

Shamsuzzoha, A., & Pelkonen, S. (2025). A robotic process automation model for order-handling optimization in supply chain management. *Supply Chain Analytics, 9*, 100102.

Sheffi, Y., & Rice, J. B., Jr. (2005). A supply chain view of the resilient enterprise. *MIT Sloan Management Review, 47*(1), 41.

Swafford, P. M., Ghosh, S., & Murthy, N. (2006). The antecedents of supply chain agility of a firm: Scale Development and model testing. *Journal of Operations Management, 24*(2), 170–188.

Tijan, E., Aksentijević, S., Ivanić, K., & Jardas, M. (2019). Blockchain technology implementation in logistics. *Sustainability, 11*(4), 1185.

Wang, G., Gunasekaran, A., Ngai, E. W., & Papadopoulos, T. (2016). Big data analytics in logistics and supply chain management: Certain investigations for research and applications. *International Journal of Production Economics, 176*, 98–110.

Westerman, G., Bonnet, D., & McAfee, A. (2014). *Leading digital: Turning technology into business transformation*. Harvard Business Press.

Winkelhaus, S., & Grosse, E. H. (2022). Smart warehouses—A sociotechnical perspective. In B. L. MacCarthy & D. Ivanov (Eds.), *The digital supply chain* (pp. 47–60). Elsevier.

Yeniyurt, S., Wu, F., Kim, D., & Cavusgil, S. T. (2019). Information Technology Resources, innovativeness, and supply chain capabilities as drivers of business performance: A retrospective and future research directions. *Industrial Marketing Management, 79*, 46–52.

Zhao, N., Hong, J., & Lau, K. H. (2023). Impact of supply chain digitalization on Supply Chain Resilience and performance: A multi-mediation model. *International Journal of Production Economics, 259*, 108817.

Chapter 2
An Invitation to Fail: Digital Transformation Without a Business Objective

Mucahit Ozden

Abstract This chapter addresses the critical importance of aligning digital transformation efforts with strategic business goals. It highlights the pitfalls of technology-driven initiatives that neglect the broader organizational vision, often leading to wasted resources, inefficiencies, and failed outcomes. The chapter emphasizes that digital transformation should serve to achieve business objectives rather than as an end-goal. Key technologies such as Artificial Intelligence (AI), blockchain, Internet of Things (IoT), and cloud computing are discussed, with an emphasis on their strategic application rather than mere adoption. By leveraging tools such as balanced scorecards (BSC), objectives and key results (OKRs), and digital maturity models, organizations can ensure that technology investments align with their long-term goals and create measurable business value. The role of leadership, change management, and continuous evaluation is explored, reinforcing the idea that a successful digital transformation requires thorough planning, cross-functional collaboration, and a focus on sustainable value creation. It concludes by addressing future trends, such as AI-driven decision-making and sustainability, urging organizations to remain agile and adaptable in their digital transformation strategies.

Keywords Digital transformation · Technology adoption · Strategic alignment · Business objectives

M. Ozden (✉)
Warwick Manufacturing Group, University of Warwick, Coventry, UK
e-mail: m.ozden.1@warwick.ac.uk

Y. Khojasteh, S.M. M. Jahan Abadi (eds.), *Digital Transformation of Supply Chains*, International Series in Operations Research & Management Science 372, https://doi.org/10.1007/978-3-032-01206-7_2

2.1 Introduction

2.1.1 Contextualizing Digital Transformation

In today's rapidly evolving business landscape, digital transformation has become a critical strategic priority for organisations seeking to maintain a competitive edge and drive sustainable growth (AlNuaimi et al., 2022; Ghosh et al., 2022). This multifaceted process involves the integration of digital technologies across various business functions, fundamentally reshaping operational processes and enhancing value creation.

Key technologies driving digital transformation include artificial intelligence (AI), machine learning, blockchain, the Internet of Things (IoT), and cloud computing. Each of these innovations presents unique opportunities to foster innovation, enhance operational efficiency, and enable the development of new business models. For instance, AI and machine learning have revolutionised data analysis and decision-making capabilities. These technologies facilitate predictive analytics, enable personalised customer experiences, and automate routine tasks, ultimately improving organisational agility and strategic decision-making (Ghobakhloo et al., 2023; Spanaki et al., 2025). Similarly, blockchain technology enhances security and transparency in transactions, making it particularly valuable in industries such as supply chain management and finance (Stroumpoulis & Kopanaki, 2022). IoT enables real-time data collection from interconnected devices, allowing for enhanced monitoring and control of business processes (Stroumpoulis & Kopanaki, 2022). Meanwhile, cloud computing provides scalable resources and storage solutions, improving organisational flexibility and cost efficiency.

Despite the considerable benefits, the rapid pace of technological advancements presents significant challenges. Many organisations face intense pressure to adopt new technologies swiftly to remain competitive. This urgency often results in hasty decision-making, where technology adoption is driven by its novelty rather than alignment with strategic objectives (Rossini et al., 2021; Scriven et al., 2024). As a result, many digital transformation initiatives fail to achieve their intended outcomes, leading to wasted resources and missed opportunities for meaningful business impact (Appio et al., 2021; Ghosh et al., 2022).

The rapid evolution of technology increases the risk of adopting solutions that may not be fully compatible with existing systems and organisational processes (Karttunen et al., 2023). A lack of comprehensive evaluation regarding the integration and long-term sustainability of new technologies can lead to operational disruptions and inefficiencies. Therefore, while digital transformation offers substantial opportunities, organisations must adopt a strategic, well-planned approach to ensure technology investments align with long-term business objectives and deliver measurable value.

2.1.2 *The Pitfalls of Technology-Driven Transformation*

Digital transformation is widely recognised as a complex process with a high failure rate (Davenport & Westerman, 2018). Estimates suggest that a significant percentage of digital transformation projects—ranging from 66 to 84%—fail to achieve their intended goals (Correani et al., 2020; Ghosh et al., 2022).These failures can result in financial losses, inefficient resource use, missed opportunities, inadequate services, and reputational damage. The complexity arises from the need to concurrently revise business and corporate strategies, redesign organisational structures, implement new digital technologies, and develop new capabilities. Despite best intentions, many organisations encounter significant challenges, leading to failed projects. These failures provide valuable insights into the pitfalls that can undermine digital transformation efforts and offer lessons on how to avoid similar issues. This section explores the common pitfalls that contribute to project success or failure.

2.1.2.1 Lack of Strategic Alignment

One of the most frequent causes of failure in digital transformation projects is misalignment with the organisation's strategic objectives. When digital initiatives are pursued in isolation or driven by technological trends without considering broader business goals, they often fail to deliver meaningful results. This misalignment leads to technologies that do not address actual business needs, resulting in wasted resources and missed opportunities. Ensuring digital transformation initiatives are closely aligned with strategic goals is therefore critical (Rossini et al., 2021). As Correani et al. (2020) highlight, companies may fail to extract value from digital transformation due to a disconnect between strategy formulation and implementation. Strategy, rather than technology alone, forms the core of digital transformation (Kane et al., 2015). A holistic approach, involving the entire organisation in digital strategy development and implementation, is essential. Without a clear vision and a detailed digital strategy that integrates corporate and business strategies, digital transformation efforts are more likely to fail. Senior management plays a crucial role in ensuring digital strategy aligns with the company's purpose and mission.

2.1.2.2 Insufficient Stakeholder Engagement

Engaging key stakeholders across departments ensures a comprehensive understanding of the broader business context and enables the integration of technology solutions that support overall objectives. Collaboration among stakeholders helps reduce cultural and knowledge barriers (Kane et al., 2015) and fosters employee commitment to change. Failure to involve all relevant stakeholders often leads to unsuccessful digital transformation projects. When decisions are made without considering the perspectives and requirements of various departments, solutions may not meet the needs of end-users or business functions. This lack of engagement can result in resistance to change, poor user adoption, and operational disruptions.

2.1.2.3 Absence of Clear Objectives and Measurable Outcomes

The lack of well-defined objectives and measurable outcomes is another common issue in failed digital transformation projects. Defining key performance indicators (KPIs) and establishing metrics to track progress is essential for evaluating success. Without clear goals, assessing project success or making informed adjustments becomes challenging. This can result in technology implementations that fail to address specific business challenges or provide tangible benefits. Moreover, without visible objectives, employees may experience insecurity regarding transformation efforts (Preindl et al., 2020).

2.1.2.4 Overemphasis on Technology

Organisations that prioritise the latest technology trends without considering their practical application risk suboptimal results (Ghobakhloo & Iranmanesh, 2021). Successful digital transformation requires strategic thinking and innovation across various business aspects rather than merely embedding new technology (Wu et al., 2023). Digital transformation projects often fail when the primary focus is on adopting new technologies rather than solving business problems. Technology should be seen as an enabler rather than the ultimate goal.

2.1.2.5 Inadequate Planning and Testing

A structured digital transformation roadmap, prioritising concrete steps over multiple years, enhances project success. Many failed digital transformation projects suffer from insufficient planning and testing. Rushing to implement new technologies without thorough testing or a well-developed implementation plan often results in system failures, integration issues, and user dissatisfaction. Proper planning and testing are essential to identify potential challenges and ensure a smooth transition. Organisations should conduct pilot tests, risk assessments, and feasibility studies to address potential issues before full-scale implementation (Preindl et al., 2020).

2.1.2.6 Neglecting Change Management

Effective change management facilitates smoother transitions and maximises project benefits. Projects that do not address the human side of change, including training, communication, and support, often encounter resistance and low adoption

Table 2.1 The issues and recommendations

Issue	Recommendation
Lack of connection to business goals	Align with strategic objectives & engage stakeholders
Poor adoption due to lack of input	Include all departments & end-users in planning
No measurable goals lead to unclear results	Set clear KPIs & track progress
Focus on technology trends over business needs	Adopt technology that solves real business problems
Rushed implementation leads to system failures	Conduct risk assessments, pilot tests & feasibility studies
Resistance due to lack of training & support	Implement training, communication & continuous support

rates. Organisations should invest in digital reskilling and upskilling initiatives to prepare employees for transformation. A digital culture—characterised by a digital mindset, innovation, data-driven decision-making, and openness to change—is fundamental to successful digital (Dubey et al., 2024)

Failed digital transformation projects offer crucial lessons for organisations aiming to navigate their transformation journeys more successfully. By focusing on strategic alignment, stakeholder engagement, clear objectives, practical technology applications, thorough planning and testing, and effective change management, organisations can mitigate failure risks and enhance the likelihood of achieving their digital transformation goals. Embracing these lessons and applying them to future projects can lead to more successful outcomes and greater value from digital transformation initiatives (Table 2.1).

2.1.3 The Lean Principle: Adopting Technology That Serves Your People and Process

Lean management principles, originally developed in the manufacturing sector, particularly within Toyota's production system, have been widely adopted across various industries to enhance efficiency, reduce waste, and optimize value creation. One of the core principles of lean management is the emphasis on using only reliable, thoroughly tested technology that serves the people and processes within an organization. The principle ensures that digital transformation initiatives remain purposeful and value driven. Rather than introducing complexity through untested solutions, this approach emphasises enhancing existing operations and empowering employees. It aligns with the broader understanding that digital transformation extends beyond the adoption of emerging technologies and must be strategically leveraged to achieve well-defined business objectives (Ghobakhloo & Iranmanesh, 2021). Large-scale digital transformation efforts demand rigorous assessment to mitigate risks associated with system failures, security vulnerabilities, and integration challenges. By prioritising technologies that align with organisational needs, firms can circumvent the pitfalls and ensure seamless implementation.

Furthermore, strategic objectives must take precedence over technological novelty. Research underscores that digital transformation should be guided by business imperatives rather than the indiscriminate adoption of new tools (Kane et al., 2015). Lean thinking reinforces this by advocating for technology that directly supports people and processes, addressing genuine operational challenges rather than serving as an end in itself (Rossini et al., 2023).

The effective integration of digital solutions necessitates careful change management. Poorly implemented technologies can disrupt workflows, create resistance among employees, and undermine transformation efforts (Scriven et al., 2024). Lean approach ensures that digital initiatives facilitate operational efficiency rather than introducing unnecessary complications. A rigorous evaluation of digital tools prior to deployment not only minimises these risks but also ensures that technological investments contribute meaningfully to organisational performance (Citybabu & Yamini, 2024). A well-integrated technology foundation enables organisations to scale their digital capabilities in a structured manner, supporting future innovation and adaptation. The lean principles provide a methodological framework for sustaining digital transformation, ensuring that technological investments remain aligned with evolving business strategies.

Research indicates that organisations committed to lean principles often exhibit a "sustaining digital transformation pattern," characterised by incremental implementation, workforce engagement, and continuous process refinement (Rossini et al., 2021). This approach allows firms to leverage technology for data-driven decision-making, performance monitoring, and waste elimination. By aligning digital initiatives with the principle, organisations can enhance productivity, improve quality, and create greater customer value. Integrating well-designed digital solutions optimises workflows, reduces inefficiencies, and enhances decision-making, thereby improving overall operational effectiveness (Wu et al., 2023). Moreover, the adoption of advanced quality control mechanisms, facilitated by digital tools, ensures higher product and service standards. The resulting improvements in process efficiency translate into superior customer experiences and increased competitive advantage.

A key advantage of this approach is its reinforcement of a culture of continuous improvement. When technology is viewed to optimise operations rather than as an end-goal, organisations cultivate a mindset of ongoing innovation and refinement. Since lean methodologies inherently prioritise iterative enhancements, their application within digital transformation strengthens this orientation. Moreover, aligning digital investments with strategic business goals ensures measurable returns. Purposeful technological integration enhances efficiency and cost-effectiveness while maximising the impact of digital transformation initiatives. Research highlights that organisations adopting structured digital strategies often experience financial gains and increased market competitiveness (Tavana et al., 2022).

Organisations that prioritise business objectives over technological trends are better positioned to avoid inefficiencies and failed initiatives. Rather than adapting innovative solutions, firms should employ digital solutions that address concrete operational needs and enhance organisational resilience. The observed "sustaining digital transformation pattern" in Lean-driven organisations demonstrates the efficacy of gradual, strategically aligned technological adoption. By embedding lean into digital strategies, firms can enhance adaptability, drive continuous improvement, and secure long-term operational excellence.

2.2 The Importance of Strategic Alignment in Digital Transformation

2.2.1 Understanding the Strategic Alignment

Strategic alignment pertains to the process of synchronizing an organization's activities, including its digital transformation initiatives, with its overarching strategic objectives, mission, and vision. It is not merely beneficial but essential for the success of digital transformation initiatives. The absence of alignment is widely recognised as a key factor behind the high failure rates of digital transformation projects. A well-defined digitalisation strategy provides a comprehensive framework that integrates supply chain objectives with the broader organisational vision. Without clear strategic direction, digital initiatives risk becoming fragmented, failing to align with overarching business goals, and ultimately failing to deliver substantial value (Aamer et al., 2023). It is therefore crucial that digital strategies are integrated holistically with both supply chain operations and broader corporate objectives.

At its core, digital transformation is about reshaping the organisation's value stream (Qiao et al., 2024). Strategic alignment ensures that all digital initiatives are deliberately structured to create, deliver, and capture value in ways that support the organisation's strategic priorities. To achieve this, companies must critically assess their existing business models through a digital lens, identify new avenues for value creation, explore alternative approaches to value appropriation, and adjust their strategic direction accordingly (Correani et al., 2020). A clearly articulated, strategically aligned digital strategy serves as a roadmap for successful implementation, ensuring consistency between digital transformation initiatives and the firm's long-term objectives. The scope of digital transformation must be explicitly defined in alignment with broader strategic goals, ensuring coherence between strategy formulation and execution.

Ultimately, strategic alignment in digital transformation is a key driver of both business performance and competitive advantage (Chawla & Goyal, 2022). When digital initiatives are effectively aligned with corporate strategy, they contribute to essential business outcomes such as enhanced customer experience, streamlined operations, and the development of innovative business models (Elia et al., 2024). Achieving alignment often requires strong strategic leadership, and in some cases, the creation of roles such as the Chief Digital Officer (CDO) to ensure that digital strategies are fully integrated with the organisation's broader mission and visibility at C-level. Additionally, the adaptability of organisational structures plays a crucial role in aligning digital transformation efforts with emerging strategic opportunities.

Conversely, the failure to align digital initiatives with corporate strategy exposes organisations to significant risks, including inefficiencies, project delays, and financial losses due to misdirected technology investments. Misalignment can also result in missed opportunities to leverage digital transformation for competitive advantage, ultimately hindering the organisation's ability to achieve its strategic goals and deliver the intended outcomes (Gartner, 2023). The disconnect between strategy formulation and implementation remains a primary reason why digital transformation efforts frequently fall short of expectations. In digital transformation, there is the tendency to adopt a technology-driven approach, where the acquisition of new digital solutions is prioritised without sufficient consideration of their alignment with existing processes and strategic imperatives (Scriven et al., 2024). This misplaced emphasis often leads to only superficial improvements in efficiency or the creation of digital products and services that fail to generate meaningful value. Digital transformation should be fundamentally strategy-led, not technology-centric, ensuring that technological investments contribute to long-term organisational objectives rather than merely serving as ends in themselves (Chawla & Goyal, 2022).

In conclusion, strategic alignment is the foundation of effective digital transformation, ensuring that digital initiatives become integral components of a broader organisational strategy rather than isolated technological upgrades (Qiao et al., 2024). By embedding digital strategies within a clearly defined business framework, organisations can drive value creation, improve operational performance, and achieve long-term strategic objectives. The development and consistent communication of a well-structured digital strategy are therefore imperative for organisations seeking to maximise the benefits of digital transformation and secure sustained competitive advantage.

2.2.2 Misalignment in Practice

Investing in technology without ensuring its alignment with organisational strategy and operational processes often leads to the failure of anticipated benefits. Consider a case of a multinational corporation that allocated significant resources to the implementation of a new enterprise resource planning (ERP) system. While the ERP system was technologically advanced and promised to enhance operational efficiency across the company, its deployment was primarily managed by the IT department with minimal input from other departments. The minimal involvement from other departments resulted in the implementation of a system that was incompatible with existing workflows. This misalignment caused operational disruptions, inefficiencies, and significant costs, while failing to deliver the expected improvements. This outcome reinforces the argument that a lack of strategic direction often leads to digital initiatives that are fragmented and disconnected from the broader business processes. The emphasis on technological advancement within the ERP system, without adequately considering its integration into core business functions and strategic objectives, ultimately undermined its potential value.

In contrast, an alternative scenario highlights the advantages of strategic alignment. Consider an organization that undertook a digital transformation initiative with a strong focus on aligning its efforts with its strategic objectives. This organization actively engaged stakeholders from various departments to ensure that the chosen digital solutions were well-integrated into the existing processes and aligned with the strategic goals. The active involvement of stakeholders from multiple departments ensured that the chosen digital solutions were effectively integrated into existing processes and aligned with overarching strategic objectives. As a result, the organisation achieved significant improvements in both operational efficiency and customer satisfaction. This outcome directly supports the argument that synchronising digital initiatives with strategic goals is essential for value creation and enhanced business performance. Ensuring that technology investments reinforce, and advance organisational strategy allows firms to fully realise the potential of digital transformation. When digital initiatives are strategically aligned, they are more likely to generate meaningful improvements in operational processes and customer experience.

In summary, these contrasting considerations underscore the fundamental principle that digital transformation must be strategy-driven rather than dictated solely by technological advancements. A well-defined digital strategy, consistently embedded within the organisation's broader objectives, is critical for ensuring the success of digital initiatives. Without alignment, organisations risk investing in technologies that fail to address core business needs or contribute to long-term goals, leading to inefficiencies, wasted resources, and unrealised opportunities.

2.3 Developing a Strategic Approach for Digital Transformation

To effectively manage the complexities of digital transformation, organisations must adopt a structured approach. This requires the creation of a comprehensive digital strategy, which serves as a guiding framework for supply chain goals and broader organisational vision (Aamer et al., 2023). Given the rapid pace of technological change, the transformation process must be cyclical in nature, ensuring continuous improvement and alignment with business objectives. Westerman et al. (2014) highlight that well-managed firms are constantly discovering new ways to operate in the digital age. This cyclical process ensures ongoing adjustment to emerging technologies and evolving business goals, enabling organisations to achieve sustained operational efficiency, improved customer engagement, and new

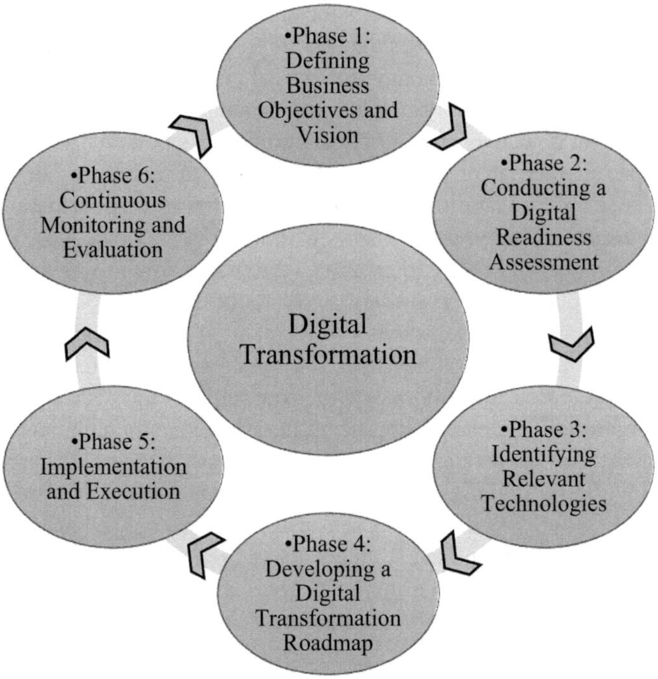

Fig. 2.1 Strategic approach for digital transformation

revenue streams (Chawla & Goyal, 2022). This section outlines an approach designed to guide organisations through the digital transformation process, structured around key phases that represent critical steps in the journey (Fig. 2.1).

Phase 1: Defining Business Objectives and Vision

The first phase of the digital transformation approach is the definition of clear business objectives and vision. Establishing the organisation's desired outcomes is critical before embarking on any digital initiative (Elia et al., 2024). This process requires a deep understanding of the organisation's mission, vision, and long-term goals (Lindquist, 2023). The strategic purpose driving digital transformation should be aligned with broader business strategies. For instance, if customer satisfaction is a core organisational goal, digital initiatives should focus on enhancing customer interactions using technologies like AI, chatbots, or personalised marketing strategies (Qiao et al., 2024). For manufacturers, the goals focused on improving visibility, enhancing efficiency, reducing lead times, and fostering collaboration, all of which can contribute to customer satisfaction (Lindquist, 2023). By articulating these objectives, organisations can ensure that digital strategies are not only focused on technology adoption but are also integrated into the core business strategy (Elia et al., 2024).

Phase 2: Conducting a Digital Readiness Assessment
Following the clear business objectives, organisations must assess their current readiness for digital transformation. This phase evaluates the organisation's current digital capabilities, including its technology infrastructure, organisational culture, and workforce skills (Aamer et al., 2023). Conducting a digital readiness assessment is essential for identifying gaps in these areas, such as outdated technology or insufficient digital skills among employees. By understanding their current capabilities, organisations can address these gaps before moving forward with digital initiatives, ensuring that the transformation is successful. For example, if the organisation lacks the necessary digital competencies in its workforce, targeted training and skill development initiatives must be implemented. This readiness assessment helps to align the organisation's existing capabilities with the demands of digital transformation.

Phase 3: Identifying Relevant Technologies
Once business objectives and readiness have been assessed, organisations must identify the technologies that are most relevant to achieving their goals. This phase involves evaluating emerging technologies based on their potential impact, feasibility, and alignment with the organisation's strategic objectives. It is essential to avoid adopting technologies simply for the sake of novelty; instead, the focus should be on selecting those that directly support business goals (Gartner, 2023). For instance, companies looking to enhance operational efficiency may prioritise automation technologies, while those aiming to improve customer engagement might focus on data analytics and personalised marketing tools. By selecting technologies that align with specific business goals, organisations can ensure that their digital initiatives have a meaningful impact to boost and sustain competitive advantage.

Phase 4: Developing a Digital Transformation Roadmap
Following the selection of appropriate technologies, organisations need to create a detailed digital transformation roadmap. This roadmap serves as a strategic plan, outlining specific initiatives, timelines, milestones, and resource allocation. A clear roadmap ensures that all digital initiatives are aligned with the broader business strategy and enables the organisation to track progress towards its goals. The roadmap should also include a multi-year plan, as digital transformation is typically a long-term, iterative process which should ensure that digitalisation strategies are aligned with supply chain goals and broader organisational objectives (Preindl et al., 2020). By incorporating clear deliverables and success metrics, organisations can track the effectiveness of their digital initiatives and ensure continuous alignment with strategic objectives.

Phase 5: Implementation and Execution
During the implementation phase, the digital transformation plan is put into action. This phase involves deploying selected technologies and implementing necessary changes to organisational processes, systems, and structures (Karttunen et al., 2023). Successful execution requires strong leadership, clear communication, and effective project management to ensure that initiatives are completed on time and

within budget. Leadership plays a critical role in guiding the organisation through this transformation, ensuring alignment among stakeholders and securing the necessary buy-in. Furthermore, organisations must remain agile and adaptable to address any challenges that arise during the transformation process, ensuring that changes are smoothly integrated into the organisation's operations (AlNuaimi et al., 2022).

Phase 6: Continuous Monitoring and Evaluation
The final phase focuses on continuous monitoring and evaluation. Digital transformation is not a one-time event but a continuous process that requires regular assessment to ensure alignment with business objectives and to track the ongoing value delivered by digital initiatives. Organisations must establish Key Performance Indicators (KPIs) that reflect both short-term outcomes and long-term objectives. These KPIs should be tied to the strategic goals defined in earlier phases, ensuring that digital transformation remains aligned with the organisation's broader vision (Gartner, 2023). Tools such as the Balanced Scorecard (BSC) or Digital Maturity Models can help organisations evaluate the performance of digital initiatives across various dimensions. These tools will be explained in Sect. 2.4.

2.4 Tools for Aligning Digital Transformation with Business Objectives

In the realm of digital transformation, aligning technology initiatives with business objectives is crucial. Several strategic alignment tools can facilitate this process, each offering unique benefits for planning, executing, and monitoring digital initiatives. This section introduces five key tools that organizations can use to bridge the gap between technology and business strategy: the Balanced Scorecard (BSC), Strategy Map, Objectives and Key Results (OKRs), Digital Maturity Models, and Value Stream Mapping.

2.4.1 Balanced Scorecard (BSC)

The BSC is a widely used strategic planning and management system designed to measure an organisation's strategic performance. Developed by Kaplan and Norton (1996), the BSC provides a framework that translates an organisation's vision and strategy into actionable goals, moving beyond financial measures, see Fig. 2.2. It balances traditional financial metrics with essential non-financial perspectives, including financial performance, customer satisfaction, internal processes, and learning and growth (Azarov et al., 2024; Frederico et al., 2021).

By using the BSC, organisations can develop specific, measurable goals for their digital initiatives. The BSC offers a structured approach for linking digital transformation efforts with critical business outcomes by expressing strategic objectives

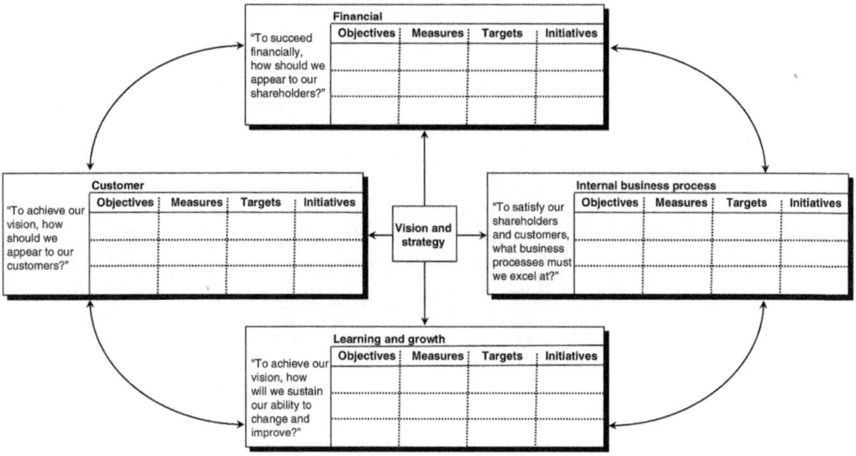

Fig. 2.2 Balanced scorecard (Kaplan, 2009)

across the four key perspectives. For example, if an organisation's strategic objectives include enhancing customer satisfaction and improving operational efficiency, the BSC translates these goals into actionable digital projects. Technology investments can then be aligned with key performance indicators such as customer response times, process automation rates, and innovation outputs. Decision-makers can leverage the BSC to identify critical drivers in digitalising the supply chain and allocate appropriate resources for the digitalisation journey.

2.4.2 Strategy Map

A Strategy Map is a visual tool designed to outline the relationships between an organisation's strategic objectives and its initiatives. It serves as a structured representation of how various digital projects contribute to achieving overarching business goals, ensuring that all digital efforts are aligned and coherent. The strategy map decomposes strategic goals, providing a clear view of how activities relate to objectives across different perspectives (Fig. 2.3).

By providing a clear visual link between digital initiatives and strategic goals, the Strategy Map enhances the coherence of digital transformation efforts. It also facilitates communication among stakeholders by visually illustrating how digital initiatives support the broader business strategy. For instance, a strategy map can help identify where digital technologies are needed most within an organisation, whether for improving customer engagement, enhancing operational efficiency, or developing innovative business models.

In the context of supply chain management, the strategy map can be used to ensure that digital technologies such as AI, big data analytics, and automation are

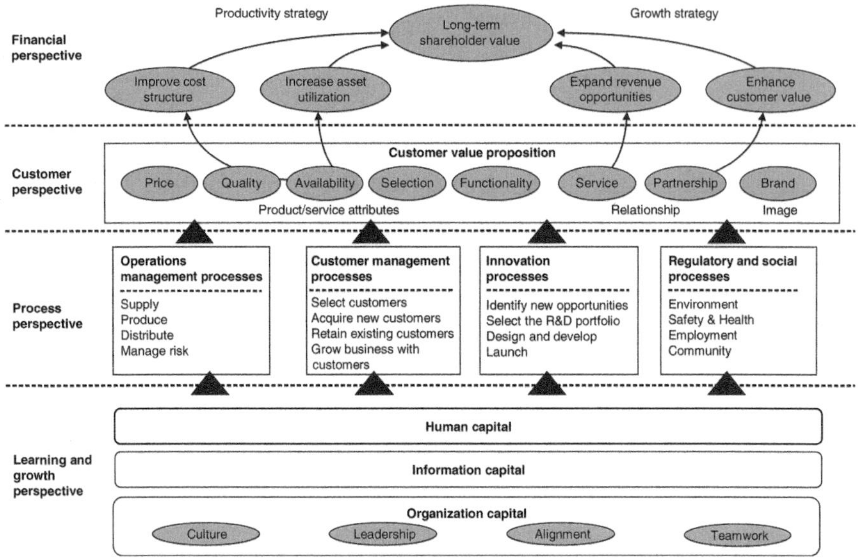

Fig. 2.3 Strategy map (Kaplan, 2009)

aligned with strategic goals like increasing operational agility or enhancing customer service. This alignment can provide clarity on the resource requirements and implementation timeline for each digital initiative.

2.4.3 Objectives and Key Results (OKRs)

OKRs is a goal-setting framework that helps organisations focus on measurable outcomes. This tool encourages transparency, alignment, and accountability by setting clear and measurable objectives that are then tracked through key results. The framework was originally developed by Andy Grove at Intel and popularised by organisations such as Google (Doerr, 2018). OKRs consist of two components: the objective, which is a qualitative description of what needs to be achieved, and the key results, which are specific, quantitative indicators that measure progress towards the objective. OKRs can be applied at various levels within an organisation, from company-wide objectives to department or team-level goals. By defining clear objectives and key results, OKRs align digital transformation initiatives with the broader business strategy, enabling organisations to assess progress and adjust their approach as needed.

For example, in a supply chain context, an organisation may set an objective to enhance the customer experience, with key results that include reducing order delivery times by 20% or increasing customer satisfaction ratings by 15%. Digital tools such as automation, cloud platforms, and AI can play a pivotal role in achieving these key results. By aligning these digital tools with OKRs, organisations can ensure that their digital transformation efforts directly support their strategic objectives.

2.4.4 Digital Maturity Models

Digital Maturity Models provide organizations with a structured approach to assessing their current digital capabilities across various dimensions. These models offer a comprehensive framework that enables organizations to evaluate their digital maturity, identify areas for improvement, and develop a roadmap for continuous digital transformation (Haryanti et al., 2024; Kalender & Žilka, 2024). By categorizing digital transformation into different stages—ranging from initial adoption to advanced, integrated digital capabilities—these models help organizations understand their progress in the digital landscape.

By utilizing a digital maturity model, organizations can benchmark their existing digital capabilities across dimensions such as data analytics, process automation, technology infrastructure, and digital strategy . Benchmarking helps identify strengths and weaknesses. For instance, an organization may excel in data analytics but lack digital strategy or automation, which are crucial for scaling operations and maintaining a competitive edge (Haryanti et al., 2024). Recognizing these gaps allows organizations to prioritize digital initiatives, such as investing in automation technologies or digital strategy development, ensuring alignment with strategic goals. Additionally, digital maturity models help organizations track their transformation efforts against established standards. The Warwick Manufacturing Group's (WMG) Industry 4.0 Readiness Assessment Tool exemplifies the value of such models in guiding organizations through digital transformation (Agca et al., 2017). This tool enables organizations to systematically assess their readiness across multiple dimensions and harness the potential of the cyber-physical age.

WMG's tool underscores the necessity of a structured approach to digital transformation, which is effectively supported by a digital maturity model. This structured methodology facilitates both the identification of an organization's current digital standing and the creation of a roadmap for progression through different maturity levels. The model comprehensively evaluates six core dimensions essential for successful digital transformation:

1. **Products and Services**: Assesses how digital technologies, such as IoT-enabled devices or digital platforms, enhance customer experience and product innovation.
2. **Manufacturing and Operations**: Evaluates the impact of digital technologies like automation, robotics, and data analytics on operational efficiency, product quality, and downtime reduction.

3. **Strategy and Organisation**: Examines how well an organization's structure and strategic objectives align with digital transformation, ensuring leadership and resources are effectively allocated to digital initiatives.
4. **Supply Chain**: Analyses the integration of digital tools and technologies in the supply chain to improve visibility, collaboration, and real-time decision-making.
5. **Business Model**: Assesses the adaptability of an organization's business model to digital opportunities, such as new revenue streams, digital customer engagement, and strategic partnerships.
6. **Legal Considerations**: Evaluates an organization's preparedness to address legal challenges related to digital transformation, including data privacy, intellectual property, and regulatory compliance.

The importance of a structured digitalization and innovation strategy has been emphasized in previous discussions, alongside the necessity of a multi-year technology investment plan. For example, in the Products and Services domain, smart and connected products with IoT, digital platforms for new value creation and the role of digital technologies in enhancing customer experience. Similarly, in Manufacturing and Operations, automation, robotics, and data analytics are highlighted as key Industry 4.0 technologies that drive efficiency and quality.

In Strategy and Organisation, successful digital transformation requires strategic alignment, top management support, and digital leadership. The Supply Chain aspect emphasizes digitalization for agility and transparency, collaboration technologies, and real-time data utilization for decision-making.

In the Business Model dimension, sources discuss digital business models, new revenue streams enabled by digital transformation, and the importance of strategic partnerships in the digital ecosystem. Finally, Legal Considerations encompass data privacy, cybersecurity, intellectual property concerns, and compliance with government regulations.

As organizations advance in digital maturity, they gain the ability to execute more complex digital strategies, integrate new technologies, and achieve greater operational efficiencies. Digital maturity models facilitate step-by-step progression, enabling organizations to handle increasingly sophisticated digital initiatives. Higher digital maturity enhances organizational agility and the ability to integrate emerging technologies effectively.

Ultimately, a digital maturity model acts as a roadmap, helping organizations continuously assess their progress, identify gaps, and prioritize initiatives in alignment with their strategic vision (Kirmizi & Kocaoglu, 2024). By providing a structured approach, these models ensure that digital transformation efforts remain sustainable, purposeful, and aligned with long-term business success. The step-by-step enhancement of digital capabilities fosters innovation in products, services, and business models. In summary, organizations that adopt digital maturity models can systematically assess, plan, and execute their digital transformation journey, ensuring sustainable growth and competitiveness in the digital era.

2.4.5 *Value Stream Mapping*

Value Stream Mapping (VSM) is a lean management tool used to analyse the flow of materials and information necessary for delivering a product or service and help organizations identify inefficiencies and areas for improvement by visually mapping out process steps. In the context of digital transformation, VSM can serve as a strategic tool for pinpointing areas where digital technologies can enhance efficiency and add value. As digital transformation leverages technology to drive significant performance improvements and operational optimization, integrating VSM into this process ensures that technology investments are effectively targeted (Westerman et al., 2014). For example, a company may use VSM to analyse its customer service processes, identifying bottlenecks and inefficiencies. Insights from this analysis can then guide the implementation of digital tools to streamline workflows and enhance overall efficiency. Aligning digital initiatives with the process improvements identified through VSM ensures that technological advancements address specific operational challenges and contribute to strategic objectives, such as improving customer satisfaction and operational efficiency. This integrated approach fosters a comprehensive and effective digital transformation strategy.

2.4.6 *The Digital Transformation Phases and the Tools*

Digital transformation is a multi-phase process that requires careful planning to ensure alignment with an organization's business objectives. To align technology initiatives with business goals, several tools can be applied at different stages of the transformation journey.

In the early stages, the BSC and Strategy Map help organizations translate strategic goals into actionable initiatives. The BSC aligns digital initiatives with KPIs across financial, customer, internal processes, and learning and growth perspectives, ensuring that technology investments support overall strategy. The Strategy Map visually connects strategic objectives to digital initiatives, helping align technology projects with long-term business goals. During the phases of identifying technologies and developing the transformation roadmap, OKRs become critical. OKRs provide measurable outcomes and define specific targets for technology adoption and project implementation, ensuring focus and accountability.

In the implementation phase, VSM and Digital Maturity Models play key roles. VSM helps identify inefficiencies and areas where digital technologies can streamline operations and enhance value delivery. By mapping value flows, VSM ensures that technology investments address critical operational challenges, contributing to efficiency improvements and waste reduction. Digital Maturity Models help assess an organization's current digital capabilities and guide the prioritization of technology adoption to align with both short-term needs and long-term strategic goals. By strategically using these tools, organizations can ensure that technology initiatives

Table 2.2 Tools and their purpose across digital transformation phases

Tool	Phase 1	Phase 2	Phase 3	Phase 4	Phase 5	Phase 6
Balanced Scorecard	Translates vision into goals	Identifies KPIs for initiatives	Aligns goals with digital projects	Checks alignment with objectives	Tracks progress against goals	Adjusts KPIs based on results
Strategy Map	Aligns goals with actions	Aligns capabilities with goals	Links tech with business objectives	Links initiatives to goals	Ensures tech alignment	Monitors impact on outcomes
OKRs	Sets measurable objectives	Identifies gaps and readiness	Turns objectives into measurable results	Defines results for each phase	Measures success with metrics	Tracks and adjusts based on outcomes
Digital Maturity Models	Assesses digital readiness	Evaluates capabilities and readiness	Identifies tech gaps	Prioritizes tech based on maturity	Tracks progress and improvements	Advances digital capabilities
Value Stream Mapping	Identifies areas to add value	Analyses processes and bottlenecks	Highlights areas for tech improvement	Integrates tech to enhance processes	Focuses on process improvement	Reviews improvements and optimizations

align with business goals while promoting continuous growth, operational improvements, and innovation. See Table 2.2 for the purpose of each tool across the phases of digital transformation.

2.5 Challenges and Solutions in Aligning with Business Objectives

2.5.1 Common Challenges

Aligning digital transformation initiatives with business objectives remains a complex task. Organizations often struggle to ensure that technology investments are both innovative and strategically aligned. Resistance to change represents one of the most significant challenges, as employees, managers, and executives may hesitate to adopt new technologies due to concerns about workflow disruptions or job security (Fabac, 2022). Resistance can manifest as passive non-compliance or active opposition, both of which can hinder transformation efforts (Faruque et al., 2024). Addressing this challenge requires transparent communication about the benefits of digital initiatives and active stakeholder engagement at all organizational levels. Leadership support is crucial in mitigating resistance and fostering a culture of digital adoption.

A lack of digital literacy among employees further complicates transformation efforts. Many organizations find that their workforce lacks the skills needed to fully utilize new technologies, resulting in inefficiencies and diminished potential gain (Faruque et al., 2024). To bridge this gap, organizations must prioritize continuous learning and development programs that equip employees with relevant digital competencies. Several studies emphasize the importance of reskilling and upskilling as essential components of successful digital transformation (Citybabu & Yamini, 2024).

Measuring the return on investment (ROI) for digital initiatives presents another major challenge. Unlike traditional investments, the outcomes of digital transformation are often long-term and intangible, encompassing aspects such as customer experience, brand reputation, and innovation (Westerman et al., 2014). This complexity makes it difficult to justify ongoing digital investments, especially when competing for finite organizational resources. The dynamic nature of digital technologies further complicates ROI measurement, as benefits may take years to materialize (Faruque et al., 2024). Factors such as customer loyalty and brand equity evolve over time, influenced by variables beyond the scope of immediate technological investments. To address this, organizations must adopt a broader evaluation framework that includes both financial and non-financial performance metrics. Tools such as the BSC offer a holistic approach to assessing ROI, integrating financial and strategic indicators (Frederico et al., 2021; Kaplan & Norton, 1996).

Integration of new digital technologies with legacy systems poses additional obstacles. Many legacy systems lack compatibility with modern digital solutions, necessitating extensive modifications or complete system overhauls (Karttunen et al., 2023). This process often results in increased costs, prolonged implementation timelines, and operational disruptions, which can undermine alignment with business objectives. Within digital supply chains, the absence of standardized technological interfaces further complicates system integration, creating inefficiencies in data exchange and workflow automation.

Addressing these challenges requires a comprehensive approach that considers both technological and strategic factors. A well-defined digital transformation roadmap, coupled with clear assessments of both tangible and intangible benefits, enables organizations to make informed decisions while managing risks and costs effectively.

2.5.2 Strategies for Overcoming Challenges

Organizations can adopt several strategic measures to enhance the success of digital transformation efforts. A commitment to continuous learning and strong leadership are critical enablers in overcoming the challenges.

Leadership plays a central role in addressing resistance to change. By establishing a clear vision for digital transformation and articulating its strategic importance, leaders can align organizational efforts toward shared business objectives.

Positioning digital transformation as a fundamental business priority fosters greater employee engagement and acceptance. Executive sponsorship is essential for securing necessary resources and providing direction throughout the transformation process. Early involvement of key stakeholders in planning and implementation phases further reduces resistance, as employees are more likely to support changes they have contributed to shaping.

Effective communication also plays a pivotal role in managing resistance to digital transformation. Leaders must clearly convey how digital initiatives will optimize processes, enhance customer value, and contribute to overall business success. Transparent discussions about potential challenges and benefits help foster trust and buy-in among employees and stakeholders.

To address the digital literacy gap, organizations must invest in comprehensive training and development programs. These initiatives should provide employees with the technical skills necessary to maximize the benefits of digital tools, ensuring a workforce that is prepared for evolving technological demands. Training content must be accessible and tailored to specific job roles to facilitate effective learning and adoption. By promoting a culture of continuous learning and upskilling, organizations empower employees to drive digital transformation efforts successfully.

Measuring the success of digital transformation requires organizations to adopt a multidimensional approach, incorporating financial and non-financial indicators. Beyond immediate financial returns, a comprehensive evaluation framework should assess long-term impacts such as customer satisfaction, operational efficiency, and competitive positioning. Broadening the scope of performance measurement enables organizations to recognize the full value of digital investments and make data-driven strategic decisions.

Technical integration challenges, organisations should adopt a phased approach to implementation, allowing for incremental integration and testing of new technologies. This incremental approach minimises the risk of widespread disruptions and allows for adjustments based on real-time feedback (Correani et al., 2020). Additionally, leveraging middleware and integration platforms can facilitate smoother connections between legacy systems and new digital tools (Mekjian, 2024; Stroumpoulis & Kopanaki, 2022). Table 2.3 provides a summary of the key challenges and strategies in digital transformation and highlights their relevance to people and process.

2.6 The Role of Leadership and Culture in Digital Transformation

Digital transformation is not merely a technological shift but a fundamental change in how organizations operate and achieve their strategic goals. As such, the role of leadership and the cultivation of a supportive organizational culture are critical components of successful digital transformation efforts.

Table 2.3 Key challenges, strategies, and their relevance to people and process in digital transformation

Challenge	Description	Strategy	Relevance to people	Relevance to process
Resistance to Change	Employees and leaders may resist new technologies, fearing disruptions or job losses	Clear communication of benefits, leadership involvement, and stakeholder engagement	Encourages buy-in and adoption by addressing fears and concerns	Ensures smoother transitions with minimal disruptions
Lack of Digital Literacy	Employees may lack the necessary skills to utilize new technologies effectively	Continuous learning and development programs, accessible training resources	Empowers employees to confidently use digital tools	Enhances operational efficiency and innovation
Difficulty in Measuring ROI	Digital transformation benefits can be intangible, making it hard to quantify success	Define clear KPIs aligned with strategic goals, including customer satisfaction and efficiency metrics	Helps employees and leaders understand the value of digital initiatives	Justifies ongoing investments and refines implementation approaches
Integration with Legacy Systems	New digital tools may be incompatible with existing infrastructure	Adopt a phased implementation approach, use middleware for smoother integration	Reduces disruptions for employees during system transitions	Ensures operational continuity and minimizes costs

2.6.1 Leadership's Role in Strategic Alignment

Effective leadership plays a pivotal role in ensuring that digital transformation initiatives align with an organization's strategic objectives. Leaders are responsible for setting a clear vision for digital transformation and embedding it within the broader organizational strategy. This responsibility extends beyond understanding technological advancements; it requires leveraging these technologies to drive meaningful business outcomes (Westerman et al., 2014). A central theme in the literature is the importance of a well-defined digital strategy, guided by strong leadership, to ensure successful digital transformation (AlNuaimi et al., 2022). Leaders must articulate this vision effectively to guarantee that digital initiatives are fully integrated into the organization's overarching strategic objectives rather than being pursued in isolation. This requires a nuanced understanding of both technology and business processes, enabling leaders to bridge gaps between technological capabilities and organizational needs.

In addition to setting strategic priorities, leaders play a critical role in selecting technologies that align with lean management principles—favouring reliable, well-tested solutions that enhance both human and operational efficiency. A carefully considered approach to technology adoption helps mitigate the risks of implementing unproven or incompatible solutions that may lead to operational inefficiencies or disruptions (Schumacher et al., 2021).

Furthermore, ongoing assessment through Digital Maturity Models, supported by leadership, enables organizations to measure progress and maintain alignment with strategic objectives. Addressing resistance to change—an essential component of change management—requires leaders to clearly communicate the transformation vision, engage employees in the process, and provide the necessary training and support to facilitate adoption. By fostering an environment that embraces change, supports innovation, and aligns digital initiatives with business goals, leadership can ensure that digital transformation becomes a sustained driver of organizational success.

2.6.2 Building a Culture of Alignment and Innovation

Beyond leadership, the success of digital transformation is heavily influenced by organizational culture. A culture that supports digital transformation embraces change, fosters innovation, and aligns with the organization's strategic objectives (Ghobakhloo & Iranmanesh, 2021). Establishing such a culture is essential for ensuring that digital initiatives contribute effectively to long-term business success. One fundamental aspect of this is fostering a digital mindset among employees. Organizations should actively promote continuous improvement and encourage employees to explore new ideas and technologies (Citybabu & Yamini, 2024). To support this shift, a structured framework should be established that prioritizes the use of reliable and thoroughly tested tools. This not only builds trust in emerging technologies but also instils confidence in their effective adoption and application.

A culture of continuous learning and adaptation is critical for navigating the evolving nature of digital transformation (Haryanti et al., 2024). Research highlights that ongoing development is integral to both lean management and digital transformation strategies (Rossini et al., 2023; Schumacher et al., 2021). Additionally, cross-functional collaboration plays a vital role in ensuring digital initiatives align with the needs of various departments and stakeholders. Open communication and knowledge sharing further reinforce a cohesive approach to digital transformation. Breaking down organizational silos is essential for optimizing the full potential of digital technologies (Lindquist, 2023). Encouraging teamwork across different functions ensures that digital transformation initiatives support business-wide objectives rather than remaining isolated within specific departments.

To reinforce this cultural shift, organizations should actively recognize and reward employees who contribute to innovation and strategic alignment (Kaplan & Norton, 2004). Embedding principles of strategic alignment and lean management into daily operations helps create an environment where digital transformation is seamlessly integrated into business processes. Moreover, cultivating a culture of innovation enables organizations to adapt to evolving digital trends, uncover new opportunities for value creation along with encouragement of encouraging experimentation, learning from failure, and embracing new technologies. By fostering an adaptive, collaborative, and innovation-driven culture, organizations can ensure that digital transformation efforts not only enhance operational efficiency but also drive sustainable growth and long-term success.

2.6.3 The Human Element in Change Management

A critical factor in the success of digital transformation is addressing the human element, as employees play a central role in adopting and sustaining new processes and technologies. Change management strategies must consider the challenges associated with altering established workflows, ensuring that employees are engaged and equipped to embrace transformation. Without employee acceptance and participation, even the most well-planned digital initiatives risk failure.

Digital transformation is not solely about implementing technology—it is equally about people and processes. Failure to prioritize the human aspect is a major reason why many transformation efforts fall short of expectations. To mitigate this risk, organizations should invest in comprehensive change management frameworks that incorporate clear communication, structured training programs, and continuous support mechanisms.

Effective communication is essential for setting expectations, articulating the benefits of digital initiatives, and addressing concerns employees may have about change. Transparent and proactive communication fosters trust and ensures that employees understand how transformation aligns with the organization's long-term goals. Training programs should be designed to equip employees with the necessary skills to navigate new digital tools and processes effectively, while ongoing support systems help sustain engagement and adaptation throughout the transition period.

Reskilling and upskilling are fundamental to preparing employees for digital transformation, particularly in areas such as supply chain digitalization and Industry 4.0. Organizations must invest in workforce development by providing targeted training, qualification programs, and user education initiatives to build digital competencies. Collaborative learning platforms, digital training factories, and emerging Industry 4.0 technologies can support these efforts, making digital education more accessible and effective.

Building a supportive and inclusive change environment is essential for minimizing resistance and ensuring successful adoption. Organizations should actively listen to employee feedback, address concerns, and involve employees in the decision-making process to foster a sense of ownership over digital transformation initiatives. Recognizing employee contributions and providing clear guidance on how changes will impact their roles further strengthens engagement and reduces opposition to transformation efforts.

Ultimately, a strong focus on the human element ensures that digital transformation efforts are not only technically sound but also socially and culturally sustainable. Organizations that cultivate a digital culture—one that encourages a digital mindset, innovation, and data-driven decision-making—are more likely to achieve successful and lasting transformation. Organizational culture plays a defining role in shaping digital transformation outcomes, reinforcing the need for strategies that integrate people, technology, and business objectives.

2.7 Future Trends and the Evolving Role of Strategic Alignment

2.7.1 Emerging Trends in Digital Transformation

A significant trend in digital transformation is the increasing integration of artificial intelligence (AI) in decision-making processes. AI-driven analytics and machine learning algorithms enable organizations to enhance decision-making by leveraging data-driven insights. These technologies contribute to strategic alignment by offering deeper insights into market trends, customer behaviour, and operational performance (Plekhanov et al., 2023; Tavana et al., 2022). As organizations increasingly rely on AI to guide business strategies, ensuring that digital initiatives align with corporate objectives becomes essential.

Another key trend in digital transformation is the growing emphasis on data governance and privacy. As organizations manage vast amounts of data across digital platforms, they must ensure compliance with regulatory requirements and implement robust data protection measures (Fabac, 2022). Effective data governance is integral to mitigating legal and reputational risks, protecting customer privacy, and maintaining organizational trust. Also, cybersecurity and data protection represent fundamental components of digital transformation readiness. Ensuring the authenticity and security of data is critical to fostering trust in digital systems.

Sustainability and corporate social responsibility (CSR) are increasingly central to digital transformation strategies (Plekhanov et al., 2023). Digital transformation contributes to sustainable development by optimizing resource utilization, improving energy efficiency, and reducing waste. As businesses increasingly recognize the importance of social responsibility, digital transformation strategies must consider their broader impact on stakeholders and society. Integrating sustainability into digital transformation initiatives strengthens an organization's commitment to social responsibility.

Another emerging trend in digital transformation is the adoption of edge computing, which facilitates real-time data processing and enhances operational responsiveness. Supply chain agility is critical consideration in digital transformation, as organizations seek to respond rapidly to changing market conditions (Dubey et al., 2024). By processing data closer to the source, edge computing reduces latency, improves decision-making efficiency, enhances transparency by enabling real-time visibility into supply chain operations, thus improving responsiveness and adaptability. The development of robust digital infrastructure that enables seamless data flow and integration is essential for achieving these objectives. Combining edge computing with cloud-based solutions can further improve efficiency and support the low-latency processing required for time-sensitive applications.

2.7.2 *The Ongoing Need for Strategic Alignment*

Despite the rapid pace of technological advancement, the necessity of strategic alignment remains a constant priority for organizations. As emerging technologies reshape industries, businesses must continuously ensure that their digital initiatives are aligned with overarching corporate objectives. Previous discussions and existing literature consistently emphasize the importance of a well-defined digitalization and innovation strategy as well as the integration with business strategy as fundamental prerequisites for successful digital transformation. Ensuring that technological advancements support broader business goals enables organizations to navigate digital change more effectively and leverage new opportunities for competitive advantage.

Given the dynamic nature of digital transformation, strategic alignment is not a one-time initiative but an ongoing process requiring continuous reassessment and adaptation. Organizations must maintain agility by regularly revisiting and refining their strategies to address new technological developments and industry trends. By fostering a strong connection between digital investments and business objectives, companies can maximize the long-term value of their transformation efforts.

As digital technologies become increasingly embedded within all aspects of business operations, the traditional boundaries between IT and other functional areas continue to dissolve (Plekhanov et al., 2023). This evolution necessitates a more holistic approach to strategic alignment, where digital transformation is regarded not as an isolated initiative but as an integral component of overall corporate strategy (Fabac, 2022). Furthermore, the growing complexity of digital ecosystems demands that organizations adopt increasingly sophisticated approaches to strategic alignment. These approaches include leveraging advanced analytics for data-driven decision-making, incorporating agile methodologies to enhance flexibility and responsiveness, and fostering a culture of continuous learning and adaptation. As technology evolves, so must the frameworks for aligning digital initiatives with business objectives, ensuring that digital transformation remains a key driver of organizational success.

2.8 Conclusions

Digital transformation presents organizations with significant opportunities to innovate, enhance operational efficiency, and maintain a competitive edge in an increasingly digital landscape. However, the full potential of these opportunities can only be realized when digital initiatives are strategically aligned with the organization's core business objectives. Without this alignment, digital investments risk becoming fragmented, leading to inefficient resource allocation, operational disruptions, and missed business opportunities.

This chapter has provided a structured framework to ensure that digital transformation efforts remain aligned with strategic business goals. A recurring theme is the necessity of clearly defined objectives, meticulous planning, and continuous evaluation. These components serve as a comprehensive roadmap to mitigate the risks associated with technology-driven approaches that prioritize technological advancements without considering their broader business implications. The integration of lean management—particularly the emphasis on leveraging proven, reliable technologies that enhance both people and processes—further underscores the critical role of strategic alignment. By positioning technology as an enabler rather than the primary driver of business objectives, organizations can achieve a more sustainable and effective transformation.

As the digital landscape continues to evolve, organizations must cultivate agility, regularly reassessing their technology investments to ensure alignment with shifting business needs. The success of digital transformation lies in maintaining a balance between technological innovation and a deep understanding of business priorities. A strategic approach that focuses on achieving specific business outcomes enables organizations to navigate the complexities of digital transformation effectively. Rather than viewing digital transformation as an isolated objective, it should be recognized as a catalyst for broader business improvements—such as enhancing supply chain efficiency, improving customer satisfaction, and driving overall organizational growth.

Lean management principles play a pivotal role in this process, ensuring that digital transformation initiatives are purposeful and generate measurable value. By adhering to these principles, organizations can avoid merely automating inefficiencies and instead leverage digital technologies to optimize processes and enhance employee engagement. This strategic alignment ensures that digital initiatives contribute not only to immediate operational improvements but also to long-term organizational success.

Furthermore, this chapter has highlighted the importance of agility in the digital transformation journey. Organizations must embrace continuous adaptation, proactively responding to technological advancements and evolving business demands. Strategic alignment is not a static goal but an ongoing process that requires regular review and refinement to maintain its effectiveness.

Ultimately, digital transformation should be regarded as a means to achieve key business objectives, such as enhancing supply chain responsiveness, improving customer experiences, and strengthening overall organizational performance. By maintaining a clear strategic focus and aligning technology investments with these goals, organizations can maximize the impact of their digital initiatives and secure a sustainable competitive advantage. Through careful planning, continuous adaptation, and an unwavering commitment to business-driven outcomes, digital transformation can serve as a powerful driver of long-term organizational success.

References

Aamer, A., Sahara, C. R., & Al-Awlaqi, M. A. (2023). Digitalization of the supply chain: transformation factors. *Journal of Science and Technology Policy Management, 14*(4), 713–733.

Agca, O., Gibson, J., Godsell, J., Ignatius, J., Davies, C. W., & Xu, O. (2017). An Industry 4 readiness assessment tool.

AlNuaimi, B. K., Singh, S. K., Ren, S., Budhwar, P., & Vorobyev, D. (2022). Mastering digital transformation: The nexus between leadership, agility, and digital strategy. *Journal of Business Research, 145*, 636–648.

Appio, F. P., Frattini, F., Petruzzelli, A. M., & Neirotti, P. (2021). Digital transformation and innovation management: A synthesis of existing research and an agenda for future studies. *Journal of Product Innovation Management, 38*(1), 4–20.

Azarov, V. N., Kuzina, E. L., Kuzina, M. A., & Chekmarev, A. V. (2024, September). Balanced scorecard and six sigma systems in the context of digital transformation. In *2024 International conference "quality management, transport and information security, information technologies" (QM&TIS&IT)* (pp. 93–98). IEEE.

Chawla, R. N., & Goyal, P. (2022). Emerging trends in digital transformation: a bibliometric analysis. *Benchmarking: An International Journal, 29*(4), 1069–1112.

Citybabu, G., & Yamini, S. (2024). Lean six sigma 4.0—A framework and review for lean six sigma practices in the digital era. *Benchmarking: An International Journal, 31*(9), 3288–3326.

Correani, A., De Massis, A., Frattini, F., Petruzzelli, A. M., & Natalicchio, A. (2020). Implementing a digital strategy: Learning from the experience of three digital transformation projects. *California Management Review, 62*(4), 37–56.

Davenport, T. H., & Westerman, G. (2018). Why so many high-profile digital transformations fail. *Harvard Business Review, 9*(4), 15.

Doerr, J. (2018). *Measure what matters: The simple idea that drives 10x growth*. Penguin.

Dubey, R., Bryde, D. J., Blome, C., Dwivedi, Y. K., Childe, S. J., & Foropon, C. (2024). Alliances and digital transformation are crucial for benefiting from dynamic supply chain capabilities during times of crisis: A multi-method study. *International Journal of Production Economics, 269*, 109166.

Elia, G., Solazzo, G., Lerro, A., Pigni, F., & Tucci, C. L. (2024). The digital transformation canvas: A conceptual framework for leading the digital transformation process. *Business Horizons, 67*(4), 381–398.

Fabac, R. (2022). Digital balanced scorecard system as a supporting strategy for digital transformation. *Sustainability, 14*(15), 9690.

Faruque, M. O., Chowdhury, S., Rabbani, G., & Nure, A. (2024). Technology adoption and digital transformation in small businesses: Trends, challenges, and opportunities. *International Journal for. Multidisciplinary Research, 6*. https://doi.org/10.36948/ijfmr.2024.v06i05.29207

Frederico, G. F., Garza-Reyes, J. A., Kumar, A., & Kumar, V. (2021). Performance measurement for supply chains in the Industry 4.0 era: a balanced scorecard approach. *International Journal of Productivity and Performance Management, 70*(4), 789–807.

Gartner. (2023). Guide to create an effective digital supply chain roadmap. Retrieved April 15, 2025, from https://emt.gartnerweb.com/ngw/globalassets/en/supply-chain/documents/topics/create-an-effective-digital-supply-chain-roadmap.pdf.

Ghobakhloo, M., & Iranmanesh, M. (2021). Digital transformation success under Industry 4.0: a strategic guideline for manufacturing SMEs. *Journal of Manufacturing Technology Management, 32*(8), 1533–1556.

Ghobakhloo, M., Iranmanesh, M., Foroughi, B., Tseng, M. L., Nikbin, D., & Khanfar, A. A. (2023). Industry 4.0 digital transformation and opportunities for supply chain resilience: a comprehensive review and a strategic roadmap. *Production Planning & Control, 36*(1), 61–91.

Ghosh, S., Hughes, M., Hodgkinson, I., & Hughes, P. (2022). Digital transformation of industrial businesses: A dynamic capability approach. *Technovation, 113*, 102414.

Haryanti, T., Rakhmawati, N. A., & Subriadi, A. P. (2024). Assessing the digital transformation landscapes of organization: The digital transformation self-assessment maturity model (DX-SAMM). *Procedia Computer Science, 234*, 1561–1569.

Kalender, Z. T., & Žilka, M. (2024). A comparative analysis of digital maturity models to determine future steps in the way of digital transformation. *Procedia Computer Science, 232*, 903–912.

Kane, G. C., Palmer, D., Phillips, A. N., Kiron, D., & Buckley, N. (2015). Strategy, not technology, drives digital transformation. *MIT Sloan Management Review*.

Kaplan, R. S. (2009). Conceptual foundations of the balanced scorecard. *Handbooks of Management Accounting Research, 3*, 1253–1269.

Kaplan, R. S., & Norton, D. P. (1996). Strategic learning & the balanced scorecard. *Strategy & Leadership, 24*(5), 18–24.

Kaplan, R. S., & Norton, D. P. (2004). The strategy map: guide to aligning intangible assets. *Strategy & Leadership, 32*(5), 10–17.

Karttunen, E., Lintukangas, K., & Hallikas, J. (2023). Digital transformation of the purchasing and supply management process. *International Journal of Physical Distribution & Logistics Management, 53*(5/6), 685–706.

Kirmizi, M., & Kocaoglu, B. (2024). Design features of digital transformation maturity models: A systematic literature analysis and future research directions. *Journal of Modelling in Management, 19*(2), 313–341.

Lindquist, M. (2023). Supply chain digital transformation: Improving performance. Retrieved April 15, 2025, from https://www.oracle.com/uk/industrial-manufacturing/supply-chain-digital-transformation/.

Mekjian, G. (2024). Supply chain collaboration: Lessons learned from best-run supply chains. Retrieved April 15, 2025, from https://www.sap.com/documents/2021/04/68de5630-dc7d-0010-87a3-c30de2ffd8ff.html.

Plekhanov, D., Franke, H., & Netland, T. H. (2023). Digital transformation: A review and research agenda. *European Management Journal, 41*(6), 821–844.

Preindl, R., Nikolopoulos, K., & Litsiou, K. (2020). January Transformation strategies for the supply chain: The impact of Industry 4.0 and digital transformation. *Supply Chain Forum: An International Journal, 21*(1), 26–34. Taylor & Francis.

Qiao, W., Ju, Y., Dong, P., & Tiong, R. L. (2024). How to realize value creation of digital transformation? A system dynamics model. *Expert Systems with Applications, 244*, 122667.

Rossini, M., Cifone, F. D., Kassem, B., Costa, F., & Portioli-Staudacher, A. (2021). Being lean: how to shape digital transformation in the manufacturing sector. *Journal of Manufacturing Technology Management, 32*(9), 239–259.

Rossini, M., Powell, D. J., & Kundu, K. (2023). Lean supply chain management and Industry 4.0: a systematic literature review. *International Journal of Lean Six Sigma, 14*(2), 253–276.

Schumacher, S., Schmid, F. A., Bildstein, A., & Bauernhansl, T. (2021). Lean production systems 4.0: the impact of the digital transformation on production system levels. *Procedia CIRP, 104*, 259–264.

Scriven, P., Ledwith, A., & Nagle, T. (2024). Towards a lean digital transformation research framework: a literature review. *Journal of Decision Systems*, 1–15.

Spanaki, K., Dennehy, D., Papadopoulos, T., & Dubey, R. (2025). Data-driven digital transformation in operations and supply chain management. *International Journal of Production Economics*. https://doi.org/10.1016/j.ijpe.2025.109599

Stroumpoulis, A., & Kopanaki, E. (2022). Theoretical perspectives on sustainable supply chain management and digital transformation: A literature review and a conceptual framework. *Sustainability, 14*(8), 4862.

Tavana, M., Shaabani, A., Raeesi Vanani, I., & Kumar Gangadhari, R. (2022). A review of digital transformation on supply chain process management using text mining. *Processes, 10*(5), 842.

Westerman, G., Bonnet, D., & McAfee, A. (2014). The nine elements of digital transformation. *MIT Sloan Management Review, 55*(3), 1–6.

Wu, Y., Li, H., Luo, R., & Yu, Y. (2023). How digital transformation helps enterprises achieve high-quality development? Empirical evidence from Chinese listed companies. *European Journal of Innovation Management, 27*(8), 2753–2779.

Chapter 3
The Impetus for Digital Supply Chain Transformation: Motivation to Change

Maria Jesus Saenz, Christoph Lennartz, and Carl Marcus Wallenburg

Abstract Motivation to change plays a significant role in digital supply chain transformation (DSCT) endeavors. In this chapter, we analyze a multidimensional framework formed along four dimensions (namely, discomfort, survival anxiety, learning anxiety, and psychological safety) in two stages (namely, momentum and continuity). As a research methodology in this study, we use a multiple case study of 17 international large case companies. We collected data from semi-structured interviews and secondary data. Psychological safety is the most important dimension of a successful DSCT. Our findings also reveal that, in the momentum stage, successful companies involve operational employees in developing a DSCT vision, providing formal training to support digital expertise, and adjusting their information technology (IT) infrastructure to connect end-to-end digital initiatives. In the continuity stage, successful companies establish a balance of realizing value-add and enabling continuous change by supporting an entrepreneurial environment, continuously updating IT infrastructure, and providing stable role models for the operational employees. This chapter provides significant actionable insights on how to successfully enable motivation and effective DSCT.

Keywords Digital supply chain transformation · Motivation to change · Transformational stages · Change theory · Digital innovations

---

M. J. Saenz (✉)
MIT Digital Supply Chain Transformation Lab, Massachusetts Institute of Technology – Center for Transportation & Logistics, Cambridge, MA, USA
e-mail: mjsaenz@mit.edu

C. Lennartz
Schunk Group, Heuchelheim, Germany

C. M. Wallenburg
WHU – Otto Beisheim School of Management, Duesseldorf, Germany

Y. Khojasteh, S.M. M. Jahan Abadi (eds.), *Digital Transformation of Supply Chains*, International Series in Operations Research & Management Science 372, https://doi.org/10.1007/978-3-032-01206-7_3

3.1 Introduction

Digital supply chain transformation (DSCT) provides companies with exclusive opportunities for radical growth and competitiveness (Gezgin et al., 2017; Saenz et al., 2022). While digital transformation has been at the epicenter of many companies' business strategies in the past several decades, their challenges illustrate that a full understanding of how to enable a successful digital transformation is missing (Danuso et al., 2022; Davenport & Westerman, 2018). Change management and personnel resistance have been identified as key challenges faced by many companies in the process of DSCT (Ellis et al., 2017). Further, employee motivation and willingness are recognized as critical factors in digital transformation context (Eden et al., 2019; Fitzgerald et al., 2014). In fact, motivational factors might be even more important in digital than in traditional organizational transformations (Schwarzmüller et al., 2018).

Despite the fact that motivation to change plays a vital role in organizational transformation, existing literature fails to identify how large established companies can build up the motivation to change for the particular context of DSCT. In the context of organizational transformation, many frameworks have been developed to address the role of motivation (Burke, 2017). However, the unique features of DSCT require a customized framework to study the role of motivation in the context of digital transformation of supply chains. We explain these features below.

First, supply chains must be adapted to the dynamics of the demand. The main goal of a supply chain is to respond to the demand of the end customers and any transformation in the supply chain must consider how the transformation impacts its responsiveness to the demand that is dynamic in nature (Lee, 2004). Second, one of the key factors in a successful DSCT is the consideration of the dynamic ecosystem including the inter-relation among suppliers, customers, logistic service providers, and other parties involved in the supply chain. Supply chain managers as the boundary spanners need to evaluate how different parties in the supply chain will react and adapt to the changes that result from the digital transformation (Reusen et al., 2020). Third, supply chains help to overcome end-to-end approaches for creating value out of digital innovations. These unique characteristics of supply chains ask for the development of frameworks that address the role of motivation in this specific context.

As mentioned before, different frameworks have been developed to study the role of motivation in the context of organizational transformation. One of the well-established frameworks in this context is the change concept of Schein and Schein (2017) that highlights the essential role of motivation for successful transformation along the dimensions of discomfort, survival anxiety, learning anxiety, and psychological safety (Schein, 2010). The objective of this research is to apply and extend the change theory via this change concept with the objective of clarifying how companies can drive motivation to change in DSCT.

The change concept as well as existing literature on DSCT divide transformations into two main stages (Akbari et al., 2024; Büyüközkan & Göçer, 2018; Dubey et al., 2024). First, momentum to change is created in companies by guiding a transformation and exploring new opportunities. Then, the continuity stage begins, where companies standardize new approaches, adapt existing organizational structures, and implement initiatives. Given the distinctive features of digital transformation in supply chain, it is necessary to view the role that motivation to change plays in DSCT in light of these specific features. In this chapter, using a multiple case study of 17 international case companies, we identify the stages of digital transformation and shed light on the role of motivation to change in DSCT of large established companies.

This research contributes to transformation literature by extending the concept of transformation stages by motivation in DSCT, providing a novel framework for motivation to change. The study contributes further to change theory by extending it via empirically-verified and enhanced themes of motivation to change and identifying the role of these themes in facilitating a successful DSCT. Multiple case study research, enhanced by a frequency analysis (Agarwal et al., 2014; Zhang et al., 2016), provides a solid foundation for this exploratory research in an emerging research area (Yin, 2014).

The remainder of the chapter is organized as follows. In the next section, the conceptual background provides a holistic overview of current research in our focus area and builds the foundation for a new conceptual framework. In Sect. 3.3, we explain the research approach and illustrate the multiple case study methodology including case and frequency analysis. Section 3.4 describes key insights including detailed information about essential themes of motivation to change. Finally, Sect. 3.5 points out theoretical and managerial implications, followed by potential limitations and suggestions for future research.

3.2 Conceptual Background

3.2.1 Digital Supply Chain Transformation

Digital transformation of companies and their supply chain operations have received a great deal of attention in the past several years (Büyüközkan & Göçer, 2018; Wang & Prajogo, 2024), which is reflective of the growing managerial importance of the subject (Hu & Monahan, 2015). The effects of digital technologies on a supply chain range from incremental adaptations of individual processes to fundamental disruptions of the overall supply chain. In both cases, digital technologies address physical and information flows of a supply chain.

Incremental adaptations relate to improvements in operational efficiency and costs in a company's supply chain. An example of the adaptation of physical processes are smart wearables, such as smart glasses (Sahu et al., 2021; Syberfeldt et al., 2017), or sensor technologies, such as RFID tracking (Abdirad & Krishnan, 2021; Wu et al., 2016), which support picking and stowing processes in warehouses.

Improvements in supply chain flows, supported by Artificial Intelligence (AI), are for example, demand planning and forecasting accuracies (Revilla et al., 2023) or IT infrastructure to enhance information exchange between companies (Dubey et al., 2024; Fawcett et al., 2011), which advances supply chain partnerships.

Fundamental disruption in supply chains relates to the reconfiguration of overall supply chain setups (Dallasega et al., 2018; Reddy et al., 2016). Physical supply chain flows are strongly affected by digital technologies such as AI technology (Sasson & Johnson, 2016). Hereby, digital technologies can even lead to new business services along the supply chain (Holmström & Partanen, 2014). Supply chain information flows are fundamentally disrupted by increasing automation and predictive analysis that can lead to a "self-thinking supply chain" (Cataldo et al., 2009). Demand forecasting and planning processes are enhanced by AI technology (Balasubramanian et al., 2020; Gartner, 2018; Huber & Stuckenschmidt, 2020). Here, digital technologies, such as internet of things and distributed ledger technologies (Seifert & Markoff, 2018), are utilized to significantly enhance the information transparency of a supply chain.

There is no commonly-accepted definition of digital transformation in the literature (Kohtamäki et al., 2024) and researchers have defined the term in varied ways (Hinings et al., 2018; Kuusisto, 2017; Reis et al., 2018). Nevertheless, there are certain elements of digital transformation that are understood as main components of the process. Digital technologies are described as technologies that enable novel utilization and application of data to generate value-add (Cerdas et al., 2017; Fitzgerald et al., 2014; Sebastian et al., 2017). Digital transformation can then refer to the change that occurs by using digital technologies (Vial, 2019) and the related adaptations of organizations in the aspects of structures, people, and processes (Chopra & Meindl, 2007; Krimpmann, 2015; Mangematin et al., 2014). The main purpose of an organizational transformation is to create and maximize value (Prahalad & Hamel, 1994). Hence, we refer to *digital supply chain transformation* as the application of digital technology to transition towards a value-driven supply chain (Meyr & Günther, 2009; Saenz et al., 2022).

3.2.2 Change Theory

The *change theory* of Schein and Schein (2017) focuses on overarching processes that take place in companies during organizational change. The change theory is based on the division of organizational change into three steps of unfreezing, moving, and freezing introduced by Lewin (1947). Schein and Schein (2017) characterize these three steps as stages, because of their potential overlap, and expand the stages by introducing four dimensions (i.e., discomfort, survival anxiety, learning anxiety, and psychological safety) that companies need to consider when pursuing change (Burke, 2017). These dimensions have negative and positive influence on the development of motivation to change (Schein, 2010).

3.2.2.1 Dimensions of Motivation to Change

Motivation or willingness is an essential element in organizational transformations and potentially even more important in digital than in traditional transformations (Schwarzmüller et al., 2018). Lack of motivation in digital transformation is a key challenge for companies (Fitzgerald et al., 2014) particularly because a high level of motivation is required throughout the entire process of digital transformation (Eden et al., 2019). In digital transformation, employees need to be willing to experiment and innovate more than before and accept to fail regularly (El Sawy et al., 2016).

In supply chain literature, motivation has been recognized as an essential element in implementing innovations, a critical component of decision-making processes (Sabri et al., 2018), and an important factor in the business performance of a company (Vanpoucke et al., 2016). In digital transformation of large organizations, "people, not technology, are the primary factor in realizing value" (Kane et al., 2019; Kohli & Johnson, 2011, p. 150). The change theory of Schein and Schein (2017) provides a framework for motivation to change in transformations and has four major dimensions, namely, discomfort, survival anxiety, learning anxiety, and psychological safety.

Discomfort refers to a disequilibrium created by information that indicates that a company might miss the targets or might not be able to accomplish expected results using the existing processes (Schein & Schein, 2017). Hence, discomfort is an internal (e.g., decreasing internal performance) or an external (e.g., changing customer requests) information that signals that a company needs to change (Schein, 2010). Other examples of such information are higher requirements of customers, improved performance of competitors (Reis et al., 2018), and failure of competitors in the same industry (Lucas & Goh, 2009).

Digital transformation is typically the consequence of significant market pressure to improve organizational efficiency and build up supply chain agility (Kohli & Johnson, 2011).

Researchers have studied the dynamic capabilities required for digital transformation of firms and supply chains (Chirumalla, 2021; Lennartz et al., 2019; Warner & Wäger, 2019). To improve performance and ultimately for business survival, companies need to identify disconfirming information and then effectively respond to new technologies (Fitzgerald et al., 2014). Further, supply chains are also digitally transformed to enable business success in anticipation of future market pressure (Büyüközkan & Göçer, 2018). Nevertheless, the transformation requires that both the leadership and the employees realize discomfort even if the transformation is initiated only by the leadership (Alos-Simo et al., 2017).

Survival anxiety refers to the anxiety or the uncomfortable impression, shared by the company employees, that without change, the company or its (individual) employees will experience negative impact (Schein & Schein, 2017). While discomfort represents the information that signals the need to change, survival anxiety is the realization of the need to change (Schein, 2010). Anxiety about missing valuable opportunities motivates employees and organizations to change. Such a need to change can range from incremental enhancements to fundamental transformations

(Cerdas et al., 2017; Xue et al., 2013). Organizational transformation, and in particular digital transformation, often leads to disruptive change (Karimi & Walter, 2015), which highlights the importance of survival anxiety to enable change.

The collective expectation from technology to provide competitive advantages leads to IT-innovations, which often create pressure for companies to adapt (Saeedikiya et al., 2024). In many cases, managers believe in the value-add of digital technologies to enable transformative business change (Fitzgerald et al., 2014). However, survival anxiety is not limited to the company's leadership. Especially in the context of digital transformation, employees also demonstrate concerns about the company's performance (Åkesson et al., 2018) or about the capabilities required for digital innovations (Svahn et al., 2017). Hence, digital transformation drives survival anxiety across different levels of the company hierarchy.

Learning anxiety is defined as the anxiety created by an expectation to learn new ways to feel, think, perceive, and behave that require giving up elements that are currently valued (Schein, 2010; Schein & Schein, 2017). A company transformation leads to significant change for individuals by adapting ways of thinking, behaving, feeling, and relating to each other (Sackmann, 1989). Employees, as well as the company's leadership, fear to be negatively impacted during or after change, which can cause resistance to change (Schein, 2010) and hamper a transformation. Resistance to change has been shown to have a significant negative impact on organizational change management (Kuipers et al., 2014).

Change causes various forms of fear for employees and companies including: (1) "loss of power and position", (2) "incompetence", (3) "loss of personal identity", and (4) "loss of group membership" (Schein, 2010: 303–304). Employees might lose influence in their job (e.g., due to new leadership) or their positions can be fully disbanded (e.g., due to automation). Employees might also feel incompetent while learning new approaches and fear being punished for incompetence (e.g., due to decreased productivity). They fear that they might not identify with new ways of working or they might lose their reference group of employees in a company due to an adapted culture and way of thinking (Schein, 2010).

One of the main challenges in digital supply chain transformation is highlighted by the fact that learning anxiety can hamper cross-functional collaboration while transforming a supply chain requires an increased level of collaboration across functions and stakeholders (Mallouf & Gammelgaard, 2016; Sáenz et al., 2014). Further, technology-induced anxiety can jeopardize the implementation of digital technologies (Kummer et al., 2017). Hence, change anxiety is a key element of organizational change that managers need to address (Valentino, 2004) because it directly influences motivation to change and thereby transformation success.

Psychological safety is defined as the perceived possibility to solve problems and learn new elements safely even though previously valued elements are given up (Schein, 2009, 2010; Schein & Schein, 2017). This perceived safety counteracts learning anxiety and positively impacts motivation to change (Schein, 2010). Psychological safety is essential for change processes that involve employees (Cataldo et al., 2009) as self-efficacy determines readiness for individual and company change (Rafferty & Simons, 2006). Employees are encouraged by psychological safety "to offer ideas, admit mistakes, ask for help, or provide feedback" (Edmondson & Lei, 2014: 36). Managers should address change anxiety by increasing psychological safety of the employees (Valentino, 2004). Further, managers should extend the within-team practices to managing the networked ecosystems in which the interactions among individual team members are also considered (Carboni et al., 2021). Transforming leadership activities (e.g., vision) increases the motivation of employees to support innovative technological change (Chen et al., 2014). For example, a digital strategy provides an overview of how to accomplish goals for a digitally-transformed supply chain via new technological opportunities (Dubey et al., 2024; Kohtamäki et al., 2024; Wang & Prajogo, 2024).

Schein (2010) provides eight activities as the major themes to increase psychological safety: (1) "compelling positive vision", i.e., communicating benefits of change; (2) "formal training", i.e., building up, via training structures, new skills and knowledge that employees require for adapted ways of working; (3) "involvement of the learner", i.e., individualizing learning of employees by involving them; (4) "informal training", i.e., imbedding new cultural assumptions by jointly building new norms and assumptions for interactions; (5) "practice fields, coaches and feedback", i.e., allowing employees to learn from mistakes without significant organizational impact and via interaction with others; (6) "positive role models", i.e., providing employees a reference of targeted behavior in form of role-models they can identify with; (7) "support groups in which learning problems can be aired and discussed", i.e., exchanging information with other learners about challenges of change; and (8) "systems and structures that are consistent with the new way of thinking and working", i.e., encouraging and rewarding targeted behavior, and punishing individuals that do not act accordingly (Schein, 2010: 305–307). This chapter transfers and extends these themes to the context of DSCT.

3.2.2.2 Transformation Stages in DSCT

A *transforming* process is typically described using either *stages* or iterative steps to change (van de Ven & Poole, 1995). The concept of differentiating two stages of preparation and implementation is well-established in the literature. For example, Sáenz et al. (2014) apply the well-known exploration and exploitation stages in a supply chain transformation and Schuchmann and Seufert (2015) describe a similar concept in the context of digital transformation. Further, the literature on DSCT recommends differentiating between preparation and implementation stages (Büyüközkan & Göçer, 2018; Dubey et al., 2024; Saenz et al., 2022; Wang & Prajogo, 2024).

Based on the reviewed literature, we summarize the transformation process into two stages: (1) *momentum:* creating impetus of change throughout an organization, identifying transformation direction, and familiarizing employees with new work approaches, and (2) *continuity:* standardizing new working approaches, structuring implementation of selected initiatives, shifting responsibility of initiatives to employees, and enforcing value creation.

The momentum stage is linked to the unfreezing stage of the change theory (Schein & Schein, 2017). This stage focuses on motivation to change, required to adapt the core structure significantly (Schein & Schein, 2017). Change requires a sufficiently large disequilibrium in an organization (Schein, 2010), which serves as a foundation for transformation. DSCT requires momentum when companies prepare to introduce digital technologies and foster awareness and culture for digital transformational change (Dremel et al., 2017; Eden et al., 2019). Companies need to experiment, innovate, and "learn how to learn" in this transformation stage (Netland & Ferdows, 2016; Schuchmann & Seufert, 2015: 32).

The continuity stage is linked to moving and refreezing of the change theory (Schein, 2010). This stage focuses on adjustment and redefinition of previous structures (e.g., roles, processes, behavior, and standards) (Schein & Schein, 2017) to enforce value-creation by transformation. Involving employees as part of the entrepreneurial culture to support digital innovation requires new standards similar to agile development (Kohtamäki et al., 2024). These actions support the desire of companies to capture value through implementation (Netland & Ferdows, 2016) in the continuity stage.

The continuity stage stabilizes new learnings, for example, by illustrating improved performance, similar to the refreezing stage of the change theory (Burke, 2017). As opposed to the refreezing stage, the continuity stage stabilizes while adjusting to quickly allow further change (Fassoula, 2006; Lewin, 1947; Schein & Schein, 2017) because of the highly dynamic nature of digital transformation. New structures need to be recommitted in the last stage, which can include rethinking core structures and thus adapting again (Eden et al., 2019) and can lead to continual change (Fassoula, 2006). Hence, while reinforcement is necessary, digital transformation does not allow a separate stage of refreezing anymore and requires continuous change.

3.2.3 Conceptual Framework

To establish additional guidance throughout our research process, we utilize the described literature to develop a novel conceptual framework displayed in Fig. 3.1. On the first level, the framework is structured along the four dimensions of motivation to change in DSCT. On the second level, the framework is structured along the transformation stages. Due to the solid theoretical foundation, this framework and its literature provide reliable guidance for our study and support our research approach, i.e., the empirical multiple case study.

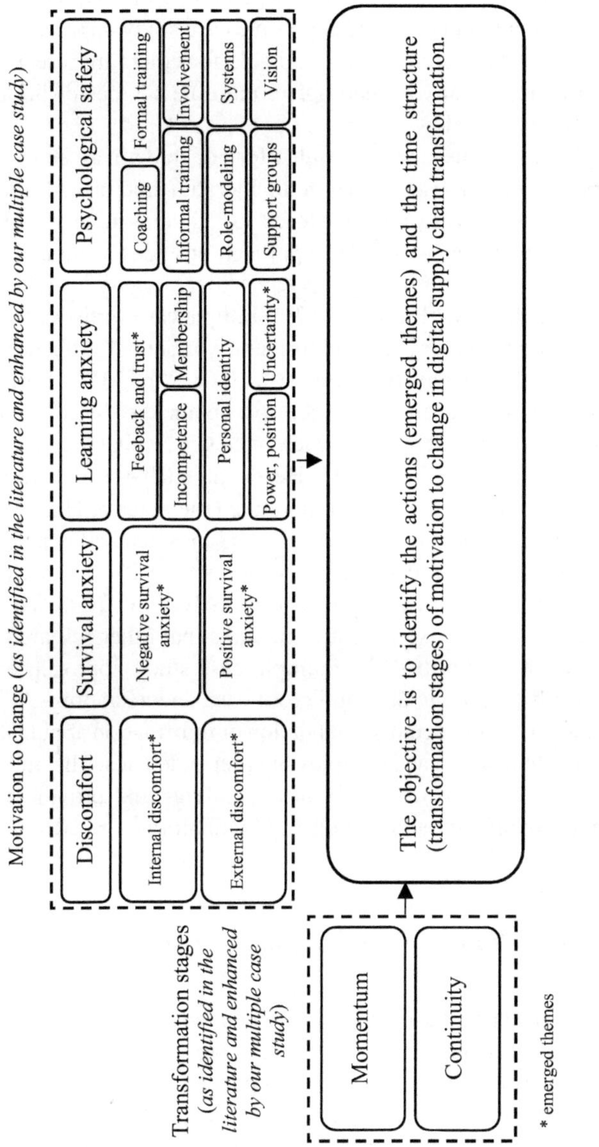

Fig. 3.1 Conceptual framework of motivation to change in DSCT

3.3 Research Approach

Our research methodology is based on a multiple case study approach, which is commonly applied to empirically develop, extend, and enhance theories (Ridder, 2017; Yin, 2014). We apply the methodology guided by the change theory to study the motivation to change in DSCT. The multiple case study approach provides two key advantages for our research: (1) it is suitable for emerging research areas, and (2) it provides the opportunity to incorporate multiple sources of information. We explain these advantages below.

First, multiple case studies are well-suited for our study in an *emerging research* area as the approach is well-established in the literature and provides reliable guidance to conduct rigorous research that leads to valuable novel findings (Ridder, 2017; Voss et al., 2002; Yin, 2014). The approach provides a framework for semi-structured interviews and continuously adapts to emerging insights throughout interviews (Eisenhardt, 1989; Ridder, 2017). Multiple case study research has often been applied to close research gaps and to identify essential influence themes and their complex interrelationships (Gibbert et al., 2008; Yin, 2014). Further, multiple case study research is an adequate choice to explain the "why" and the "how" (Bansal & Corley, 2011) in research, which makes it a suitable approach to study digital transformation (Ba & Nault, 2017; Hess et al., 2016) as a subject that has not been sufficiently covered by the current literature (Bienhaus & Haddud, 2018).

Second, the multiple case study approach combines data collected in case company interaction (e.g., semi-structured interviews, site visits) and multiple enhancing sources of information (e.g., annual reports) (Alvesson, 2003; Yin, 2014). The company interactions allow us to identify findings that other research methodologies would not uncover. Further, the multiple case study approach enables us to develop an in-depth case-individual and cross-case understanding. As a result, we are able to create a solid and detailed foundation of information about DSCT aspects while considering their subtle and complex interdependencies. The multiple sources of information are essential to identify novel findings and provide a richness of information that supports research validity and reliability (Yin, 2014).

3.3.1 Case Selection and the Unit of Analysis

We selected company cases based on a clearly defined process of identifying potential case companies, screening cases based on fit, and conducting the research process. The unit of analysis in our study is the supply chain function within a company. We first identified potential case companies based on publicly available information (e.g., annual reports) about their DSCT experience. Then, we screened the companies based on selection criteria. The criteria were based on literature (e.g., Davenport & Westerman, 2018; Kohli & Johnson, 2011; Sebastian et al., 2017) and a pre-interview workshop on DSCT with 27 supply chain representatives of 20 companies. The resulting information highlighted three criteria: (1) Large company size: Large companies face the challenge of complexity due to interwoven internal organizational structures and the pressure of international competition. Participating companies had to employ at least 5000 employees, create more than EUR 1 billion in revenue per year, and pursue international business activities; (2) long-established company: Over time companies build up legacy structures and processes which hamper change. Hence, case companies had to exist at least 20 years to be included; (3) first-hand experience in DSCT: First-hand experience ensures a fact-based and reliable data source for this research. Hence, involved companies had to be actively involved in DSCT activities. The criteria ensured case comparability of our research findings, while differentiation in industry, duration of transformation, and within large company size, supported the generalizability of our research (Yin, 2014).

We then contacted potential companies out of which 31 showed interest to participate in the study. After we ran up-front conversations and reviewed additional case material, we selected 17 case companies, all of which had not financial entanglement with the research team in order to ensure research integrity. Table 3.1 provides an overview of the case companies and their key characteristics.

3.3.2 Data Collection Process

In our data collection process, we considered the criteria of Yin (2014) to ensure rigorous case study research and increase findings reliability and validity (i.e., construct, internal, and external validity). We followed a three-step process of initiation phase, interview conduction, and data review. First, we reviewed the existing literature and conducted discussions with peer researchers (Yin, 2014). To enhance this foundation, we also conducted expert discussions and workshops, reviewed company reports, and ran a pilot interview.

Second, we conducted case study interviews that represented our key source of case data. All interviews utilized a semi-structured interview protocol, which was continuously enhanced as recommended by literature (Ridder, 2017). These interviews followed established ethics standards and did not involve deception. First interviews were conducted with multiple researchers and case data was frequently reviewed with research scholars to enhance reliability and support the research process.

The first interviewee of each case company represented a high hierarchical level (e.g., senior manager, vice-president, director) to ensure an overarching perspective and enable a snowball sampling tactic (i.e., the first interviewee nominates next suitable interviewee(s)). Table 3.1 outlines for each of the 17 case companies, interviewee role, and interview duration. In total, we interviewed 49 company representatives. Thereby, each case interview was scheduled for 60–90 min. Following the established process in the literature, we transcribed case interview records via the denaturalistic transcription method (Saunders et al., 2015). The resulting transcripts, excluding additional field notes, covered ~1000 pages of text

Table 3.1 Company case overview

Case	#FTEs (in ,000)	Start of transformation	Revenue EUR bn.	Founded	Interview: number/ duration	Position
AerospaceCo	>50	2016	>30	1951–1995	2/205	Director Digitalization; Internal IT-Advisor
AgricultureCo	5–20	2014	5–30	1901–1950	3/188	Head SC Planning; Head SC IT; Head of Logistics
ChemicalsCo	20–50	2015	5–30	≤1900	2/137	Head of Digital SC; Director SC Digitalization
ChemicalsInc	5–20	2016	5–30	1951–1995	2/106	Head of Digital; SC Manager
ComputerCo	>50	2016	>30	1951–1995	2/119	Sr. Director Operations; Director Operations
EquipmentCo	5–20	2016	5–30	≤1900	3/198	SC Manager; CDO Manufacturing; CDO S&M
EquipmentInc	20–50	2015	1–5	1901–1950	5/342	4x Head of SC area; Head of SC
FoodCo	20–50	2015	5–30	≤1900	2/118	SC Director; Logistics Manager
MachineryCo	5–20	2015	5–30	≤1900	5/384	2x GM Operations; Direct Global Supply; Planning manager; Director Finance
MiningCo	>50	2015	5–30	1901–1950	2/117	SC analytics Manager; Digitalization Manager
PharmaCo	5–20	2015	1–5	1901–1950	2/123	VP Supply; Director SC development
PharmaInc	20–50	2016	5–30	≤1900	1/69	Head of SC Innovation
RetailCo	5–20	2015	1–5	1951–1995	2/96	Head SC; SC Executive
RetailGroup	>50	2015	>30	≤1900	3/183	SC Director; SC Manager; CI Manager

(continued)

Table 3.1 (continued)

Case	#FTEs (in ,000)	Start of transformation	Revenue EUR bn.	Founded	Interview: number/ duration	Position
RetailInc	>50	2016	>30	≤1900	3/165	VP SC; Manager SC; Director SC
TeleCo	>50	2016	>30	1951–1995	2/104	Head Digital SC; Head SC (Digital) Operations
TextileCo	5–20	2017	1–5	≤1900	6/472	SC Expert; Head of SC; Head of Operations; SC Manager; Sr. VP Global SC; COO

by ~408,000 words. This interview data is a solid foundation for cross- and within-case coding and frequency analysis.

We collected secondary data to ensure a comprehensive case company overview, enable triangulation, and support external, internal, and construct validity (Yin, 2014). The secondary data included related literature, case company websites and annual reports, internal documents, a case company workshop, and field note documentation for each case company (e.g., photos, and walkthrough documents).

Third, we conducted further extensive iterations with research scholars. These included international research conferences and company workshops. One 2-day workshop about DSCT involved large case and non-case companies.

3.3.3 *Multiple Case Study and Data Analysis*

As part of the multiple case study data analysis, we first conducted the coding process and frequency analysis. The conceptual framework provided us with research guidance and supported the consolidation of case information. We conducted within- and cross-case analyses based on the resulting information and in-depth case company understanding. This enabled us to extend the theory-based framework to the context of DSCT.

We conducted open-, axial-, and selective coding within the framework structure, supported by the coding software NVivo (Corbin & Strauss, 2008; Locke, 2001). In total, we identified 1789 codes, related to research framework dimensions. This process was accompanied by continuous review of case study data, related to relevant literature and theory. Furthermore, we enhanced the validity and reliability of our study through research discussions, conference presentations, and workshops (Yin, 2014), as mentioned before.

To utilize established processes, we applied the systematic approach of Gioia et al. (2013), integrating codes into concepts. Iterative case coding and analyses enhanced the conceptual framework depicted in Fig. 3.1. We identified themes and evaluated their importance based on the case study approach described by literature (Ridder, 2017; Yin, 2014). Furthermore, we conducted a frequency analysis to enhance coding and create an additional indication for their relevance (Agarwal et al., 2014; Zhang et al., 2016). The frequency analysis counts the number of codes identified within an emerged theme.

We segmented cases into three performance groups (i.e., low, medium, and high) depending on their success in implementing digital technologies, adjusting their IT infrastructure and master data, and adapting organizational processes. The research findings illustrated that these elements are core elements of a successful DSCT. In particular, we measure three criteria: (1) number of identified, tested, and implemented technologies, (2) existence and the role of a team dedicated to DSCT, and (3) existence and the form of a vision including a supply chain vision for digital transformation. Case segmentation is well-established in digital transformation and supply chain literature (Busse et al., 2017; Knoppen et al., 2011; Singh & Hess, 2017).

The motivation to change is divided into transformation stages (i.e., momentum and continuity), and grouped (i.e., low, medium, high), based on coding frequencies and in-depth case analysis.

3.4 Case Findings

Our case analysis shows that psychological safety is the most important motivational factor in the success of DSCT. This is the result of our detailed within- and cross-case analysis and is reflected in 748 of the 1789 total codes (i.e., ~42% of all codes) referring to psychological safety. Overall, companies with a higher motivation to change, and especially a more pronounced focus on higher psychological safety, perform superior in their DSCT.

Our analysis also confirms that the process of DSCT consists of two stages of creating momentum and establishing continuity. Interestingly, the highly dynamic business context causes the actions of refreezing (mainly reinforcement) to be essentially embedded into the continuity stage. We explain this below.

Companies used the momentum stage to establish a transformation mindset and develop impetus throughout the organization, for example, by encouraging the initiation of a diverse digital initiative portfolio and adapting organizational structures. This stage focuses on developing a need to change due to DSCT, which is expressed in survival anxiety and a high level of discomfort. The actual financial return out of digital initiatives is not the key focus of this stage.

Once the impetus throughout the organization is created, companies need to establish more control of DSCT activities while, at the same time, encouraging further development. Hence, the continuity stage focuses on a balance of capturing the value of initiatives and providing a relatively high level of freedom to experiment with digital initiatives. This balance needs to be held in place by continuous adaptation of organizational structures, for example, in the form of an adapted transformation vision or training.

Companies that do not shift from the momentum to the continuity stage, do not realize sufficient value-add for their supply chain out of digital initiatives and will not be able to conduct a successful DSCT over an extended period of time. Also, companies cannot skip the momentum stage and move directly to the continuity stage, as several important themes, such as the involvement of operational employees, would not be possible without an established momentum stage. This is because, in the absence of a momentum stage, employees are less likely to experience the need to change and, consequently, may not support transformation initiatives. Further, organizational structures, such as vision, that are critical for encouraging digital initiatives in the continuity stage, may not be sufficiently developed unless a momentum stage is fully established.

Figure 3.2 provides a comparative overview of themes per motivational dimension, DSCT performance, and transformation stages. Each bar summarizes the coding statement by aggregating the maturity levels for each theme and average per company. In the remainder of this section, we describe the momentum stage and the subsequent continuity stage in detail.

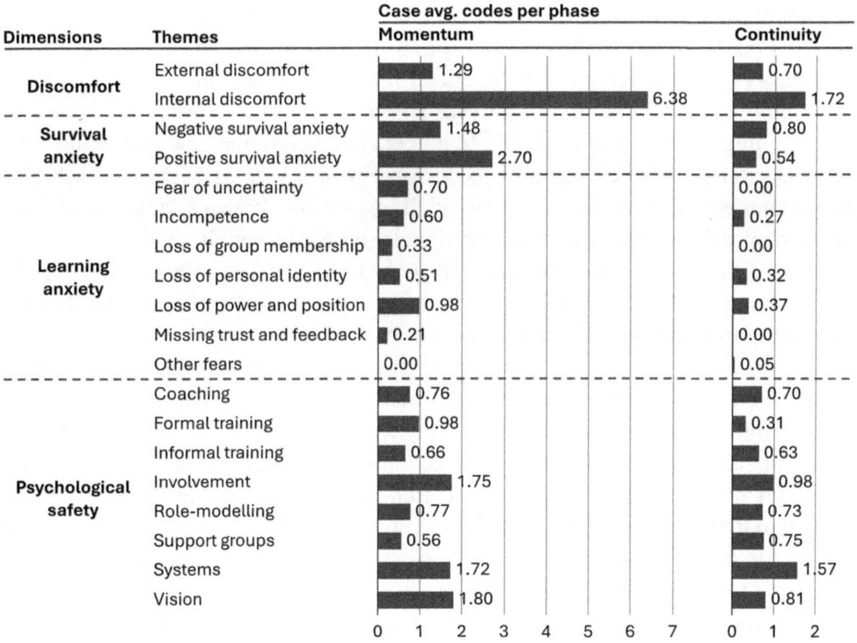

Fig. 3.2 Transformation performance by motivational dimensions and themes for the momentum and continuity stage

3.4.1 Momentum Stage: Getting the Transformation Started

At the core of the momentum stage is the development of an understructure that facilitates initiating digital transformation and learning to innovate in an environment of rapid technological progress. This includes building a modified cultural foundation in the workforce (Eden et al., 2019) and establishing a mindset that is well-adapted to ever-emerging technological opportunities. Consequently, the workforce needs to realize that failures will occur much more frequently than before, that learning now occurs differently (e.g., by employing more trial and error in short cycles), and that "change is the new normal".

In the momentum stage, *psychological safety* was the most important motivational dimension. Out of the eight themes for psychological safety identified by Schein (2010), five proved to be particularly relevant in the DSCT context. Two themes (namely, *vision* and *involvement of employees*) are foundational elements, which over time even the less successful companies in the transformation realized their importance and incorporated them. The three themes (namely, *role-modeling*, *formal training*, and *adjustment of systems*) are differentiating elements as they facilitate superior performance.

Vision and involvement of employees were among the most frequently mentioned elements of providing psychological safety. Both mechanisms in combination prove to be most beneficial for DSCT success. A clear *vision* regarding DSCT provides guidance to employees and allows them to understand why (and potentially how) the transformation is beneficial for the company and why it can be an opportunity for the individual employee. This addresses the uncertainty of employees regarding changes due to DSCT. Vision creates a common goal across the organization, and establishes common needs among the various regions in which a global company is operating and among all employees. It also enables structuring the employees' involvement cross-functionally along the supply chain.

Involving employees in creating a transformation vision significantly impacts the success of digital transformation by helping employees develop ownership for transformation activities and identify their individual processes of change within the overall transformation (Schein, 2010). It also allows employees to understand the requirements and benefits of transformation from their own perspective and, consequently, gives them a feeling of safety within the disruptive transformation of the supply chain. For example, at TeleCo, after a company vision was established, digital channels for internal communication about digital initiatives were implemented to increase the involvement of operational employees. If employees are not involved in creating the vision, they only develop an abstract and aggregated understanding of the overall benefits to the company, while being uncertain about the actual changes and not understanding the impact of DSCT on them as individuals.

The involvement of employees also fosters the creation of diverse and cross-functional groups of employees, which is not only beneficial for enhancing creativity in developing digital initiatives but also creates new communities and groups of employees within the organization. Thus, employees develop a sense of belonging, feel safe working together in these setups, and share their knowledge about the actual supply chain challenges. For example, DSCT-coordinating staff at PharmaInc introduced an annual employee survey to regularly involve employees in the progress of DSCT. Also, the leadership of EquipmentInc organized annual multi-day company offsites with the company employees where they jointly discuss and further develop a transformation vision of the supply chain.

The case of ChemicalsCo illustrates the importance of the involvement of employees for a successful transformation. At ChemicalsCo, the top management initially designed and started to execute digital initiatives via a top-down approach where employees were not directly involved in the decision-making process. Consequently, operational employees did not develop a sense of ownership over the process of digital transformation, which led to a low level of perceived psychological safety. The top management of ChemicalsCo identified this issue and started to encourage active involvement of employees throughout the organization structured via particular supply chain processes. This led to a substantial increase in motivation and, consequently, higher transformation performance.

Among the three differentiating elements, role-modeling and formal training are both directly related to the individual employee's career in the company. *Role-modeling* fosters psychological safety and is beneficial in the momentum stage by illustrating the direct benefits of DSCT. Role-modeling facilitates the use of successful cases of digital initiatives (e.g., Robotic Process Automation (RPA) in order management) from within the company or from other companies. Employees view role-models as a guide that they can follow their behavior, which helps them feel safe in a transformational environment. In the case of AerospaceCo, top management explicitly and positively highlighted employees as role models both with successful and failed digital initiatives to show that failure is part of any digital transformation. To apply role-modeling, ChemicalsCo provided employees with career opportunities as change ambassadors that facilitated a change management program in support of DSCT. Change ambassadors enhance their skills in the digital context, develop a mindset that supports digital transformation approaches (e.g., agile working), and receive direct top-management exposure.

Formal training is directly related to role-modeling as it provides employees with skills to follow or even become role-models. Formal training significantly contributes to transformation success by providing psychological safety and reducing negative anxiety. It sets standards that provide a basic and common understanding of DSCT elements (e.g., terminologies). The skills gained through these trainings make the employees feel more self-assured in discussions about digital transformation and enable employees to better understand the implications of DSCT. For example, the top management of ChemicalsCo initiated training programs focusing on employees who showed interest in supporting DSCT. Employees who participated in the programs better understood the company-related and individual benefits of DSCT. Hence, employees identified their individual opportunities for change and actively contributed to a successful transformation.

The *adjustment of systems* and structures supports psychological safety and, consequently, transformation success by providing the settings that help the company to successfully utilize role-modeling and formal training. In the momentum stage, the adjustment of systems and structures focuses on improving the IT infrastructure (e.g., integrating IT systems) and the creation of a risk-free development environment in which operational employees can experiment with digital initiatives. An IT infrastructure that consists of an integrated IT systems and utilizes the correct master data is especially important for DSCT because this transformation intensively uses the IT infrastructure and particularly the data it manages. In the absence of a well-developed IT infrastructure, the system and the data might lead to higher levels of uncertainty. In other words, an IT infrastructure provides a feeling of safety to employees if they can rely on the data and the IT systems.

A well-adjusted infrastructure includes the introduction of formal training (e.g., via training courses) and supports the development of digital initiatives (e.g., via accurate data) that can turn the operational employees into powerful role-models. Examples are EquipmentInc and ChemicalsCo, which improved the transparency in the supply chain information flows and benefited from a more end-to-end data-driven decision-making process by utilizing a more integrated IT infrastructure that relied on accurate master data across the supply chain. If IT infrastructure is not integrated or data is not accurate, employees may find the experience frustrating, which in turn negatively impacts the transformation process. Employees at MachineryCo ran digital initiatives, estimated their impact based on available master data and presented impact estimations to the executive management. The employees progressed with the digital initiative and identified severe errors in master data they utilized, which caused failures in impact of demand forecasting and the design of the initiative. This experience was frustrating for the employees and led to mistrust regarding the value-add of not only this single digital initiative but also the overall digital transformation, considering the core role of demand forecasting for their digital vision. This had a direct negative impact on transformation performance.

Companies can use different approaches to adjust an IT infrastructure and its data. ChemicalsCo utilized an "80/20 approach" focusing only on the data that was directly required by selected digital initiatives. This contributed to transformation success because the data was cleaned after a short period of time and employees were able to immediately start their digital initiatives. Employees felt safe to utilize the IT infrastructure for digital initiatives while being aware of possible but minimal inaccuracies. The opportunity for further scalability of the initiative, upstream or downstream in the supply chain, created additional motivation. In contrast, EquipmentInc and PharmaCo first conducted a very rigorous adjustment of the master data and the IT systems (e.g., through a global integration of IT systems). This process can provide a significant level of psychological safety. However, due to an extremely long preparation process and extended scope for the end-to-end approach, such safety might be achieved too late to help generate motivation in the momentum stage.

Another structural adjustment to support psychological safety is the creation of a development environment that contributes to the success of DSCT via a risk-free environment for employees to develop digital initiatives. Such an environment enhances creativity, diversity, and collaboration among the employees during the development of the initiatives. Employees feel safe in the environment as they can learn about digital technologies, combine the skills of the other employees, and actively contribute as members of the DSCT success. MiningCo and AerospaceCo created such an environment in the form of a center of excellence and a collaboration platform.

Discomfort is the main source of *survival anxiety*, which is the reason why these two aspects of motivation go hand in hand. Internal discomfort has a more positive impact on the transformation success, as it creates mainly positive survival anxiety. The supply chain management of ChemicalsInc expected improved forecasting, business opportunities, and customer experience, due to digital technologies. Hence, companies are more proactive, starting DSCT, before external discomfort increases. Companies usually initiate the transformation more structured and with a holistic perspective. The external discomfort that high-performing companies mainly receive, is in the form of suggestions by supply chain partners. An external digital technology provider and previous lateral partner approached AerospaceCo and suggested starting digital initiatives as a free test period of joint development, illustrating a positive impact on success.

In comparison, external discomfort can drive negative survival anxiety. This has a negative impact on transformation success, as companies are more reactive, starting DSCT partly in a panic due to external discomfort (e.g., customer complaints), which leads to less efficient change. This impact on transformation success is also reflected in high-performing companies, that have a higher internal to external discomfort ratio than a low-performing companies. TextileCo, a company with lower transformation performance, is an example that faced external pressure of customers to improve by a digital supply chain. RetailCo described the resulting external pressure as: "If you do not adapt, you will die".

Learning anxiety has a negative effect on the motivation to change and can hamper the success of DSCT. Psychological safety has a detrimental impact on learning anxiety. Hence, companies with high psychological safety can better control learning anxiety, which then has a less negative impact on the success of digital transformation.

In the momentum stage, the two most important themes are fears of loss of power or position and uncertainty, whose negative impact on the transformation can be limited. The fear of employees losing their power or position (e.g., being replaced by automation) is the strongest fear in low- and high-performing companies. Employees negatively impact the transformation success, if they do not support DSCT due to their fear of digital technologies and adjusted organizational structures. Companies can positively impact their transformation success by shifting this fear into the perception of an opportunity. For this, companies apply role-modeling, building up digital technology skills of employees and illustrating opportunities of digital technologies. Employees start to translate their fear into actions to improve their individual position, encouraged by company management that continuously points out opportunities of a DSCT.

Fear of uncertainty is significantly higher in DSCT than in a traditional organizational transformation, due to the linked chain effects. For example, digital technologies require often experimentation along different functions and supply chain actors to understand their actual value-add. Therefore, holistic vision and diverse employee involvement are even more important. They support transformation success, as they reduce the uncertainty of employees by providing the described guidance for and understanding of digital transformation activities.

3.4.2 Continuity Stage: Standardizing and Controlling the Change

For the continuity stage, we focus on themes to successfully manage the ambidextrous challenge of enforcing continuous change simultaneously with impact and value-creation that is typical for this stage. DSCT is structured along stages, not phases. While phases have very specific points of ending and beginning, stages can overlap in their ending and beginning, also as a change process involves case individual aspects (e.g., being in continuity stage and realizing that further momentum should be established). The continuity stage begins when companies have created a portfolio of digital initiatives and start to focus on their selected implementation, shifting the responsibilities for initiatives progressively to operational employees. Companies start to standardize new working approaches that are part of DSCT (e.g., more agile development culture) and intensify their inter-organizational collaboration activities. Also, companies have defined their digital transformation strategy or vision and adjusted their organizational structures, especially regarding financial funding processes of digital initiatives and the organization of employees that have extensive digital expertise.

Out of these changes, the continuity stage faces different challenges than the momentum stage. The momentum stage mobilized employees' motivation for DSCT, adjusting organizational structures and initiating digital initiatives. The continuity stage needs to enable continuous change and simultaneously ensure realized value-add via transformation changes. Hence, companies need to keep their employees motivated, ensure that existing digital initiatives are continuously improved, and facilitate to identification and develop new digital initiatives, despite more restrictive development and implementation processes. Although there is a risk of employees falling back into pre-transformation behaviors (e.g., rejecting digital working approaches again), due to less focus on building momentum, the risk of lower employee excitement about DSCT is more important. For example, do employees lose excitement for required tasks (e.g., improvement of master data). Also, employees want to continue working in a quite unrestricted environment (e.g., having access to "play budgets") and not follow comparably restrictive processes, although these processes still provide more freedom than given by structures before DSCT.

To address *psychological safety*, companies can conduct several activities including guidance via a more specified digital transformation vision that is essential for transformation success. *Adjustment of systems* and a combination of *role-modeling and involvement* contribute significantly to the success of the continuity stage. Successful companies that already identified the contribution of adjusting systems and structures in the momentum stage, focus more on role-modeling and involvement in continuity stage. Thus, less successful companies focus slightly more on adjusting their systems in the continuity stage than highly successful case companies. In the continuity stage, companies focus on implementing prototypes and pilots of digital technologies that are scaled throughout a supply chain. These are based on an established IT infrastructure and the data it stores, which, if IT systems are integrated and data is accurate, provide safety to progress. Simultaneous to this focus and structure, it is important to empower employees in developing and implementing digital initiatives via a risk-free and cross-functional environment that utilizes the improved IT infrastructure and data, to keep a continuous momentum that enables transformation success.

Related to the adjustment of structures, training supports the realization of value. Companies should utilize less formal training (e.g., standardized training courses) but keep the informal training (e.g., learning by testing and communication also via support groups) for transformation success. In the momentum stage, formal training provided employees with a basic understanding of DSCT. In the continuity stage, resources are better invested elsewhere, as standardized training cannot provide employees with the expertise required at this stage, due to a variety of digital-technology-specific expertise that is utilized. PharmaInc illustrated how employees utilized informal training and specialized in selected areas of expertise, supporting initiatives cross-functionally. Hence, not every employee needs to understand every digital technology but could focus her or his capacity on one technological area. This balance of focus and freedom to develop contributes to success in the continuity stage.

Successful companies can utilize more used cases and *role-models* in the continuity stage, which contributes to their transformation success. Role-modeling illustrates the benefits of supporting DSCT and encourages employees who are increasingly *involved* and take over responsibility for digital initiatives in this stage. PharmaInc established key users in functions to develop role-models who are peers of operational employees. The "peer-to-peer type" of role-modeling (e.g., at ChemicalsCo) is a valuable mechanism, as employees feel even more secure if their functional colleagues successfully participate and support change processes. This often facilitates an entrepreneurial environment which helps employees identify, develop, and implement digital initiatives.

Discomfort and, consequently, *survival anxiety* reduce significantly as the company moves from the momentum stage to the continuity stage. The continuity stage does not have the focus of creating momentum anymore but should ensure ongoing change. To successfully keep momentum, especially internal discomfort (usually linked to positive survival anxiety) is important, as it provides a more positive source of motivation to change (e.g., employees expecting further benefits out of digital technologies) than external discomfort (usually linked to negative survival anxiety). A notion of missing opportunities without digital transformation is established, supporting continuous development.

With a decreasing focus on momentum, also *learning anxiety* decreases, while several anxieties even disappear (e.g., loss of uncertainty), as people get used to a new mindset and behavior. MachineryCo provided an example of a special source of learning anxiety that companies should be aware of to not risk their previous transformation success. A team dedicated to DSCT feared losing power or position if digital initiatives were not developing successfully. In the continuity stage, the expectation of delivering value-add via digital initiatives increased, but the limited impact of digital initiatives frustrated operational employees of MachineryCo. Hence, previously supportive employees built up resistance against digital transformation actions and the dedicated team feared to be blamed for failures of digital initiatives. To keep the team resources highly utilized in front of the top management, the dedicated team started digital initiatives, even if direct value-add was questionable. This behavior had a negative impact on role-modeling, as the team was not representing a "risk-free trial-and-error"-behavior anymore. This negatively impacted the success of MachineryCo's DSCT. Companies should consider this aspect of learning anxiety when building up psychological safety by introducing a dedicated team as part of adapting company structures.

3.5 Discussion and Conclusion

3.5.1 Theoretical Contribution

This research contributes to the change theory in the context of operations/supply chain literature, specifically focusing on motivation to change in DSCT. The findings extend the theory by describing by what (i.e., motivational dimensions) and how (i.e., themes) companies enable a successful DSCT. Hereby, this study addresses the increasing relevance of the people dimension in the context of digital transformations, currently emphasized by literature (e.g., Dubey et al., 2024; Kane et al., 2019; Westerman, 2016).

First, our research extends the change theory, as it applies the dimensions of motivation to change (i.e., discomfort, survival anxiety, learning anxiety, and psychological safety) in the context of DSCT, uncovering their role in this highly dynamic change process. Although previous literature described psychological safety as the most important dimension for change (Schein, 2010), this study illustrates the actual importance of psychological safety in the digital domain and more in particular in DSCT. Moreover, our research reveals that psychological safety is directly related to success in a DSCT, enhancing motivation to change; a relation that is not described by current literature (e.g., Cataldo et al., 2009). Insights of this study also show how psychological safety has a direct influence on other motivational dimensions besides learning anxiety. The interrelations illustrate that levels of discomfort and anxiety should be viewed in relation to the level of psychological safety established in a company.

Second, this study extends the change theory by enhancing and describing themes about how companies can increase motivation to change via specific actions. Schein (2010) provides themes for learning anxiety and psychological safety, which are selectively utilized by literature to describe organizational change (e.g., Valentino, 2004). Our findings enhance this literature, by identifying themes for discomfort and survival anxiety while adding themes of "uncertainty" and "missing trust and feedback" to the dimension of learning anxiety. Hereby, especially uncertainty shows a high relevance for motivation in DSCT. Based on in-depth case understanding and frequency analysis, the research findings describe the themes that are most essential for transformation success (e.g., vision) in more detail, specified for DSCT. In comparison to current literature (e.g., Dubey et al., 2024; Schein & Schein, 2017), the insights point out specific interrelations of key themes, for example vision and employee involvement, which have an essential impact on the success of a transformation.

Third, the research contributes to the digital transformation literature in the context of motivation in DSCT, by identifying and describing novel transformation stages. Change theory described three stages of unfreezing, moving, and refreezing in the context of motivation to change (Akbari et al., 2024; Schein, 2010). Existing literature also points out the importance of stages in DSCT and introduced own constructs (Büyüközkan & Göçer, 2018; Wang & Prajogo, 2024), but does not illustrate the influence on motivation to change nor describes in-depth themes related to establishing this motivation. In comparison, our findings identified and describe the two transformation stages of momentum and continuity in the context of motivation to change, enhanced by detailed descriptions of related themes, pointing out interrelations between the stages. Hereby, the insights show how the importance of themes for DSCT success shifts between stages and which impact this has on the role and combination of themes in the corresponding stage.

Fourth, we add to the supply chain literature by identifying supply chain challenges, and cross-functional as well as inter-organizational aspects in the context of motivation in DSCT. Previous literature points out supply-chain-specific challenges (e.g., missing capabilities (Saenz et al., 2022)) that digital technologies create but also solve (e.g., Calatayud et al., 2019). Nevertheless, the literature misses supply-chain-specific challenges in the context of motivation in a DSCT. Our research describes these, including the cross-functional collaboration that is increasingly required for DSCT, the role of lateral and vertical supply chain partners in the context of discomfort and psychological safety, and IT infrastructure obstacles for the development and implementation of digital initiatives along the end-to-end supply chain.

Finally, this study contributes to digital transformation and supply chain research by integrating relevant literature in the context of motivation in DSCT. Current literature emphasizes limited consideration of supply chain aspects in digital transformation research (e.g., Hu & Monahan, 2015). Although researchers conducted literature reviews to consolidate existing literature on DSCT (e.g., Büyüközkan & Göçer, 2018; Wang & Prajogo, 2024), the focus on digital transformation and supply chain literature in the context of motivation is still missing. Our research integrates literature of both fields and brings them into the context of motivation to change. Hereby, we develop a research framework that future research can utilize to structure their investigation of motivational DSCT aspects, as well as their empirical validation.

3.5.2 Managerial Implications

In order to translate our theoretical findings into practical actions, it is useful to map the key motivational dimensions onto the distinct stages of a successful digital transformation journey—specifically, the momentum and the continuity stages. This mapping helps managers identify which actions and strategies are most pertinent at each phase, thereby facilitating a targeted and effective DSCT implementation.

Companies need to actively consider and engage their employees for a successful digital transformation, as the "people-dimension" is of increasing importance in the digital context (Chirumalla, 2021; Grover & Kohli, 2013; Kane et al., 2019; Kohtamäki et al., 2024; Warner & Wäger, 2019). Our research insights confirm this, by emphasizing the essential role of motivation to change for DSCT success, especially in large established companies. The consequences of failing a digital transformation are severe for these companies as demonstrated by the experiences of General Electric and Kodak (Davenport & Westerman, 2018; Schoemaker et al., 2018). Hence, our study provides significant value-add for companies by providing actionable insights on how to successfully establish motivation and enable a fruitful DSCT.

First, this research provides company leadership with a structured framework to assess their current DSCT motivation and stage throughout their supply chain, detailed out with actionable themes. Leadership should identify their motivational state, down to the described themes, to identify gaps that need to be addressed. For example, leadership can take the framework as a frequently reviewed reference to identify decreasing psychological safety when prototyping and scaling up digital supply chain processes.

Second, our study describes themes for each of the motivational dimensions and thereby concrete actions to increase psychological safety, understand discomfort, and address survival and learning anxiety. This provides companies with the opportunity to adjust the intensity of themes according to the two foundational transformation stages (i.e., momentum and continuity) , supporting the continuity of DSCT success.

In the momentum stage, where initiating change is critical, our findings clearly indicate that establishing psychological safety—through a well-defined vision, active employee involvement, and structured role-modeling—is essential to create an environment that embraces trial and error and rapid learning. Companies should at least develop a DSCT vision, involving operational employees to encourage ownership and benefit from their perspective. To further support transformation success, company leadership needs to establish role-models that illustrate the targeted behavior and provide formal training that enables operational employees to build up basic digital expertise. Simultaneously, companies need to adjust their IT infrastructure and master data to provide a foundation for digital initiative development and implementation using an end-to-end approach.

As organizations transition into the continuity stage, the focus shifts from generating initial momentum to sustaining continuous innovation and improvement. Here, our research highlights that companies need to establish a balance between realizing value-add and enabling continuous change. For this, company leadership needs to support an entrepreneurial environment—a culture facilitated by organizational structures that allows employees to identify, develop, and implement digital initiatives. In addition, continuous updates of IT infrastructure and data are required. Role-models provide stable and encouraging references for the operational employees while involving operational employees enables that these are increasingly responsible for digital initiatives.

Third, the research insights describe the importance of balancing the efforts in the context of DSCT, focused on technology-driven change management with proactive behavior. Companies that proactively approach digital technologies as opportunities for improvement, creating an internally initiated need for change, can start their DSCT before external partners create anxiety that leads to less efficient, reactive change. This also applies to individual operational employees who can utilize DSCT as an opportunity for their individual development in a company. The insights illustrate essential roles that operational employees can take in DSCT, for example in the form of role-models that are supported by top management.

3.5.3 Potential Limitations and Suggestions for Future Research

This chapter created insightful findings by focusing on motivation to change in DSCT. Throughout the research process, we identified the following potential research limitations that are of interest to future research. First, our multiple case study approach included companies out of different industries to enable overarching findings. Research can investigate the influence on the research findings, if companies out of a selected industry such as the retail industry, are in focus. Based on our cases positioned in the retail industry, we would expect an on average higher level of digital maturity than in companies of the manufacturing industry. Scholars who utilize this research opportunity should consider the insights (especially themes) of this study to point out interesting industry-specific characteristics.

Second, this research focused on large, established, and internationally acting companies in DSCT. Digital transformation is also influencing small size companies (Saeedikiya et al., 2024), such as small manufacturing companies that have established a long company history. Based on discussions in our research process, these companies are also under competitive pressure in the context of DSCT. Hereby, we would expect a lower level of current discomfort regarding DSCT at companies that have focused their business model on local, rural customers (vom Berg et al., 2017). Future research can identify the implications of the research findings in this context of small-size companies, which would provide a direct value-add to these businesses.

Third, this chapter focuses on a topic of increasing relevance and continuous change. We encourage future research to replicate this study as more companies progress into their DSCT to provide further valuable research insights. Further, our focus in this chapter was on the role of motivation in DSCT from an organizational perspective while considering the implications for the supply chain. We expect that our research will pursue future work with a focus on the role of motivation in coordinating digital transformation initiatives among upstream and downstream supply-chain partners.

References

Abdirad, M., & Krishnan, K. (2021). Industry 4.0 in logistics and supply chain management: A systematic literature review. *Engineering Management Journal, 33*(3), 187–201.

Agarwal, R., Selen, W., Sajib, S., & Scerri, M. (2014). Dynamic capability building in service networks: An exploratory case study. *Journal of New Business Ideas & Trends, 12*(1), 27–41.

Akbari, M., Kok, S. K., Hopkins, J., Frederico, G. F., Nguyen, H., & Alonso, A. D. (2024). The changing landscape of digital transformation in supply chains: Impacts of industry 4.0 in Vietnam. *The International Journal of Logistics Management, 35*(4), 1040–1072.

Åkesson, M., Sørensen, C., & Eriksson, C. I. (2018). Ambidexterity under digitalization: A tale of two decades of new media at a Swedish newspaper. *Scandinavian Journal of Management, 34*, 276–288.

Alos-Simo, L., Verdu-Jover, A. J., & Gomez-Gras, J.-M. (2017). How transformational leadership facilitates e-business adoption. *Industrial Management & Data Systems, 117*(2), 382–397.

Alvesson, M. (2003). Beyond neopositivists, romantics, and localists: A reflexive approach to interviews in organizational research. *Academy of Management Review, 28*(1), 13–33.

Ba, S., & Nault, B. R. (2017). Emergent themes in the interface between economics of information systems and management of technology. *Production and Operations Management, 26*(4), 652–666.

Balasubramanian, N., Ye, Y., & Xu, M. (2020). Substituting human decision-making with machine learning: Implications for organizational learning. *Academy of Management Review*. https://doi.org/10.2139/ssrn.4543450

Bansal, P., & Corley, K. (2011). The coming age for qualitative research: Embracing the diversity of qualitative methods. *Academy of Management Journal, 54*(2), 233–237.

Bienhaus, F., & Haddud, A. (2018). Procurement 4.0: factors influencing the digitisation of procurement and supply chains. *Business Process Management Journal, 24*(4), 965–984.

Burke, W. W. (2017). *Organization change: Theory and practice*. Sage.

Busse, C., Meinlschmidt, J., & Foerstl, K. (2017). Managing information processing needs in global supply chains: A prerequisite to sustainable supply chain management. *Journal of Supply Chain Management, 53*(1), 87–113.
Büyüközkan, G., & Göçer, F. (2018). Digital supply chain: Literature review and a proposed framework for future research. *Computers in Industry, 97*, 157–177.
Calatayud, A., Mangan, J., & Christopher, M. (2019). The self-thinking supply chain. *Supply Chain Management: An International Journal, 1*, 22.
Carboni, I., Cross, R., & Edmondson, A. C. (2021). No team is an Island: How leaders shape networked ecosystems for team success. *California Management Review, 64*(1), 5–28.
Cataldo, C. G., Raelin, J. D., & Lambert, M. (2009). Reinvigorating the struggling organization: The unification of Schein's oeuvre into a diagnostic model. *The Journal of Applied Behavioral Science, 45*(1), 122–140.
Cerdas, F., Juraschek, M., Thiede, S., & Herrmann, C. (2017). Life cycle assessment of 3D printed products in a distributed manufacturing system. *Journal of Industrial Ecology, 21*, 80–93.
Chen, Y., Tang, G., Jin, J., Xie, Q., & Li, J. (2014). CEOs' transformational leadership and product innovation performance: The roles of corporate entrepreneurship and technology orientation. *Journal of Product Innovation Management, 31*, 2–17.
Chirumalla, K. (2021). Building digitally-enabled process innovation in the process industries: A dynamic capabilities approach. *Technovation, 105*, 1–24.
Chopra, S., & Meindl, P. (2007). Supply chain management. In C. Boersch & R. Elschen (Eds.), *Strategy, planning & operation, Das summa summarum des management* (pp. 265–275). Gabler.
Corbin, J. M., & Strauss, A. L. (2008). *Basics of qualitative research: techniques and procedures for developing grounded theory* (3rd ed.). Sage.
Dallasega, P., Rauch, E., & Linder, C. (2018). Industry 4.0 as an enabler of proximity for construction supply chains: A systematic literature review. *Computers in Industry, 99*, 205–225.
Danuso, A., Giones, F., & Ribeiro da Silva, E. (2022). The digital transformation of industrial players: A guide. *Business Horizons, 65*(3), 341–349.
Davenport, T. H., & Westerman, G. (2018). Why so many high-profile digital transformations fail. *Harvard Business Review Digital Articles, 9*, 2–5.
Dremel, C., Wulf, J., Herterich, M. M., Waizmann, J.-C., & Brenner, W. (2017). How AUDI AG established big data analytics in its digital transformation. *MIS Quarterly Executive, 16*(2), 81–100.
Dubey, R., Bryde, D. J., Blome, C., Dwivedi, Y. K., Childe, S. J., & Foropon, C. (2024). Alliances and digital transformation are crucial for benefiting from dynamic supply chain capabilities during times of crisis: A multi-method study. *International Journal of Production Economics, 269*, 109166.
Eden, R., Burton-Jones, A., Casey, V., & Draheim, M. (2019). Digital transformation requires workforce transformation. *MIS Quarterly Executive, 18*(1), 1–17.
Edmondson, A. C., & Lei, Z. (2014). Psychological safety: The history, renaissance, and future of an interpersonal construct. *Annual Review of Organizational Psychology and Organizational Behavior, 1*(1), 23–43.
Eisenhardt, K. M. (1989). Building theories from case study research. *Academy of Management Review, 14*(4), 532–550.
Ellis, S., Knickle, K., Veronesi, L., & Brown, V. (2017). *Digital transformation drives supply chain restructuring imperative*. IDC: International Data Corporation.
El Sawy, O. A., Kræmmergaard, P., Amsinck, H., & Vinther, A. L. (2016). How LEGO built the foundations and enterprise capabilities for digital leadership. *MIS Quarterly Executive, 15*(2), 141–166.
Fassoula, E. D. (2006). Transforming the supply chain. *Journal of Manufacturing Technology Management, 17*(6), 848–860.
Fawcett, S. E., Wallin, C., Allred, C., Fawcett, A. M., & Magnan, G. M. (2011). Information technology as an enabler of supply chain collaboration: a dynamic-capabilities perspective. *Journal of Supply Chain Management, 47*(1), 38–59.

Fitzgerald, M., Kruschwitz, N., Bonnet, D., & Welch, M. (2014). Embracing digital technology: A new strategic imperative. *MIT Sloan Management Review, 55*(2), 1–12.

Gartner. (2018). Current use cases for machine learning in supply chain solutions. Retrieved September 7, 2020, from https://www.gartner.com/en/documents/3875876

Gezgin, E., Huang, X., Samal, P., & Silva, I. (2017, November 17). Digital transformation: Raising supply-chain performance to new levels. https://www.mckinsey.com/business-functions/operations/our-insights/digital-transformation-raising-supply-chain-performance-to-new-levels.

Gibbert, M., Ruigrok, W., & Wicki, B. (2008). What passes as a rigorous case study? *Strategic Management Journal, 29*(13), 1465–1474.

Gioia, D. A., Corley, K. G., & Hamilton, A. L. (2013). Seeking qualitative rigor in inductive research: Notes on the Gioia methodology. *Organizational Research Methods, 16*(1), 15–31.

Grover, V., & Kohli, R. (2013). Revealing your hand: Caveats in implementing digital business strategy. *MIS Quarterly, 37*(2), 1–8.

Hess, T., Matt, C., Benlian, A., & Wiesböck, F. (2016). Options for formulating a digital transformation strategy. *MIS Quarterly Executive, 15*(2), 123–139.

Hinings, B., Gegenhuber, T., & Greenwood, R. (2018). Digital innovation and transformation: An institutional perspective. *Information and Organization, 28*(1), 52–61.

Holmström, J., & Partanen, J. (2014). Digital manufacturing-driven transformations of service supply chains for complex products. *Supply Chain Management, 19*(4), 421–430.

Hu, M., & Monahan, S. T. (2015). Sharing supply chain data in the digital era. *MIT Sloan Management Review, 57*(1), 95–96.

Huber, J., & Stuckenschmidt, H. (2020). Daily retail demand forecasting using machine learning with emphasis on calendric special days. *International Journal of Forecasting, 36*(4), 1420–1438.

Kane, G. C., Michelman, P., Copulsky, J. R., Phillips, A. N., & Andrus, G. R. (2019). *The technology fallacy: How people are the real key to digital transformation*. MIT Press.

Karimi, J., & Walter, Z. (2015). The role of dynamic capabilities in responding to digital disruption: A factor-based study of the newspaper industry. *Journal of Management Information Systems, 32*(1), 39–81.

Knoppen, D., Sáenz, M. J., & Johnston, D. A. (2011). Innovations in a relational context: mechanisms to connect learning processes of absorptive capacity. *Development and Learning in Organizations: An International Journal, 2*, 419–438.

Kohli, R., & Johnson, S. (2011). Digital transformation in latecomer industries: CIO and CEO leadership lessons from Encana Oil & Gas (USA) Inc. *MIS Quarterly Executive, 10*(4), 141–156.

Kohtamäki, M., Leminen, S., & Parida, V. (2024). Conceptualizing digital business models (DBM): Framing the interplay between digitalization and business models. *Technovation, 133*, 103013.

Krimpmann, D. (2015). IT/IS organisation design in the digital age—A literature review. *International Journal of Social, Behavioral, Educational, Economic, Business and Industrial Engineering, 9*(4), 1208–1218.

Kuipers, B. S., Kickert, W., Tummers, L., Grandia, J., Van Der Voet, J., & Higgs, M. (2014). The management of change in public organizations: A literature review. *Public Administration, 92*(1), 1–20.

Kummer, T.-F., Recker, J., & Bick, M. (2017). Technology-induced anxiety: Manifestations, cultural influences, and its effect on the adoption of sensor-based technology in German and Australian hospitals. *Information & Management, 54*, 73–89.

Kuusisto, M. (2017). Organizational effects of digitalization: A literature review. *International Journal of Organization Theory & Behavior, 20*(3), 341–362.

Lee, H. L. (2004). The triple-A supply chain. *Harvard Business Review, 82*(10), 102–113.

Lennartz, C., Saenz, M. J., & Wallenburg, C. M., (2019, July). Facilitating digital supply chain transformations: The influence of dynamic capabilities. In *Academy of Management proceedings* (Vol. 2019, No. 1, p. 10202). Academy of Management.

Lewin, K. (1947). Group decision and social change. *Readings in Social Psychology, 3*, 197–211.

Locke, K. (2001). *Grounded theory in management research*. Sage.

Lucas, J. H. C., & Goh, J. M. (2009). Disruptive technology: How Kodak missed the digital photography revolution. *Journal of Strategic Information Systems, 18*(1), 46–55.

Mallouf, M., & Gammelgaard, B. (2016). Managing paradoxical tensions during the implementation of lean capabilities for improvement. *International Journal of Operations & Production Management, 6*, 687–709.

Mangematin, V., Sapsed, J., & Schussler, E. (2014). Disassembly and reassembly: An introduction to the Special Issue on digital technology and creative industries. *Technological Forecasting and Social Change, 83*, 1–9.

Meyr, H., & Günther, H.-O. (2009). *Supply chain planning*. Springer.

Netland, T. H., & Ferdows, K. (2016). The S-curve effect of lean implementation. *Production & Operations Management, 25*(6), 1106–1120.

Prahalad, C. K., & Hamel, G. (1994). Strategy as a field of study: Why search for a new paradigm? *Strategic Management Journal, 15*(S2), 5–16.

Rafferty, A., & Simons, R. (2006). An examination of the antecedents of readiness for fine-tuning and corporate transformation changes. *Journal of Business & Psychology, 20*(3), 325–350.

Reddy, G. R. K., Singh, H., & Hariharan, S. (2016). Supply chain wide transformation of traditional industry to industry 4.0. *ARPN Journal of Engineering and Applied Sciences, 11*(18), 11089–11097.

Reis, J., Amorim, M., Melao, N., & Matos, P. (2018). Digital transformation: A literature review and guidelines for future research. In Á. Rocha, H. Adeli, L. P. Reis, & S. Costanzo (Eds.), *Trends and advances in information systems and technologies* (pp. 411–421). Springer.

Reusen, E., Stouthuysen, K., Roodhooft, F., Van den Abbeele, A., & Slabbinck, H. (2020). Imitation of management practices in supply networks: relational and environmental effects. *Journal of Supply Chain Management, 56*(1), 54–72.

Revilla, E., Saenz, M. J., Seifert, M., & Ma, Y. (2023). Human–AI collaboration in prediction: A field experiment in the retail industry. *Journal of Information and Management Systems, 40*(4).

Ridder, H.-G. (2017). The theory contribution of case study research designs. *Business Research, 10*(2), 281–305.

Sabri, Y., Micheli, G. J. L., & Nuur, C. (2018). Exploring the impact of innovation implementation on supply chain configuration. *Journal of Engineering and Technology Management, 49*, 60–75.

Sackmann, S. (1989). The role of metaphors in organization transformation. *Human Relations, 42*(6), 463–485.

Saeedikiya, M., Salunke, S., & Kowalkiewicz, M. (2024). Toward a dynamic capability perspective of digital transformation in SMEs: A study of the mobility sector. *Journal of Cleaner Production, 439*, 140718.

Saenz, M. J., Borrella, I., & Revilla, E. (2022, March/April). Digital supply chain transformation: Aligning operations and strategy. *Supply Chain Management Review*.

Sáenz, M. J., Revilla, E., & Knoppen, D. (2014). Absorptive capacity in buyer–supplier relationships: Empirical evidence of its mediating role. *Journal of Supply Chain Management, 50*(2), 18–40.

Sahu, C. K., Young, C., & Rai, R. (2021). Artificial intelligence (AI) in augmented reality (AR)-assisted manufacturing applications: a review. *International Journal of Production Research, 59*(16), 4903–4959.

Sasson, A., & Johnson, J. C. (2016). The 3D printing order: variability, supercenters and supply chain reconfigurations. *International Journal of Physical Distribution & Logistics Management, 46*(1), 82.

Saunders, M., Lewis, P., & Thornhill, A. (2015). *Research methods for business students* (7th ed.). Pearson.

Schein, E. H. (2009). *The corporate culture survival guide*. Wiley.

Schein, E. H. (2010). *Organizational culture and leadership* (4th ed.). Wiley.

Schein, E. H., & Schein, P. (2017). *Organizational culture and leadership* (5th ed.). Wiley.

Schoemaker, P. J. H., Heaton, S., & Teece, D. (2018). Innovation, dynamic capabilities, and leadership. *California Management Review, 61*(1), 15–42.
Schuchmann, D., & Seufert, S. (2015). Corporate learning in times of digital transformation: A conceptual framework and service portfolio for the learning function in banking organisations. *International Journal of Advanced Corporate Learning, 8*(1), 31–39.
Schwarzmüller, T., Brosi, P., Duman, D., & Welpe, I. M. (2018). How does the digital transformation affect organizations? Key themes of change in work design and leadership. *Management Revue, 29*(2), 114–138.
Sebastian, I. M., Ross, J. W., Beath, C., Mocker, M., Moloney, K. G., & Fonstad, N. O. (2017). How big old companies navigate digital transformation. *MIS Quarterly Executive, 16*(3), 197–213.
Seifert, R. W., & Markoff, R. (2018). Tetra Pak: A digitally enabled supply chain as a competitive advantage. https://www.imd.org/research-knowledge/for-educators/case-studies/tetra-pak-a-digitally-enabled-supply-chain-as-a-competitive-advantage/; 2019/06/16.
Singh, A., & Hess, T. (2017). How chief digital officers promote the digital transformation of their companies. *MIS Quarterly Executive, 16*(1), 1–17.
Svahn, F., Mathiassen, L., & Lindgren, R. (2017). Embracing digital innovation in incumbent firms: How Volvo cars managed competing concerns. *MIS Quarterly, 41*(1), 239–253.
Syberfeldt, A., Danielsson, O., & Gustavsson, P. (2017). Augmented reality smart glasses in the smart factory: Product evaluation guidelines and review of available products. *IEEE Access, 5*, 9118–9130.
Valentino, C. L. (2004). The role of middle managers in the transmission and integration of organizational culture. *Journal of Healthcare Management, 49*(6), 393–404.
van de Ven, A. H., & Poole, M. S. (1995). Explaining development and change in organizations. *The Academy of Management Review, 20*(3), 510–540.
Vanpoucke, E., Quintens, L., & Van Engelshoven, M. (2016). The role of motivation in relating green supply chain management to performance. *Supply Chain Management: An International Journal, 21*(6), 732–742.
Vial, G. (2019). Understanding digital transformation: A review and a research agenda. *Journal of Strategic Information Systems, 28*(2), 118–144.
vom Berg, B. W., Gómez, J. M., & Sandau, A. (2017). ICT-Platform to transform car dealerships to regional providers of sustainable mobility services. *Interdisciplinary Journal of Information, Knowledge & Management, 12*, 37–51.
Voss, C., Tsikriktsis, N., & Frohlich, M. (2002). Case research in operations management. *International Journal of Operations & Production Management, 22*(2), 195–219.
Wang, M., & Prajogo, D. (2024). The effect of supply chain digitalisation on a firm's performance. *Industrial Management & Data Systems, 124*(5), 1725–1745.
Warner, K. S., & Wäger, M. (2019). Building dynamic capabilities for digital transformation: An ongoing process of strategic renewal. *Long Range Planning, 52*(3), 326–349.
Westerman, G. (2016). Why digital transformation needs a heart. *MIT Sloan Management Review, 58*(1), 19–21.
Wu, L., Yue, X., Jin, A., & Yen, D. C. (2016). Smart supply chain management: a review and implications for future research. *The International Journal of Logistics Management, 27*(2), 395–417.
Xue, L., Zhang, C., Ling, H., & Zhao, X. (2013). Risk mitigation in supply chain digitization: System modularity and information technology governance. *Journal of Management Information Systems, 30*(1), 325–352.
Yin, R. K. (2014). *Case study research: design and methods*. Sage.
Zhang, Y. F., Gregory, M., & Neely, A. (2016). Global engineering services: Shedding light on network capabilities. *Journal of Operations Management, 42*, 80–94.

Chapter 4
The Impact of Industry 4.0 Technologies on Efficiency, Transparency, and Resilience in Global Supply Chains

Alexander Samuels and Blessing Takawira

Abstract Currently, the rapid advancement of Industry 4.0 technologies, including AI, blockchain, IoT, and big data analytics, is causing a substantial transformation in global supply chains. These advanced technologies are transforming supply chain operations by improving efficiency, transparency, and resilience, as this chapter examines. For instance, the implementation of blockchain technology improves traceability and accountability in intricate supply chains, while the Internet of Things (IoT) enables the continuous collection and monitoring of data in real time, thereby enabling proactive decision-making and predictive analytics. Furthermore, operational optimisation and cost reduction are facilitated by AI and machine learning. The COVID-19 pandemic has emphasised the urgent necessity of supply chains that are both adaptable and strong, which has led to a wider adoption of these technologies to mitigate disruptions and adjust to changing demands. Employing a systematic literature review, the chapter investigates the effects of Industry 4.0 on global supply chain management. The potential for digital transformation is immense, providing supply chain managers with valuable insights to encourage more sustainable and resilient networks, even though challenges remain, including the need for skilled professionals and high implementation costs.

Keywords Blockchain technology · Global supply chain · Industry 4.0 · Supply chain efficiency · Supply chain resilience

---

A. Samuels (✉)
North West University Business School, Potchefstroom, South Africa
e-mail: Alexander.Samuels@nwu.ac.za

B. Takawira
Department of Business Management, University of Johannesburg, Auckland Park, Kingsway Campus, South Africa

Y. Khojasteh, S.M. M. Jahan Abadi (eds.), *Digital Transformation of Supply Chains*, International Series in Operations Research & Management Science 372, https://doi.org/10.1007/978-3-032-01206-7_4

4.1 Introduction

Global supply chains are currently facing a myriad of challenges that are impacting their efficiency, transparency, and resilience. The complexity of supply chain management in a globalised context has led to difficulties in ensuring transparency, traceability, and efficiency (Lalitha, 2024). Conventional systems are facing difficulties in handling the growing complexity of worldwide supply chains, which are worsened by concerns including insufficient transparency regulations resulting in corporate accountability problems associated to contemporary slavery and unethical labour practices (Islam, 2021). The absence of standardised measures and economic considerations in achieving transparency in supply chains worsens these difficulties (Holloway, 2024). Technological progress, specifically the implementation of blockchain technology, has been recognised as a possible remedy for tackling these difficulties. Blockchain presents prospects for improving efficiency, transparency, and innovation in worldwide supply chains (Oriekhoe, 2024a). Blockchain technology has been recognised for its potential to improve traceability, accountability, and information flow in supply chains, specifically in businesses such as food supply chains that strive to accomplish sustainable objectives (Chandan et al., 2023). The utilisation of blockchain technology has the potential to improve the integration of supply chains and promote sustainable performance, as demonstrated in the automobile industry (Kamble et al., 2021). The COVID-19 pandemic has underlined the significance of resilience in worldwide supply systems. The global pandemic has caused significant disruptions to logistics processes, highlighting the importance for supply chain leaders to consistently assess their strategic positions in order to meet key supply requirements (Forehand et al., 2021). The aftermath of the pandemic has shown obstacles to transparency in supply chains, including unclear short-term return on investment and limited adoption of technology. These issues must be resolved in order to improve the efficiency of supply systems (Chatterjee & Chatterjee, 2022).

The technologies encompassed in this list are the IoT, blockchain, AI, and big data analytics. By using these technologies, organisations may optimise their operations, improve transparency throughout the supply chain, and efficiently adapt to disruptions and fluctuations in demand. Blockchain technology has become a powerful tool for addressing supply chain difficulties. The framework improves transparency, traceability, and efficiency by utilising a decentralised and secure system (Oriekhoe, 2024a).

Companies are able to authenticate products, monitor commodities across the supply chain, and establish confidence among stakeholders by leveraging blockchain technology (Badhwar et al., 2023). Technology also facilitates the adoption of ethical sourcing procedures and improves transparency in areas such as agriculture and textiles (Ibrahim, 2024). The integration of Industry 4.0 technologies such as the IoT and AI allows for the continuous monitoring of supply chain activities in real-time, the use of predictive analytics to make proactive decisions, and the automation of repetitive tasks (Olutimehin, 2024). These improvements improve operational efficiency and mitigate risks by providing data-driven insights and visibility into the whole supply chain network. The implementation of digital technologies such as RFID, robots, business intelligence systems, and cloud technology within the industry 4.0 framework guarantees long-lasting organisational effectiveness (Kamble et al., 2021). These technologies enable the integration of processes, sharing of data, and collaboration across partners in the supply chain, resulting in improved performance and adaptability to market changes. Drawing from this background, the chapter objectives are

- To assess the impact of industry 4.0 technologies on operational efficiency in global supply chains
- To investigate the role of blockchain technology in improving transparency and trust among supply chain participants
- To explore the Contribution of industry 4.0 technologies to building resilience in global supply chains

4.2 Theoretical Background

4.2.1 An Overview of Industry 4.0 Technologies

Global supply chains have been profoundly impacted by Industry 4.0, which is defined by the integration of cutting-edge technologies into numerous industries. The IoT, AI, Blockchain, and Big Data Analytics are the key technologies fuelling this transformation. The IoT is a fusion of sensor, embedded, computer, and communication technologies that allows for the interconnection of devices and systems (Swamy & Raju, 2020). AI, utilising its machine learning capabilities, improves operational efficiency by automating procedures and making predictions about outcomes (Zhou et al., 2023). Blockchain technology, renowned for its decentralised and immutable ledger, plays a pivotal role in improving transparency and encouraging trust among actors in the supply chain. The utilisation of blockchain technology in manufacturing guarantees the execution of secure and transparent transactions, hence improving the management of supply chains and the tracing of products.

Big Data Analytics utilises extensive quantities of data to extract insights, improve processes, and make well-informed judgements. Analysing data from IoT devices and other sources allows for the optimisation of supply chain activities, resulting in improved efficiency (Zhou et al., 2023). Integrating blockchain technology with IoT and AI provides a comprehensive solution for constructing strong supply networks. Traceability, security, and reliability are key for maintaining continuity in the presence of disturbances (Park & Li, 2021). However, the implementation of Industry 4.0 technologies such as Blockchain in the supply chains of agriculture and construction has demonstrated encouraging outcomes in bolstering resilience, particularly in times of crises like as the COVID-19 pandemic (Khan, Malik, et al., 2022).

Significant milestones have characterised the progression of these technologies throughout the years. Blockchain technology has significantly improved multiple elements of supply chain management. The use of information and communication technology (ICT) into sectors such as agriculture and food initially facilitated developments in production and transportation (Chandan et al., 2023). The advent of blockchain technology has revolutionised supply chains by offering transparency, traceability, and security (Revathi, 2024). Blockchain has played a key role in connecting stakeholders, guaranteeing the accuracy of data, and encouraging confidence among participants (Santhi & Muthuswamy, 2022). Supply chains have become more transparent and efficient as a result of the use of blockchain, especially in the luxury goods and agricultural industries (Boissieu et al., 2021). Research has demonstrated that blockchain technology improves trust among individuals or groups involved, diminishes the likelihood of fraudulent activities, and improves the ability to track and trace things (Oriekhoe, 2024c). Blockchain technology has been recognised as a disruptive force having the ability to reshape the movement of goods and information in supply chains (Alqarni et al., 2023).With Industry 4.0 still developing, the emphasis now is on using blockchain to increase supply chain flexibility and encourage trust (Wang & Yang, 2022). Businesses are urged to allocate resources towards blockchain technology in order to adjust to ever-changing market conditions and improve operational flexibility (Meidutė-Kavaliauskienė et al., 2021).

Global supply chains have been profoundly impacted by Industry 4.0 technologies, which have increased operational resilience, efficiency, and transparency. The integration of blockchain technology is a key element of Industry 4.0 technologies. Studies have demonstrated that the implementation of blockchain technology improves the level of transparency and confidence among actors in supply chains (Ibrahim, 2024). By implementing blockchain technology, companies are able to achieve greater transparency in their supply chains and adopt ethical sourcing practices, leading to improved integrity of the entire supply chain (Ibrahim, 2024). Blockchain technology facilitates instantaneous transparency throughout the complete supply chain network, promoting cooperation, streamlining inventory control, and improving operational effectiveness (Revathi, 2024). Furthermore, blockchain technology offers a distributed, unchangeable, and open ledger system, which effectively tackles issues like fraud, counterfeiting, and inefficiencies in the worldwide supply chain industry (Sachan, 2024).

Industry 4.0 technologies are key for strengthening the strength of global supply networks, alongside blockchain. Implementing Industry 4.0 technologies, such as blockchain, can improve the ability of organisations to bounce back from disruptions and potentially achieve a more favourable condition than before, thus improving supply chain resilience. These technologies improve supply chains by guaranteeing the dependability, trackability, and genuineness of information, which are key for efficient supply chain management (Jiang, 2023). However, the integration of blockchain technology in supply chain finance can improve transparency, decrease regulatory expenses, mitigate fraud risks, and improve operational efficiency by means of digitisation and smart contracts (Wang, 2024).

4.2.2 Impact of Industry 4.0 on Supply Chain Efficiency

Supply chain processes have been profoundly impacted by Industry 4.0's real-time monitoring, data collection, and predictive maintenance capabilities. These technologies facilitate operational efficiency improvements by providing greater visibility and control over processes (Revathi, 2024). Through the utilisation of blockchain technology, actors in the supply chain can get improved levels of transparency and trust (Huzaifa, 2024). Blockchain technology guarantees the authenticity of transactions and information, minimising the possibility of deceit and imitation, therefore improving effectiveness and openness in supply chains (Oriekhoe, 2024b).

Adoption of blockchain technology also improves supply chain resilience by facilitating improved stakeholder cooperation and information sharing (Reynolds, 2024c). The decentralised structure of blockchain enables direct transactions between peers, minimising the need for intermediaries and simplifying trade processes (Reynolds, 2024a). Blockchain technology improves resilience by improving traceability and guaranteeing the genuineness of items (Hurduzeu, 2024). Additionally, it expedites the process of identifying contaminated products, resulting in focused recalls and minimising the consequences of foodborne illness outbreaks (Oriekhoe, 2024b). However, the utilisation of blockchain technology in supply chain management has advantages such as improved visibility, strength, and effectiveness (Priyadarshi & Asif, 2023).

AI and machine learning applications play a key role in improving operational efficiency by utilising predictive analytics, demand forecasting, and automated decision-making (Reynolds, 2024c). These technologies enable organisations to streamline procedures, minimise expenses, and improve overall efficiency (Atadoga, 2024a). AI greatly improves the efficiency of supply chains by improving decision-making, decreasing expenses, and optimising the allocation of resources (Eyo-Udo, 2024). Blockchain technology is a key factor that is transforming supply chains by improving transparency and encouraging trust among players. The system provides immediate visibility, insights based on data, and proactive methods to mitigate interruptions, hence improving the resilience of supply chains (Odimarha, 2024).

Blockchain technology improves the accuracy, ability to track, and environmental friendliness of supply chains (Chen, 2023).The characteristics and advantages of this solution effectively tackle common industrial difficulties such as inadequate information sharing and a lack of trust among stakeholders, leading to significant improvements in supply chain processes (Xiao, 2024). Industry 4.0 technologies, specifically AI and machine learning, play a key role in establishing resilience in global supply chains. Organisations can improve their ability to identify, assess, and respond to risks in real-time by using AI technologies like as machine learning, predictive analytics, and optimisation algorithms. This proactive approach helps mitigate disruptions and improves resilience (Bidin, 2024). These technologies aid companies in increasing their profit margins, minimising supply chain expenses, delivering exceptional customer service, and establishing intelligent supply networks, therefore strengthening overall resilience (Rana & Daultani, 2022).

The IoT, AI, and Blockchain have facilitated instantaneous exchange of information, collaboration among parties involved, and improved visibility and confidence in supply chains (Odimarha, 2024). Integrating Industry 4.0 technologies has been proven to improve operational efficiency in global supply chains by improving visibility, predictive analytics, and secure information sharing. This enables organisations to proactively identify risks and quickly respond to disruptions (Atadoga, 2024b).

Blockchain technology is key concept for improving transparency and trust among actors in the supply chain. It facilitates safe and transparent transactions, allows for product traceability, and enables effective record management, hence promoting confidence and accountability in supply chain activities (Remko, 2020). However, the implementation of Industry 4.0 technologies such as IoT, AI, and machine learning presents a considerable opportunity to improve the resilience of the supply chain. This is achieved through the provision of immediate and up-to-date visibility, insights derived from data analysis, and pre-emptive measures to mitigate possible issues (Reynolds, 2024b).

Furthermore, the impact of Industry 4.0 technologies on improving the ability of global supply chains to withstand and recover from crises, such as the COVID-19 pandemic, has been clearly demonstrated. Companies with digital capabilities demonstrated superior readiness and efficacy in addressing supply chain disturbances, underlining the significance of digital technology in strengthening resilience (Nakandala et al., 2023). The relevance of resilience capacities in moderating the influence of Industry 4.0 technologies on supply chain performance, particularly in vital areas such as healthcare, has been recognised as a key aspect (Tortorella et al., 2023).

4.2.3 *Building Resilience with Industry 4.0 Technologies*

Organisations are able to proactively identify and manage supply chain risks with the use of AI-driven risk assessment and mitigation approaches, which guarantees operational continuity in the face of uncertainties and interruptions (Chukwu, 2024). By utilising AI technology like machine learning and predictive analytics, organisations are able to strengthen their supply networks against unexpected disruptions, thus improving their resilience (Bidin, 2024).

In addition, AI has the capability to automate and improve risk management procedures, resulting in improved operational efficiency within worldwide supply chains (Cui et al., 2024). Furthermore, by enabling more efficient resource management and quick reactions to shocks, the combination of AI and big data analytics dramatically increases the resilience of supply chains (Zamani et al., 2022). The utilisation of AI in supply chains increases their resilience by enabling proactive decision-making and risk reduction through real-time information and rapid responses to shocks (Singh, 2023). However, AI systems play a role in improving resilience and mitigating supply chain hazards, demonstrating its distinctive utility in constructing strong supply chains (Gupta et al., 2024).

Industry 4.0 technologies, such as the IoT, Cyber-Physical Systems (CPS), AI, and Blockchain, are key for improving the resilience of supply chains (Marinagi et al., 2023). These technologies promote improved cooperation, confidence, flexibility, adaptation, and integration within supply chains (Brookbanks & Parry, 2024). By employing IT capabilities, companies can develop and sustain their ability to withstand and recover from many threats, including natural disasters, geopolitical tensions, and pandemics (Atadoga, 2024b).The ability of supply chains to withstand and recover from disruptions, such as supply chain failures or cyber-attacks, is of utmost importance. These disruptions can have profound consequences for society as a whole (Fowler, 2023). To further on this concept, resilience pertains to the ability to recover from component failures, attacks, and service disruptions (Mwangi, 2024).

Furthermore, the incorporation of key supporting technologies of Industry 4.0 improves the degree of transparency and durability in supply chains, hence strengthening their ability to withstand challenges (Patidar et al., 2022). To improve the resilience of global supply networks against disruptions, it is key to prioritise actions such as advancing scientific research, stimulating innovation, extending coverage of disaster insurance, and strengthening emergency protection systems (Yuan, 2023). The cooperation between the manufacturing and information communication service sectors can significantly strengthen economic resilience, highlighting the importance of a coordinated strategy (Xu et al., 2022).

4.2.4 Challenges in Implementing Industry 4.0 Technologies

Technological constraints include high costs and complexity of implementation, as well as concerns regarding data privacy and security (Raj et al., 2020). On the other hand, organisational and cultural barriers encompass resistance to change, lack of skilled personnel, and the need for strategic alignment and collaboration (Ghadge et al., 2020). Research has indicated that the adoption of Industry 4.0 technology may not result in quick financial gains, which is a difficulty for management (Raj et al., 2020). However, the integration of Industry 4.0 principles into supply chains requires the resolution of multiple challenges, underlining the importance of a proficient adaptation framework (Ghadge et al., 2020).

The importance of organisational culture is key in the effective implementation of Industry 4.0. A favourable culture and suitable incentives are vital for securing employee dedication (Liu et al., 2022). However, the utilisation of blockchain technology to improve transparency and trust among participants in the supply chain is of utmost importance. However, its deployment encounters challenges such as issues with data dependability and complexity in tracking (Zhang et al., 2020). Furthermore, it is key for organisations to have strategic alignment and coordination in order to effectively integrate Industry 4.0 technologies into global supply chains (Ghadge et al., 2020).

4.2.5 Impact of Industry 4.0 on Supply Chain Transparency

Supply chain transparency has been significantly altered by the emergence of Industry 4.0, which has facilitated a paradigm shift that improves operational efficiency and resilience. Technologies such as the IoT and blockchain are essential to these transformations, as they enable organisations to achieve unparalleled levels of visibility and traceability throughout their supply chains. The operational dynamics of supply chains are significantly altered to improve transparency, thereby contributing to increased trust among supply chain participants, as organisations adopt these technologies at an increasing rate.

The transparency of the supply chain is significantly improved by blockchain technology. Blockchain ensures that all parties involved in the supply chain have access to verifiable information regarding product provenance and status at all stages of the supply chain by providing a decentralised, immutable record of transactions. The integration of IoT devices with blockchain technology enables organisations to collect and monitor real-time data, which is essential for capturing information about inventory levels, condition, and location throughout the supply chain. This information can be used to improve transparency and efficiency by enabling predictive analytics and decision-making.

Industry 4.0 technologies have significantly increased the importance of operational efficiency, which is not limited to improving supply chain performance metrics. Firms' proactive risk management, which is essential for ensuring resilience against unforeseen disturbances, is facilitated by their capacity to acquire precise and real-time information about supply chain disruptions and demand fluctuations (Aslam et al., 2024; Reynolds, 2024b). Organisations can improve efficiency and reduce risk by more accurately forecasting demand and adjusting their strategies to effectively mitigate potential disruptions, due to the advanced analytics and machine learning capabilities (Aslam et al., 2024; Reynolds, 2024b). These technologies converge to allow firms to respond dynamically to real-time challenges, thereby improving their operational resilience (Mukucha & Chari, 2022).

Providing visibility to the supply chain is essential in the context of resilience building, which is facilitated by Industry 4.0. Encouraging a stronger supply chain architecture, increased visibility enables firms to identify vulnerabilities within their supply chains and develop strategies to address them (Aslam et al., 2024). Recently, the significance of establishing resilient supply chain frameworks has been emphasised in the literature, particularly in the context of global disruptions like the COVID-19 pandemic, which have emphasised the vulnerability of current supply chain structures (Golan et al., 2020). The integration of Industry 4.0 technologies has resulted in increased transparency, which has enabled organisations to develop adaptive strategies. This has enabled them to navigate uncertainties and maintain continuous operations in the face of supply chain volatility (Hobbs, 2021). The data-sharing capabilities of Industry 4.0 technologies facilitate collaboration among supply chain partners. This collaborative ecosystem encourages collective risk management practices by providing stakeholders with a comprehensive understanding of their interdependencies, thereby increasing transparency and trust (Aslam et al., 2024; Reynolds, 2024b).

Table 4.1 provides an analysis of the implementation of Industry 4.0 technologies in various sectors, emphasising their influence on the efficiency, transparency, and resilience of the supply chain. Blockchain technology improves the transparency and trust of humanitarian, logistics, and manufacturing supply chains.

Table 4.1 Industry technology implementation and benefits

Industry	Technology	Implementation & benefits	Reference
Manufacturing	Blockchain	Improved supply chain transparency and trust. Improved traceability and product authenticity verification	Khan, Malik, et al. (2022)
Food & Agriculture	IoT, Blockchain	Improved food traceability, safety, and sustainability. Improved coordination in agri-food supply chains	Bhat et al. (2021)
Healthcare	AI, IoT	Improved process integration and risk resilience. Improved supplier relationships and green supply chain practices	Junaid et al. (2023)
E-commerce	AI, Big Data	Improved supply chain flexibility and last-mile delivery. Personalized solutions for upstream and downstream stakeholders	Andrei et al. (2024)
Automotive	Blockchain, AI	Improved supply chain integration and sustainable performance. Improved traceability and transparency	Kamble et al. (2021)
Logistics	Blockchain	Improved efficiency, transparency, and innovation in freight logistics. Improved decision-making for blockchain adoption	Koh et al. (2020)
Humanitarian	Blockchain	Improved swift-trust, collaboration, and resilience in humanitarian supply chains. Improved transparency and traceability of disaster relief materials	Dubey et al. (2020)
Electrical & Electronics	Blockchain, IoT	Improved supply chain mapping and integration. Improved sustainability practices across the supply chain	Khan, Mubarik, et al. (2022)
Perishable Goods	Blockchain, IoT	Improved operational excellence in perishable food supply chains during outbreaks. Improved responsiveness and flexibility	Kayikci et al. (2022)
Garment Manufacturing	AI, Big Data	Improved supply chain resilience and performance. Improved visibility and integration across the supply chain	Huang et al. (2023); Rafid et al. (2024)

Traceability, risk resilience, and process integration are enhanced in the food, healthcare, and perishable goods industries by IoT and AI. Big data analytics enhances decision-making and adaptability in e-commerce and garment manufacturing. The automotive sector receives advantages from AI and blockchain technology in terms of sustainable performance and integration. These developments collectively improve supply chain optimisation, thereby encouraging adaptability, sustainability, and efficiency in global industries.

4.3 Methodology

In our comprehensive examination of the impact of Industry 4.0 technologies on global supply chains, we employed a systematic literature review methodology, drawing inspiration from the seminal work of Denyer and Tranfield (2009), This approach allowed us to analyse and synthesise the current body of knowledge on supply chain efficiency, transparency, and resilience in the context of digital transformation meticulously. We primarily utilised the Scopus database, recognised for its extensive coverage of high-quality, peer-reviewed literature in the fields of technology and supply chain management. Our search strategy was carefully crafted to capture the most relevant and recent publications, focusing on the period from 2020 to 2024 to ensure the inclusion of cutting-edge research. This methodological rigor enabled us to identify key trends, challenges, and best practices in the implementation of Industry 4.0 technologies such as IoT, AI, blockchain, and big data analytics in global supply chains (Chatterjee et al., 2023; Xu et al., 2021). By adopting this systematic approach, we were able to provide a comprehensive and up-to-date analysis of how these technologies are revolutionizing supply chain operations, improving transparency, and building resilience in an increasingly complex global business environment (Belhadi et al., 2022; Ivanov & Dolgui, 2020).

To ensure a comprehensive and targeted review of the literature on Industry 4.0 technologies and their impact on global supply chains, we employed a sophisticated search strategy within the Scopus database. Our search parameters were meticulously crafted to capture the most relevant and recent scholarship in this rapidly evolving field. We focused on publications from 2020 to 2024, reflecting the cutting-edge nature of Industry 4.0 applications in supply chain management. Our keyword combination included “global supply chains”, “industry 4.0 technologies”, “efficiency”, “transparency”, and “resilience”, allowing us to pinpoint studies that specifically address the transformative effects of digital technologies on supply chain operations. To maintain academic rigor, we limited our search to peer-reviewed journal articles and reviews within the Business, Management and Accounting subject area. This approach ensured that we captured high-quality scholarly contributions to the field. The search yielded 76 publications, providing a strong foundation for our analysis. This carefully curated collection of literature represents the forefront of research on how Industry 4.0 technologies are reshaping global supply chains, offering insights into improved efficiency, improved transparency, and strengthened resilience. By focusing on open-access publications, they aimed to promote accessibility and encourage wider engagement with these critical developments in supply chain management. Our systematic approach to literature selection enables us to present a comprehensive, up-to-date overview of the intersection between Industry 4.0 technologies and global supply chain dynamics, reflecting the most current thinking and empirical findings in this rapidly advancing field.

In the subsequent sections of this chapter, we present a comprehensive analysis of the transformative impact of Industry 4.0 technologies on global supply chains. We begin by examining the theoretical underpinnings of how these advanced technologies improve supply chain efficiency, transparency, and resilience. Our discussion then investigates into a detailed review of cutting-edge research and industry applications, highlighting how IoT, AI, blockchain, and big data analytics are revolutionising supply chain operations. We explored case studies and empirical evidence that demonstrate the practical implementation of these technologies across various industries, showcasing their potential to optimise processes, improve decision-making, and encourage greater collaboration among supply chain partners. Our analysis also addresses the challenges and opportunities associated with adopting Industry 4.0 technologies in supply chain management, providing insights into best practices and strategic considerations for organisations looking to leverage these innovations. Furthermore, we examine the interplay between technological adoption and organisational capabilities, illustrating how companies can develop and nurture the necessary skills and infrastructure to fully capitalise on the benefits of Industry 4.0. This exploration aims to provide a roadmap for supply chain managers and executives to navigate the digital transformation of their operations, ultimately leading to more resilient, efficient, and transparent global supply networks.

4.4 Operational Efficiency in Global SC

It has been demonstrated that the integration of advanced technologies, particularly blockchain, can significantly improve the resilience, transparency, and operational efficiency of supply chains. To clarify the influence of these technologies on operational efficiency in global supply chains, this discussion synthesises the results of a variety of studies. The IoT, artificial intelligence, and blockchain are among the industry 4.0 technologies that improve operational efficiency by streamlining processes and improving data visibility across supply chains. For example, the utilisation of blockchain technology enables the real-time tracking and monitoring of goods, thereby optimising inventory management and minimising the delays and errors that are associated with conventional supply chain processes (Oriekhoe, 2024b; Revathi, 2024). This ability is especially critical in the current fast-paced market, where agility and responsiveness are of the utmost importance. Ivanov and Dolgui (2020) emphasise that the utilisation of digital twins further improves operational efficiency by offering predictive analytics and real-time data visualisation, which allows companies to effectively manage disruptions and ensure business continuity.

Blockchain technology is instrumental in improving transparency among supply chain participants. Trust and collaboration among stakeholders are encouraged by blockchain, which provides a secure and immutable ledger, which is crucial for operational efficiency. Blockchain's transparency has been demonstrated to facilitate improved decision-making and decreased transaction costs by eliminating the necessity for intermediaries and reducing disputes (Dutta et al., 2020; Rijanto, 2021). Furthermore, the capacity to track products throughout the supply chain is becoming increasingly critical in sectors such as pharmaceuticals and food, as it improves accountability and adherence to regulatory standards (Kraft & Kellner, 2022). The adoption of Industry 4.0 technologies also significantly improves the resilience of global supply chains. Traditional supply chains were exposed as vulnerable to the COVID-19 pandemic, which prompted organisations to pursue innovative solutions to improve their resilience (Ivanov & Dolgui, 2020). Blockchain technology, in particular, has been recognised as a transformative tool that can reduce the risks associated with supply chain disruptions by facilitating the rapid response to changes in demand or supply and providing real-time data (Oriekhoe, 2024b). Additionally, the integration of blockchain technology with IoT devices can improve the monitoring of supply chain conditions, thereby improving the capacity to predict and respond to potential disruptions (Xu et al., 2021). The integration of AI-driven predictive analytics and Big Data monitoring in a single e-commerce firm resulted in a 25% reduction in inventory errors and a reduction in last-mile delivery times. This empirical success emphasises the importance of Industry 4.0 technologies in improving operational efficiency across intricate supply networks.

4.5 Transparency in Global SC

In the context of Industry 4.0 technologies, which are reshaping operational frameworks and improving efficiency, resilience, and trust among supply chain participants, transparency in global supply chains (SC) has become increasingly critical. This transparency is fundamental for sustainable supply chain management, and it is achieved through the integration of technologies such as blockchain, IoT, and data analytics. Industry 4.0 technologies significantly improve operational efficiency by facilitating real-time data sharing and visibility throughout the supply chain. For example, Kamble et al. emphasise that improvement in ICT is essential for the effective management of the intricacies of contemporary supply chains (Kamble et al., 2021). Ghadge et al. (2020) further substantiate this integration by observing that digital networks improve overall supply chain performance by improving forecasting, planning, and material traceability. Increased operational efficiency and responsiveness to market demands are the results of the capacity to monitor and control supply chain processes in real time.

Blockchain technology has the potential to significantly improve the transparency and trust of supply chain participants. Tezel et al. contend that blockchain has the potential to transform procurement processes by establishing a secure, immutable ledger that documents all transactions, thereby promoting collaboration and oversight among stakeholders (Tezel et al., 2020). This is consistent with Oriekhoe's assertion that blockchain technology facilitates end-to-end visibility, which enables stakeholders to monitor the origin and movement of products. This, in turn, improves accountability and compliance with regulatory standards (Oriekhoe, 2024a). Additionally, Zilin (2023) has observed that the automation of processes and the reduction of the necessity for intermediaries can be achieved through the integration of smart contracts into blockchain systems, thereby streamlining operations.

The impact of Industry 4.0 technologies on the development of resilience in global supply chains is also a significant factor. Mubarik et al. (2021) examine the critical role of supply chain mapping and visibility in the development of resilience and cleaner production. These technologies facilitate these processes. Proactive and reactive strategies that capitalise on real-time data and analytics improve the capacity to adjust to disruptions, as emphasised by Özdemir et al. (2022). This adaptability is essential in the context of global challenges, such as the COVID-19 pandemic, which has emphasised the necessity of resilient supply chains that can maintain operational continuity in the face of stress. Furthermore, the narrative of transparency encompasses more than just operational efficiency; it also embraces brand reputation and ethical considerations. In industries where ethical sourcing and sustainability are of the utmost importance, Holloway emphasises the necessity of transparency in supply chains to preserve consumer trust (Holloway, 2024). This landscape is further complicated by the cultural and regional differences in consumer expectations regarding transparency, which requires that brands develop customised communication strategies (Holloway, 2024). End-to-end product tracking was enabled by blockchain implementation in a significant pharmaceutical company, resulting in a 40% reduction in counterfeit incidents and regulatory compliance. This transparency significantly increased audit accuracy and encouraged stakeholder trust, demonstrating the significant impact of digital ledgers on supply chain integrity and accountability.

4.6 Resiliency in Global SC

The IoT, AI, and blockchain are among the advanced technologies that collectively enable real-time data sharing, improved decision-making, and improved operational flexibility, collectively referred to as Industry 4.0. These technologies facilitate the rapid adaptation of supply chains to evolving conditions, thereby increasing their resilience to unforeseen disruptions (Madrid-Guijarro, 2024; Patidar et al., 2022). The vulnerabilities that are inherent in global supply chains have been emphasised by the COVID-19 pandemic. The sudden disruptions resulted in significant operational challenges for numerous organisations, as they lacked the requisite strategies to manage them (Reynolds, 2024c). Nevertheless, organisations that had implemented risk mitigation strategies and invested in digitalisation were more adept at withstanding these shocks. Investments in data-driven monitoring systems and adaptable supply chain configurations enabled companies to respond more effectively to the challenges posed by the pandemic (Ivanov & Das, 2020). Resistant supply chains are distinguished by their capacity to anticipate risks and promptly mobilise resources, underscoring the significance of cultivating capabilities such as agility and absorptive capacity (Ali et al., 2022; Madrid-Guijarro, 2024).

The significance of blockchain technology in encouraging trust and transparency among supply chain participants is not overstated. The integrity and traceability of transactions are paramount to the preservation of trust in supply chain operations, and blockchain technology offers a decentralised ledger that guarantees this (Cherian et al., 2022; Um & Han, 2020). This transparency is especially important during periods of crisis, as it enables stakeholders to confirm the reliability of suppliers and the authenticity of products, thereby minimising the likelihood of disruptions caused by misinformation or fraud (Shi et al., 2022). The integration of blockchain with other Industry 4.0 technologies further improves its impact, resulting in a more interconnected and resilient supply chain ecosystem (Tortorella et al., 2023). AI and Big Data analytics were employed by a prominent garment manufacturer to accurately predict demand fluctuations, thereby reducing supply disruptions during market shifts. Industry 4.0's ability to improve resilience in volatile supply environments was illustrated by the 30% increase in response times that their logistics strategy achieved.

4.7 Challenges and Opportunities

One of the primary challenges in adopting Industry 4.0 technologies is the high implementation cost and complexity. As noted by Queiroz et al. (2021), the initial investment required for technologies such as blockchain can be substantial, potentially deterring smaller organisations or those with limited resources. This is further complicated by the rapid pace of technological advancement, which may lead to concerns about the longevity and relevance of current investments. Organisations must carefully evaluate the return on investment and consider phased implementation strategies to mitigate these financial risks. Another significant challenge is the lack of standardization across Industry 4.0 technologies. Bienhaus and Haddud (2018) highlight that the absence of universal standards can lead to interoperability issues, particularly when integrating systems across different supply chain partners. This challenge emphasises the importance of collaborative efforts within industries to establish common standards and protocols. The work of Kayikci et al. (2022), in perishable food supply chains demonstrates how standardized approaches to blockchain and IoT implementation can improve operational excellence across the entire supply chain ecosystem.

Data privacy and security concerns represent another critical challenge. As supply chains become increasingly digitized, the volume of sensitive data being collected, processed, and shared grows exponentially. Kshetri (2018) argues that while blockchain technology offers improved security features, it is not immune to cyber threats. Organisations must invest in strong cybersecurity measures and ensure compliance with data protection regulations, which can vary significantly across different jurisdictions. Despite these challenges, the opportunities presented by Industry 4.0 technologies are substantial. The potential for improved operational efficiency is evident across various industries. For instance, Modgil et al. (2022) demonstrate how AI and big data analytics in e-commerce supply chains can significantly improve last-mile delivery and enable personalized solutions for stakeholders. Similarly, Yadav et al. (2023) illustrate how IoT and blockchain technologies in agri-food supply chains can dramatically improve traceability and sustainability.

The ability to improve transparency and trust among supply chain participants represents another significant opportunity. Khan, Malik, et al. (2022) show how blockchain implementation in manufacturing supply chains has improved product authenticity verification and overall supply chain transparency. This increased transparency can lead to improved stakeholder relationships, reduced fraud, and improved brand reputation. Perhaps most critically, Industry 4.0 technologies offer the opportunity to build more resilient supply chains. The work of Huang et al. (2023) in the garment manufacturing industry demonstrates how AI and big data analytics can improve supply chain resilience and performance. This is particularly relevant in the context of global disruptions, such as the COVID-19 pandemic, where supply chain resilience has proven to be a critical factor in organisational survival and success.

In terms of best practices and strategic considerations, several key points emerge from the literature and empirical evidence. Firstly, organisations should adopt a holistic approach to technology implementation. Bag et al. (2021) emphasize the importance of considering organisational culture, employee skills, and management support alongside technological factors. A perishable food supply chain project implemented blockchain technology in conjunction with IoT sensors to facilitate real-time monitoring and traceability. The system demonstrated the ability of effective solutions to overcome implementation barriers by reducing spoilage by 20% and improving collaboration among partners, despite the high costs and integration challenges.

Figure 4.1 labelled "Improving Global Supply Chains with Industry 4.0" illustrates the impact of Industry 4.0 on the transformation of global supply chains, emphasising operational efficiency, transparency, resilience, and significant emerging opportunities.

The transformative impact of Industry 4.0 on global supply chains is emphasised by Fig. 4.1, examining four key dimensions: operational efficiency, transparency, resilience, and challenges/opportunities. It emphasises the significance of utilising data-driven insights to streamline processes, which improve decision-making and alleviate operational bottlenecks. Companies can more effectively anticipate and adapt to market disruptions by utilising real-time data, automation, and

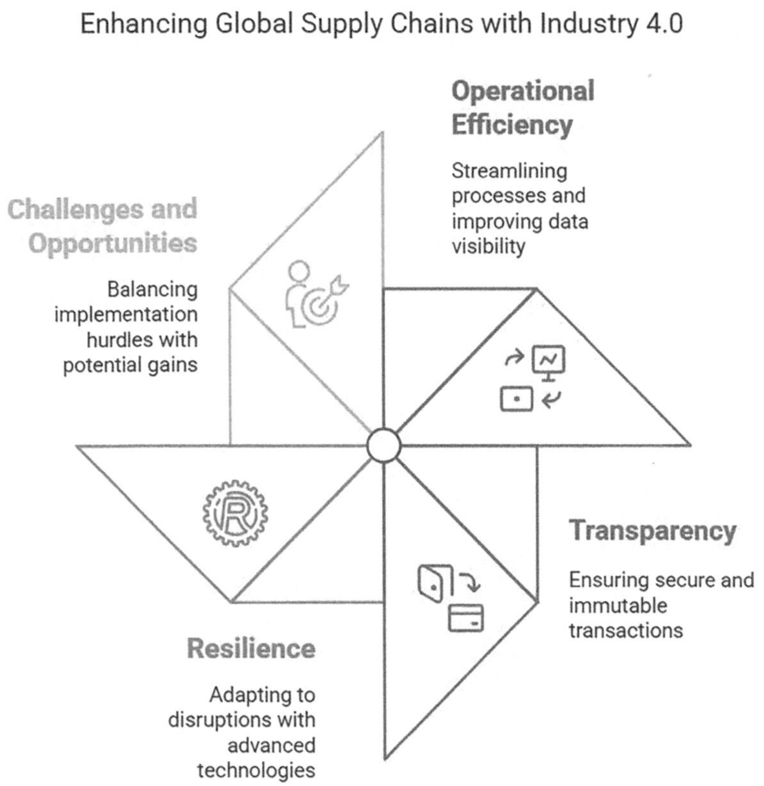

Fig. 4.1 Improving global supply chains with Industry 4.0

interconnected systems. Transparency in transactions is further emphasised by the figure, which emphasises the importance of secure digital records in encouraging accountability and trust. Despite the potential for regulatory complexities, talent shortages, or high initial costs to impede adoption, Industry 4.0 is necessary due to the potential gains in productivity, innovation, and competitive advantage.

4.8 Discussion and Conclusions

This chapter aimed to explore the transformative impact of Industry 4.0 technologies on global supply chains, focusing specifically on how these technologies improve efficiency, transparency, and resilience. Our analysis demonstrates that integrating advanced technologies such as blockchain, IoT, AI, and big data analytics fundamentally reshapes supply chain operations across various sectors. The empirical evidence presented in this chapter confirms that these technologies are not merely incremental improvements to existing systems but represent a paradigm shift in how global supply chains are managed and optimised.

Our findings indicate that blockchain technology plays a pivotal role in improving transparency and trust among supply chain participants. By providing a secure, immutable, and decentralised ledger, blockchain enables real-time tracking of goods, authentication of products, and verification of transactions (Kumar et al., 2024; Oriekhoe, 2024b). This capability is particularly crucial in pharmaceuticals, food, and luxury goods industries, where provenance and authenticity are paramount concerns for consumers and regulators alike. As demonstrated by Khan, Malik, et al. (2022), the implementation of blockchain in manufacturing supply chains has significantly improved product traceability and authenticity verification, directly addressing issues of counterfeiting and fraud that have long plagued global supply networks.

IoT technology, through its network of sensors and connected devices, has revolutionised data collection and monitoring capabilities in supply chains. Our analysis reveals that IoT enables continuous, real-time tracking of inventory, environmental conditions, and equipment status, providing unprecedented visibility into supply chain operations. This improved visibility allows for proactive decision-making and predictive maintenance, reducing downtime and optimising resource allocation. The work of Yadav et al. (2023) in agri-food supply chains demonstrates how IoT-based coordination systems improve not only traceability but also sustainability and overall performance, highlighting the multifaceted benefits of this technology.

The integration of AI and machine learning into supply chain operations has significantly improved predictive capabilities and decision-making processes. Our findings show that AI-driven analytics enable more accurate demand forecasting, risk assessment, and strategic planning (Modgil et al., 2022; Reynolds, 2024a). This predictive capability is crucial for building resilient supply chains that can anticipate and mitigate disruptions before they occur. The empirical evidence from the garment manufacturing industry, as presented by (Huang et al., 2023), illustrates how AI and big data analytics have improved supply chain responsiveness and flexibility in the face of market volatility.

Big data analytics emerges from our analysis as a foundational technology that underpins the effectiveness of other Industry 4.0 technologies. By processing and analysing vast amounts of structured and unstructured data, big data analytics provides the insights necessary for strategic decision-making and operational optimisation (Queiroz et al., 2021). The ability to derive meaningful patterns and correlations from complex datasets enables supply chain managers to identify inefficiencies, predict trends, and implement targeted improvements. As demonstrated in Table 4.1, the synergistic implementation of these technologies across various industries has led to significant improvements in operational efficiency, transparency, and resilience.

Our research also highlights the critical role of Industry 4.0 technologies in building supply chain resilience, particularly in the context of global disruptions such as the COVID-19 pandemic. The pandemic exposed vulnerabilities in traditional supply chain models and accelerated the adoption of digital technologies as a means of improving resilience (Ivanov & Dolgui, 2020; Nakandala et al., 2023). Organisations that had implemented digital capabilities demonstrated superior preparedness and effectiveness in addressing supply chain disruptions, underscoring the importance of technological adoption in strengthening resilience. The work of Tortorella et al. (2023) further emphasises the significance of resilience capacities in moderating the influence of Industry 4.0 technologies on supply chain performance, particularly in critical sectors such as healthcare.

Despite the clear benefits, our analysis also identifies significant challenges in implementing Industry 4.0 technologies in global supply chains. These include high implementation costs, complexity, data privacy concerns, and organisational resistance to change (Ghadge et al., 2020; Raj et al., 2020; Tortorella et al., 2023). The research of Bag et al. (2021) emphasises that successful implementation requires more than just technological investment; it necessitates appropriate organisational culture, skilled personnel, and strategic alignment. This suggests that a holistic approach, considering both technological and organisational factors, is essential for the effective adoption of Industry 4.0 technologies in supply chains.

Looking ahead, our proposed model for Technology-Driven Supply Chain Transformation (Fig. 4.2) offers a framework for organisations to systematically implement Industry 4.0 technologies to improve efficiency, transparency, and resilience. This model emphasises the interconnected nature of these technologies and their collective impact on supply chain operations. By adopting this integrated approach, organisations can leverage the full potential of Industry 4.0 technologies to build supply chains that are not only more efficient and transparent but also more resilient in the face of disruptions. Therefore, our findings demonstrate that Industry

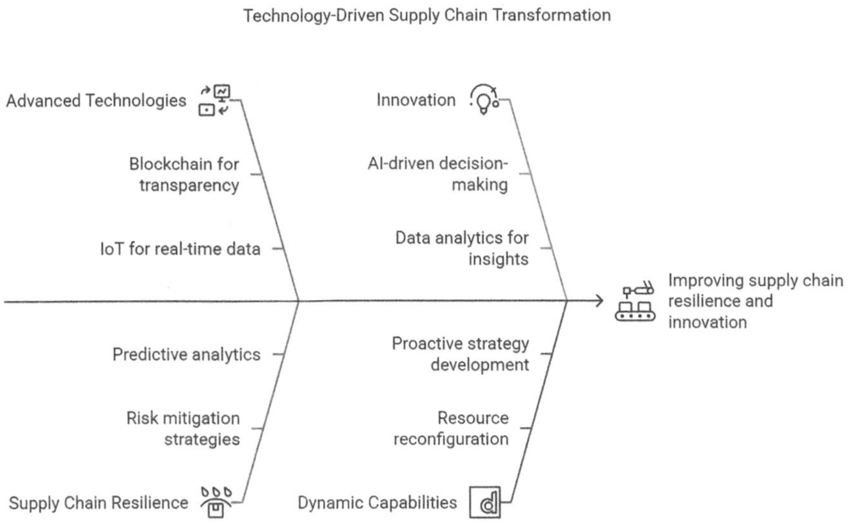

Fig. 4.2 Technology-driven supply chain transformation

4.0 technologies are fundamentally transforming global supply chains, enabling unprecedented levels of efficiency, transparency, and resilience. While challenges remain, the potential benefits of these technologies are substantial, particularly in terms of improved operational performance, improved stakeholder trust, and increased supply chain agility. As these technologies continue to evolve and mature, the impact on global supply chains is likely to become even more profound, potentially leading to more sustainable, resilient, and efficient supply chain ecosystems.

Concluding Remarks

Based on the comprehensive analysis presented in this chapter, we emphasise the strategic imperative for organisations to develop systematic approaches to Industry 4.0 technology adoption in global supply chains. The evidence clearly demonstrates that blockchain, IoT, AI, and big data analytics are not merely optional improvements but essential tools for building supply chains that can withstand contemporary challenges while delivering improved efficiency and transparency. Supply chain executives, technology strategists, and policymakers should collaborate to establish frameworks that facilitate the widespread implementation of these technologies across sectors.

Furthermore, it is recommended that to accelerate the digital transformation of global supply chains, educational institutions and industry bodies should develop specialised training programs focused on building the necessary technical and managerial capabilities. Organisations such as the World Economic Forum, in partnership with technology providers and supply chain associations, should establish standards and best practices for Industry 4.0 implementation that address interoperability, data security, and ethical considerations. Additionally, governments should consider incentive structures and regulatory frameworks that encourage technology adoption while safeguarding against potential risks related to data privacy and cybersecurity.

The integration of blockchain technology deserves particular attention, as our research indicates its transformative potential for supply chain transparency and trust-building. Companies should prioritize pilot projects that demonstrate blockchain's effectiveness in their specific operational contexts, with special consideration for industries where traceability and provenance are paramount concerns, such as pharmaceuticals, food, and high-value manufacturing. These initiatives can play a crucial role in developing more resilient, efficient, and transparent supply chain ecosystems capable of meeting the challenges of an increasingly complex global business environment.

As the transition from Industry 4.0 to Industry 5.0 begins, future research should explore the evolving relationship between technological systems and human operators in supply chain contexts. The focus should shift toward collaborative intelligence that combines the analytical power of advanced technologies with human creativity, ethical judgment, and contextual understanding. This human-centric approach to technology implementation will be essential for developing supply chains that are efficient, resilient, sustainable, and aligned with broader societal goals.

References

Ali, I., Arslan, A., Chowdhury, M. M. H., Khan, Z., & Tarba, S. (2022). Reimagining global food value chains through effective resilience to COVID-19 shocks and similar future events: A dynamic capability perspective. *Journal of Business Research, 141*, 1–12. https://doi.org/10.1016/j.jbusres.2021.12.006

Alqarni, M. A., Alkatheiri, M. S., Chauhdary, S. H., & Saleem, S. (2023). Use of blockchain-based smart contracts in logistics and supply chains. *Electronics, 12*(6), 1340. https://doi.org/10.3390/electronics12061340

Andrei, N., Scarlat, C., & Ioanid, A. (2024). Transforming E-commerce logistics: sustainable practices through autonomous maritime and last-mile transportation solutions. *Logistics, 8*(3), 71.

Aslam, H., Syed, T. A., Blome, C., Ramish, A., & Ayaz, K. (2024). The multifaceted role of social capital for achieving organizational ambidexterity and supply chain resilience. *IEEE Transactions on Engineering Management, 71*, 10571–10584. https://doi.org/10.1109/tem.2022.3174069

Atadoga, A. (2024a). AI in supply chain optimization: A comparative review of USA and African trends. *International Journal of Science and Research Archive, 11*(1), 896–903. https://doi.org/10.30574/ijsra.2024.11.1.0156

Atadoga, A. (2024b). The role of it in enhancing supply chain resilience: A global review. *International Journal of Management & Entrepreneurship Research, 6*(2), 336–351. https://doi.org/10.51594/ijmer.v6i2.774

Badhwar, A., Islam, S., & Tan, C. S. L. (2023). Exploring the potential of blockchain technology within the fashion and textile supply chain with a focus on traceability, transparency, and product authenticity: A systematic review. *Frontiers in Blockchain, 6*. https://doi.org/10.3389/fbloc.2023.1044723

Bag, S., Gupta, S., & Kumar, S. (2021). Industry 4.0 adoption and 10R advance manufacturing capabilities for sustainable development. *International journal of production economics, 231*, 107844.

Belhadi, A., Kamble, S. S., Venkatesh, M., Jabbour, C. J. C., & Benkhati, I. (2022). Building supply chain resilience and efficiency through additive manufacturing: An ambidextrous perspective on the dynamic capability view. *International Journal of Production Economics, 249*, 108516.

Bhat, S. A., Huang, N.-F., Sofi, I. B., & Sultan, M. (2021). Agriculture-food supply chain management based on blockchain and IoT: A narrative on enterprise blockchain interoperability. *Agriculture, 12*(1), 40.

Bidin, M. F. (2024). Adaptive supply chain risk management using AI mitigating disruptions and enhancing resilience in the post-pandemic era. *TJJPT, 45*(02), 2655–2664. https://doi.org/10.52783/tjjpt.v45.i02.6294

Bienhaus, F., & Haddud, A. (2018). Procurement 4.0: factors influencing the digitisation of procurement and supply chains. *Business Process Management Journal, 24*(4), 965–984.

Boissieu, E. D., Kondrateva, G., Baudier, P., & Ammi, C. (2021). The use of blockchain in the luxury industry: Supply chains and the traceability of goods. *Journal of Enterprise Information Management, 34*(5), 1318–1338. https://doi.org/10.1108/jeim-11-2020-0471

Brookbanks, M. & Parry, G.C. (2024). The impact of Industry 4.0 technologies on the resilience of established cross-border supply chains. *Supply Chain Management an International Journal, 29*(4), 731–754. https://doi.org/10.1108/scm-07-2023-0333

Chandan, A., John, M., & Potdar, V. (2023). Achieving UN SDGs in food supply chain using blockchain technology. *Sustainability, 15*(3), 2109. https://doi.org/10.3390/su15032109

Chatterjee, S., & Chatterjee, D. (2022). Modelling the barriers of supply chain transparency in the post-Covid-19 scenario. *International Journal of Logistics Economics and Globalisation, 9*(3), 277. https://doi.org/10.1504/ijleg.2022.120812

Chatterjee, S., Chaudhuri, R., Gupta, S., Sivarajah, U., & Bag, S. (2023). Assessing the impact of big data analytics on decision-making processes, forecasting, and performance of a firm. *Technological Forecasting and Social Change, 196*, 122824.

Chen, Y. (2023). How blockchain adoption affects supply chain sustainability in the fashion industry: A systematic review and case studies. *International Transactions in Operational Research, 31*(6), 3592–3620. https://doi.org/10.1111/itor.13273

Cherian, T. M., Mathivathanan, D., Sj, C. J. A., Ramasubramaniam, M., & Alathur, S. (2022). Influence of supply chain resilience, information technology capabilities and agility on cost and delivery performance in construction supply chains: An Indian perspective. *The International Journal of Logistics Management, 34*(4), 1050–1076. https://doi.org/10.1108/ijlm-04-2022-0164

Chukwu, N. (2024). Resilient chain: AI-enhanced supply chain security and efficiency integration. *IJSMR, 07*(03), 46–65. https://doi.org/10.37502/ijsmr.2024.7306

Cui, H., Xu, C., & Sun, K. (2024). Unveiling the future of engineering management: The role of artificial intelligence and big data. In *Proceedings of the first international conference on science, engineering and technology practices for sustainable development, ICSETPSD 2023, 17th–18th November 2023, Coimbatore, Tamilnadu, India*. https://doi.org/10.4108/eai.17-11-2023.2342763

Denyer, D., & Tranfield, D. (2009). Producing a systematic review. In D. A. Buchanan & A. Bryman (Eds.), *The Sage handbook of organizational research methods* (pp. 671–689). Sage.

Dubey, R., Gunasekaran, A., Bryde, D. J., Dwivedi, Y. K., & Papadopoulos, T. (2020). Blockchain technology for enhancing swift-trust, collaboration and resilience within a humanitarian supply chain setting. *International Journal of Production Research, 58*(11), 3381–3398. https://doi.org/10.1080/00207543.2020.1722860

Dutta, P., Choi, T. M., Somani, S., & Butala, R. (2020). Blockchain technology in supply chain operations: Applications, challenges and research opportunities. *Transportation Research Part E Logistics and Transportation Review, 142*, 102067. https://doi.org/10.1016/j.tre.2020.102067

Eyo-Udo, N. L. (2024). Leveraging artificial intelligence for enhanced supply chain optimization. *Open Access Research Journal of Multidisciplinary Studies, 7*(2), 001–015. https://doi.org/10.53022/oarjms.2024.7.2.0044

Forehand, K., Román, J. B., & Schaefer, T. E. (2021). Supply chain efficiency in the discount store industry post Covid-19: Applying the supply chain efficiency ratio. *Operations and Supply Chain Management an International Journal, 14*(4), 423–430. https://doi.org/10.31387/oscm0470313

Fowler, D. S. (2023). Aspects of resilience for smart manufacturing systems. *Strategic Change, 32*(6), 183–193. https://doi.org/10.1002/jsc.2555

Ghadge, A., Kara, M. E., Moradlou, H., & Goswami, M. (2020). The impact of Industry 4.0 implementation on supply chains. *Journal of Manufacturing Technology Management, 31*(4), 669–686. https://doi.org/10.1108/jmtm-10-2019-0368

Golan, M. S., Jernegan, L. H., & Linkov, I. (2020). Trends and applications of resilience analytics in supply chain modeling: Systematic literature review in the context of the COVID-19 pandemic. *Environment Systems & Decisions, 40*(2), 222–243. https://doi.org/10.1007/s10669-020-09777-w

Gupta, S., Modgil, S., Meissonier, R., & Dwivedi, Y. K. (2024). Artificial intelligence and information system resilience to cope with supply chain disruption. *IEEE Transactions on Engineering Management, 71*, 10496–10506. https://doi.org/10.1109/tem.2021.3116770

Hobbs, J. E. (2021). Food supply chain resilience and the COVID-19 pandemic: What have we learned? *Canadian Journal of Agricultural Economics/Revue Canadienne D Agroeconomie, 69*(2), 189–196. https://doi.org/10.1111/cjag.12279

Holloway, S. (2024). Narratives of transparency: How supply chain communication shapes brand image in marketing. https://doi.org/10.20944/preprints202406.1124.v1.

Huang, Y.-F., Phan, V.-D.-V., & Do, M.-H. (2023). The impacts of supply chain capabilities, visibility, resilience on supply chain performance and firm performance. *Administrative Sciences, 13*(10), 225.

Hurduzeu, G. (2024). Blockchain in agriculture: Transforming the food supply chain for transparency and efficiency in the European Union. *Aerd, 2023*(1), 145–153. https://doi.org/10.59277/aerd.2023.2.04

Huzaifa, I. (2024). Impact of blockchain technology on supply chain management efficiency and transparency in Pakistan. *International Journal of Supply Chain Management, 9*(1), 16–26. https://doi.org/10.47604/ijscm.2320

Ibrahim, M. (2024). The effect of blockchain technology in enhancing ethical sourcing and supply chain transparency: Evidence from the cocoa and agricultural sectors in Ghana. *African Journal of Empirical Research, 5*(2), 55–64. https://doi.org/10.51867/ajernet.5.2.6

Islam, M. A. (2021). Modern slavery disclosure regulation and global supply chains: Insights from stakeholder narratives on the UK Modern Slavery Act. *Journal of Business Ethics, 180*(2), 455–479. https://doi.org/10.1007/s10551-021-04878-1

Ivanov, D., & Das, A. (2020). Coronavirus (COVID-19/SARS-CoV-2) and supply chain resilience: A research note. *International Journal of Integrated Supply Management, 13*(1), 90. https://doi.org/10.1504/ijism.2020.107780

Ivanov, D., & Dolgui, A. (2020). A digital supply chain twin for managing the disruption risks and resilience in the era of Industry 4.0. *Production Planning & Control, 32*(9), 775–788. https://doi.org/10.1080/09537287.2020.1768450

Jiang, F. (2023). Analyzing price and profit dynamics in free trade port supply chains: A blockchain-centric approach under consumer sensitivity. *Journal of Intelligent Management Decision, 2*(4), 202–214. https://doi.org/10.56578/jimd020405

Junaid, M., Zhang, Q., Cao, M., & Luqman, A. (2023). Nexus between technology enabled supply chain dynamic capabilities, integration, resilience, and sustainable performance: An empirical examination of healthcare organizations. *Technological Forecasting and Social Change, 196*, 122828.

Kamble, S. S., Gunasekaran, A., Subramanian, N., Ghadge, A., Belhadi, A., & Mani, V. (2021). Blockchain technology's impact on supply chain integration and sustainable supply chain performance: Evidence from the automotive industry. *Annals of Operations Research, 327*(1), 575–600. https://doi.org/10.1007/s10479-021-04129-6

Kayikci, Y., Gozacan-Chase, N., Rejeb, A., & Mathiyazhagan, K. (2022). Critical success factors for implementing blockchain-based circular supply chain. *Business Strategy and the Environment, 31*(7), 3595–3615.

Khan, H. H., Malik, M. F., Konečná, Z., Chofreh, A. G., Goni, F. A., & Klemeš, J. J. (2022). Blockchain technology for agricultural supply chains during the COVID-19 pandemic: Benefits and cleaner solutions. *Journal of Cleaner Production, 347*, 131268. https://doi.org/10.1016/j.jclepro.2022.131268

Khan, S. A., Mubarik, M. S., Kusi-Sarpong, S., Gupta, H., Zaman, S. I., & Mubarik, M. (2022). Blockchain technologies as enablers of supply chain mapping for sustainable supply chains. *Business Strategy and the Environment, 31*(8), 3742–3756.

Koh, L., Dolgui, A., & Sarkis, J. (2020). Blockchain in transport and logistics—paradigms and transitions. *International Journal of Production Research, 58*, 2054–2062. Taylor & Francis.

Kraft, S. K., & Kellner, F. (2022). Can blockchain be a basis to ensure transparency in an agricultural supply chain? *Sustainability, 14*(13), 8044. https://doi.org/10.3390/su14138044

Kshetri, N. (2018). 1 Blockchain's roles in meeting key supply chain management objectives. *International Journal of Information Management, 39*, 80–89.

Kumar, R., Jain, V., Yie, L. W., & Teyarachakul, S. (2024). *Convergence of IoT, blockchain, and computational intelligence in smart cities*. CRC Press, Taylor & Francis.

Lalitha, G. (2024). Blockchain-enabled simulation and optimization for supply chain transparency. *E3s Web of Conferences, 472*, 02007. https://doi.org/10.1051/e3sconf/202447202007

Liu, S.-F., Fan, Y.-J., Luh, D.-B., & Teng, P.-S. (2022). Organizational culture: The key to improving service management in Industry 4.0. *Applied Sciences, 12*(1), 437. https://doi.org/10.3390/app12010437

Madrid-Guijarro, A. (2024). Unlocking resilience: The impact of Industry 4.0 technologies on manufacturing firms' response to the COVID-19 pandemic. *Management Decision, 63*(1), 126–154. https://doi.org/10.1108/md-02-2024-0262

Marinagi, C., Reklitis, P., Trivellas, P., & Sakas, D. (2023). The impact of Industry 4.0 technologies on key performance indicators for a resilient supply chain 4.0. *Sustainability, 15*(6), 5185. https://doi.org/10.3390/su15065185

Meidutė-Kavaliauskienė, I., Yıldız, B., Çiğdem, Ş., & Činčikaitė, R. (2021). An integrated impact of blockchain on supply chain applications. *Logistics, 5*(2), 33. https://doi.org/10.3390/logistics5020033

Modgil, S., Singh, R. K., & Hannibal, C. (2022). Artificial intelligence for supply chain resilience: learning from Covid-19. *The International Journal of Logistics Management, 33*(4), 1246–1268.

Mubarik, M. S., Naghavi, N., Mubarik, M., Kusi-Sarpong, S., Khan, S. A., Zaman, S. I., & Kazmi, S. H. A. (2021). Resilience and cleaner production in Industry 4.0: Role of supply chain mapping and visibility. *Journal of Cleaner Production, 292*, 126058. https://doi.org/10.1016/j.jclepro.2021.126058

Mukucha, P., & Chari, F. (2022). Supply chain resilience: The role of supplier development in the form of contract farming in fast-food outlets in Zimbabwe. *Continuity & Resilience Review, 4*(3), 280–299. https://doi.org/10.1108/crr-03-2022-0006

Mwangi, A. (2024). Towards a software-defined industrial IoT-edge network for next-generation offshore wind farms: State of the art, resilience, and self-X network and service management. *Energies, 17*(12), 2897. https://doi.org/10.3390/en17122897

Nakandala, D., Yang, C., Lau, H. C. W., & Weerabahu, S. K. (2023). Industry 4.0 technology capabilities, resilience and incremental innovation in Australian manufacturing firms: A serial mediation model. *Supply Chain Management an International Journal, 28*(4), 760–772. https://doi.org/10.1108/scm-08-2022-0325

Odimarha, A. C. (2024). The role of technology in supply chain risk management: Innovations and challenges in logistics. *Magna Scientia Advanced Research and Reviews, 10*(2), 138–145. https://doi.org/10.30574/msarr.2024.10.2.0052

Olutimehin, D. O. (2024). The role of technology in supply chain risk management: Innovations and challenges in logistics. *International Journal of Management & Entrepreneurship Research, 6*(3), 878–889. https://doi.org/10.51594/ijmer.v6i3.941

Oriekhoe, O. I. (2024a). Blockchain in supply chain management: A review of efficiency, transparency, and innovation. *International Journal of Science and Research Archive, 11*(1), 173–181. https://doi.org/10.30574/ijsra.2024.11.1.0028

Oriekhoe, O. I. (2024b). Blockchain in supply chain management: A systematic review: Evaluating the implementation, challenges, and future prospects of blockchain technology in supply chains. *Engineering Science & Technology Journal, 5*(1), 128–151. https://doi.org/10.51594/estj.v5i1.732

Oriekhoe, O. I. (2024c). Ensuring global food safety: Integrating blockchain technology into food supply chains. *Engineering Science & Technology Journal, 5*(3), 811–820. https://doi.org/10.51594/estj.v5i3.905

Özdemir, D., Sharma, M., Dhir, A., & Daim, T. U. (2022). Supply chain resilience during the COVID-19 pandemic. *Technology in Society, 68*, 101847. https://doi.org/10.1016/j.techsoc.2021.101847

Park, A., & Li, H. (2021). The effect of blockchain technology on supply chain sustainability performances. *Sustainability, 13*(4), 1726. https://doi.org/10.3390/su13041726

Patidar, A., Sharma, M., Agrawal, R., & Sangwan, K. S. (2022). Supply chain resilience and its key performance indicators: An evaluation under Industry 4.0 and sustainability perspective. *Management of Environmental Quality an International Journal, 34*(4), 962–980. https://doi.org/10.1108/meq-03-2022-0091

Priyadarshi, A., & Asif, M. (2023). Dynamics of blockchain in supply chain management. In *Social development and governance innovations in education, technology and management* (pp. 221–234). QTanalytics. https://doi.org/10.48001/978-81-966500-9-4_18

Queiroz, M. M., Fosso Wamba, S., De Bourmont, M., & Telles, R. (2021). Blockchain adoption in operations and supply chain management: empirical evidence from an emerging economy. *International Journal of Production Research, 59*(20), 6087–6103.

Rafid, M., Mostafa, T. B., Wasi, R. H., & Avi, R. I. (2024). Harnessing big data and AI for supply chain resilience in the garments industry: A comparative analysis of Bangladesh and China. In *7th IEOM Bangladesh international conference on industrial engineering and operations management*.

Raj, A., Dwivedi, G., Sharma, A., Lopes de Sousa Jabbour, A. B., & Rajak, S. (2020). Barriers to the adoption of Industry 4.0 technologies in the manufacturing sector: An inter-country comparative perspective. *International Journal of Production Economics, 224*, 107546. https://doi.org/10.1016/j.ijpe.2019.107546

Rana, J., & Daultani, Y. (2022). Mapping the role and impact of artificial intelligence and machine learning applications in supply chain digital transformation: A bibliometric analysis. *Operations Management Research, 16*(4), 1641–1666. https://doi.org/10.1007/s12063-022-00335-y

Remko, v. H. (2020). Research opportunities for a more resilient post-Covid-19 supply chain—Closing the gap between research findings and industry practice. *International Journal of Operations & Production Management, 40*(4), 341–355. https://doi.org/10.1108/ijopm-03-2020-0165

Revathi, M. K. L. (2024). Impact of blockchain technology in supply chain system. *International Research Journal on Advanced Engineering and Management, 2*(05), 1670–1672. https://doi.org/10.47392/irjaem.2024.0238

Reynolds, S. (2024a). Delving into the adoption of blockchain technology in supply chain management. https://doi.org/10.20944/preprints202406.0481.v1.

Reynolds, S. (2024b). Exploring supply chain resilience strategies in the face of price inflation in renewable energy markets. https://doi.org/10.20944/preprints202405.1662.v1.

Reynolds, S. (2024c). Investigating the impact of information sharing on supply chain resilience. https://doi.org/10.20944/preprints202406.0557.v1

Rijanto, A. (2021). Blockchain technology adoption in supply chain finance. *Journal of Theoretical and Applied Electronic Commerce Research, 16*(7), 3078–3098. https://doi.org/10.3390/jtaer16070168

Sachan, R. (2024). Investigating the use of blockchain for secure and efficient supply chain management. https://doi.org/10.55524/csistw.2024.12.1.39.

Santhi, A. R., & Muthuswamy, P. (2022). Influence of blockchain technology in manufacturing supply chain and logistics. *Logistics, 6*(1), 15. https://doi.org/10.3390/logistics6010015

Shi, Y., Zheng, X., Venkatesh, V. G., Humdan, E. A., & Paul, S. K. (2022). The impact of digitalization on supply chain resilience: An empirical study of the Chinese manufacturing industry. *Journal of Business and Industrial Marketing, 38*(1), 1–11. https://doi.org/10.1108/jbim-09-2021-0456

Singh, P. K. (2023). Digital transformation in supply chain management: Artificial intelligence (AI) and machine learning (ML) as catalysts for value creation. *International Journal of Supply Chain Management, 12*(6), 57–63. https://doi.org/10.59160/ijscm.v12i6.6216

Swamy, S., & Raju, K. S. (2020). An empirical study on system level aspects of internet of things (IoT). *IEEE Access, 8*, 188082–188134. https://doi.org/10.1109/access.2020.3029847

Tezel, A., Papadonikolaki, E., Yitmen, İ., & Hilletofth, P. (2020). Preparing construction supply chains for blockchain technology: An investigation of its potential and future directions. *Frontiers of Engineering Management, 7*(4), 547–563. https://doi.org/10.1007/s42524-020-0110-8

Tortorella, G. L., Prashar, A., Antony, J., Fogliatto, F. S., González, V. A., & Filho, M. G. (2023). Industry 4.0 adoption for healthcare supply chain performance during COVID-19 pandemic in Brazil and India: The mediating role of resilience abilities development. *Operations Management Research, 17*(2), 389–405. https://doi.org/10.1007/s12063-023-00366-z

Um, J., & Han, N.-h. (2020). Understanding the relationships between global supply chain risk and supply chain resilience: The role of mitigating strategies. *Supply Chain Management an International Journal, 26*(2), 240–255. https://doi.org/10.1108/scm-06-2020-0248

Wang, C. (2024). Research on receivables financing model in supply chain finance based on blockchain technology. https://doi.org/10.4108/eai.8-12-2023.2344789

Wang, M., & Yang, Y. (2022). An empirical analysis of the supply chain flexibility using blockchain technology. *Frontiers in Psychology, 13*. https://doi.org/10.3389/fpsyg.2022.1004007

Xiao, M. (2024). Application decision model of blockchain technology in construction supply chain. https://doi.org/10.21203/rs.3.rs-4133660/v1.

Xu, P., Lee, J., Barth, J. R., & Richey, R. G. (2021). Blockchain as supply chain technology: Considering transparency and security. *International Journal of Physical Distribution & Logistics Management, 51*(3), 305–324. https://doi.org/10.1108/ijpdlm-08-2019-0234

Xu, Y., Li, J., Yan, Y., Gao, P., & Xie, H. (2022). Can coordinated development of manufacturing and information communication service industries boost economic resilience? An empirical study based on China's provinces. *Sustainability, 14*(17), 10758. https://doi.org/10.3390/su141710758

Yadav, S., Kumar, A., Mehlawat, M. K., Gupta, P., & Charles, V. (2023). A multi-objective sustainable financial portfolio selection approach under an intuitionistic fuzzy framework. *Information Sciences, 646*, 119379.

Yuan, Y. (2023). Measurements and influencing factors of new rural collective economies' resilience toward mountain disasters in indigent areas: A case study of Liangshan Yi Autonomous Prefecture, China. *Land, 12*(11), 2042. https://doi.org/10.3390/land12112042

Zamani, E. D., Smyth, C., Gupta, S., & Dennehy, D. (2022). Artificial intelligence and big data analytics for supply chain resilience: A systematic literature review. *Annals of Operations Research, 327*(2), 605–632. https://doi.org/10.1007/s10479-022-04983-y

Zhang, A., Zhong, R. Y., Farooque, M., Kang, K., & Venkatesh, V. G. (2020). Blockchain-based life cycle assessment: An implementation framework and system architecture. *Resources Conservation and Recycling, 152*, 104512. https://doi.org/10.1016/j.resconrec.2019.104512

Zhou, M., Qiao, Y., Liu, B., Vogel-Heuser, B., & Kim, H. (2023). Machine learning for Industry 4.0 [from the guest editors]. *IEEE Robotics & Automation Magazine, 30*(2), 8–9. https://doi.org/10.1109/mra.2023.3266618

Zilin, Z. (2023). Blockchain technology in construction supply chain management: Enhance transaction speed, cost effectiveness and security. *Journal of Advanced Research in Applied Sciences and Engineering Technology, 32*(3), 400–420. https://doi.org/10.37934/araset.32.3.400420

Chapter 5
Developing a Customer Value Oriented Digital Servitization Strategy: A Case Study of Komatsu

Junichi Tomita

Abstract This chapter examines a customer value oriented digital servitization strategy for smart, connected products. While Japanese companies may appear to lag behind their foreign counterparts, some effectively leverage their technological strengths, smart product development, and digital transformation (DX) to gain a competitive edge. Using Komatsu Ltd. as a case study, this chapter explores a DX-driven digital servitization strategy. Komatsu emphasizes customer touch-points, collects extensive data through monitoring, uploads it to the cloud, analyzes it with AI, and proposes manufacturing and service solutions. Through Komtrax and ICT-enabled (ICT) construction machineries, Komatsu promotes Smart Construction, optimizing processes with IoT and big data to create customer value. By prioritizing customer touchpoints, data collection, solution optimization, and customer satisfaction, Komatsu exemplifies a robust value creation strategy.

Keywords Digital servitization · Smart · Connected product · Customer value creation

5.1 Introduction

We are surrounded by Internet-connected products such as smartphones, smart-watches, and home appliances. These are referred to as "smart, connected products" (Porter & Heppelmann, 2014, 2015). As consumers of such products, our satisfaction depends on the effectiveness of the products and services provided.

This chapter outlines the concept of digital servitization strategies for smart, connected products. Japanese companies are often perceived as lagging behind their foreign competitors in digital servitization. However, this perception is not entirely

J. Tomita (✉)
Department of Business Administration, Toyo University, Tokyo, Japan
e-mail: tomita@toyo.jp

Y. Khojasteh, S.M. M. Jahan Abadi (eds.), *Digital Transformation of Supply Chains*, International Series in Operations Research & Management Science 372, https://doi.org/10.1007/978-3-032-01206-7_5

accurate. Some companies have gained a competitive advantage by leveraging their technological strengths and developing smart, connected products through digital transformation (DX).

Komatsu Ltd. serves as an example of a company leading advanced DX initiatives. By valuing customer touchpoints, collecting and analyzing vast amounts of data, and proposing tailored solutions, Komatsu integrates smart, connected products into a broader value creation framework.

This chapter aims to answer key questions about solutions incorporating smart, connected products and the digital servitization strategies and processes that enable them. A case study of Komatsu's approach is used to provide insights.

5.2 Literature Review

The concept of "digital servitization" builds upon the foundation of "servitization". Vandermerwe and Rada (1988) defined servitization as "the addition of services to product offerings to provide additional customer value", marking a shift from selling standalone products to providing integrated bundles of products, services, support, and knowledge (Paschou et al., 2020). Servitization offers financial (Neely, 2008), strategic (Baines & Lightfoot, 2013), and marketing (Gebauer et al., 2017) benefits by adding value to existing offerings.

With advancements in digital technologies including IoT, Industry 4.0, blockchain, big data, cloud computing, and robotics—the sophistication of services has increased, enabling new business models. Digital servitization integrates servitization with technological advancements, offering two key perspectives (Shen et al., 2023). The broad perspective considers digital servitization as the provision of digital services through digital components embedded in physical products (Holmström & Partanen, 2014). In contrast, the narrow perspective defines it as the use of digital technology to transition from a product-centric to a service-centric business model (Coreynen et al., 2017). These perspectives highlight how digital servitization enables companies to enhance product functionality while simultaneously transforming their business models to focus on service-oriented value creation.

Various researchers have proposed definitions of digital servitization. In this chapter, we modified the definition provided by Paschou et al. (2020): "The process of leveraging digital technologies to develop or enhance services, enabling new business models, value co-creation, data-driven insights, and improved operational and environmental performance, ultimately achieving a competitive advantage."

This definition emphasizes the transformative role of digital technologies in enabling service-oriented business models, fostering collaborative value creation, and enhancing operational efficiency and sustainability to strengthen competitive positioning.

Research on digital servitization is believed to have originated with Allmendinger and Lombreglia (2005), who highlighted the importance of leveraging technology to embed intelligence into products. They argued that technological advancements are essential for enabling smart services.

Subsequent research in this field, up until around 2017, primarily focused on foundational discussions regarding the role of digital technology in servitization, particularly in facilitating the shift from product-centric to service-centric business models. Following this period, the research landscape evolved, with increasing attention directed toward digital servitization as a form of business model innovation, platform strategy, and business ecosystem development (Shen et al., 2023).

Despite significant advancements, gaps remain in understanding the causal relationships between digital technology and business model innovation. Research on the digital servitization process—the shift from product-centric to service-centric models—has yet to fully explore its impact on customer relationships and competitive advantage.

This chapter seeks to address these gaps by focusing on customer value—one of the key benefits of digital servitization (Paschou et al., 2020). Komatsu's case is used to examine the digital servitization process and its implications for customer value creation.

5.3 Digital Serivtization Strategy with a Focus on Smart, Connected Products

5.3.1 Digital Servitization Strategy

A digital servitization strategy leverages evolving digital technologies to achieve competitive advantages. Porter and Heppelmann (2014, 2015) described such strategies as "business models for smart solutions that integrate products, services, software, and analytics." By collecting real-time data from smart, connected products, companies can gain valuable insights that enable enhanced customer solutions. The strategic integration of these products into business models significantly expands opportunities, facilitating greater efficiency, innovation, and customer value creation.

5.3.2 Smart, Connected Product Elements

A smart, connected product consists of three key elements: physical components, smart components, and connectivity elements.

The physical components include mechanical and electrical elements that form the core structure and functionality of the product. The smart components consist of sensors, microprocessors, data storage units, control devices, software, embedded operating systems, and user interfaces, all of which enable data processing, automation, and enhanced functionality. Lastly, the connectivity elements encompass ports, antennas, and communication protocols that facilitate wired or wireless Internet connections, enabling remote monitoring, data exchange, and integration with other digital systems.

By integrating these three elements, smart, connected products enhance operational efficiency, optimize performance, and enable new service-oriented business models. Through real-time data collection and analysis, companies can develop innovative, customer-focused AI solutions, further strengthening their competitive positioning in an increasingly digitalized market.

5.3.3 Smart, Connected Product Features

Smart, connected products are characterized by four key functions: monitoring, control, optimization, and autonomy (Porter & Heppelmann, 2014, 2015).

The monitoring function utilizes sensors to measure both the internal state of the product and its external environment. These sensors collect data such as position, status, and usage conditions, which are then stored in a cloud-based data lake for further analysis.

The control function enables remote operation of the product through cloud-based commands. When specific environmental conditions are met, the system autonomously adjusts the product's behavior to ensure optimal performance.

The optimization function integrates control mechanisms with big data analytics to enhance product performance and service delivery. AI-driven tools analyze real-time and historical data to assess the product's operational status, diagnose performance issues, predict potential failures, and prescribe corrective measures. This process improves product availability, enhances productivity, prevents breakdowns, and informs users when maintenance or component replacements are necessary.

The autonomy function is achieved when the monitoring, control, and optimization functions are seamlessly integrated. Autonomy enables self-diagnosis and automated responses tailored to customer preferences. By leveraging Internet connectivity, smart, connected products can enhance performance, reliability, operational efficiency, and processing speed. For example, an autonomous system can assess its operational environment, self-diagnose faults, initiate repairs, and respond proactively to user requirements. Moreover, highly autonomous products can interact and collaborate with other smart, connected products and systems, forming integrated smart ecosystems that provide optimal solutions for customers.

Given these capabilities, some scholars argue that the true essence of the Internet of Things (IoT) lies not in the "Internet of Things" but rather in "taking information from things (IfT)" (Fujimoto, 2017). In this perspective, IoT's primary function is to extract and process data from smart, connected products to optimize both product performance and service delivery.

However, simply developing and connecting smart, connected products to the Internet is insufficient for achieving comprehensive value creation. A narrow focus on individual product optimization may lead to isolated improvements rather than holistic, system-wide enhancements. To maximize value, companies must first identify customer needs and then develop smart, connected products, IoT-enabled solutions, and service models that align with these needs. By adopting this customer-centric approach, firms can establish innovative business models and scalable digital platforms that support long-term value creation.

5.4 Digital Servitization Process

5.4.1 Customer Value Creation Process in B2B Business

To examine the digital servitization strategy, it is essential to consider it from the perspective of customer value creation (Fig. 5.1). This discussion focuses on two primary business models: Business-to-Business (B2B), where the customer is another company, and Business-to-Consumer (B2C), where the customer is an individual consumer.

In the B2B value creation process, a client company formulates its business strategy, designs products and services, procures and prepares raw materials and components, and subsequently engages in production, inspection, sales, and after-sales service. Applying this framework to a security sensor company, the customer in this case is a security firm, and its value creation process follows a structured approach. First, the security company develops a security plan for a given building, outlining the necessary measures to ensure safety and security. Next, the company conducts site surveys and measurements, performs structural calculations, designs security systems, and estimates the associated costs. Once the design phase is complete, the company procures security sensors and other necessary components, assesses existing site conditions, and refines the detailed security design before proceeding to the installation phase.

During the production phase, the security company installs and configures the security system for the end user. Once installation is complete, the system undergoes performance monitoring and validation, ensuring that all security components function as intended. Following the inspection, if necessary, repairs or system upgrades are conducted to maintain continued operational efficiency. In the sales and after-sales service phase, the security company ensures long-term reliability by providing maintenance, technical support, and upgrades to enhance customer satisfaction.

Through this structured sequence of activities, security companies create value by integrating and optimizing the entire security process, ultimately enhancing the effectiveness and reliability of their security solutions. The digital servitization process further strengthens this value creation by leveraging data-driven insights, predictive maintenance, and automated monitoring to ensure a more efficient, scalable, and customer-centric approach.

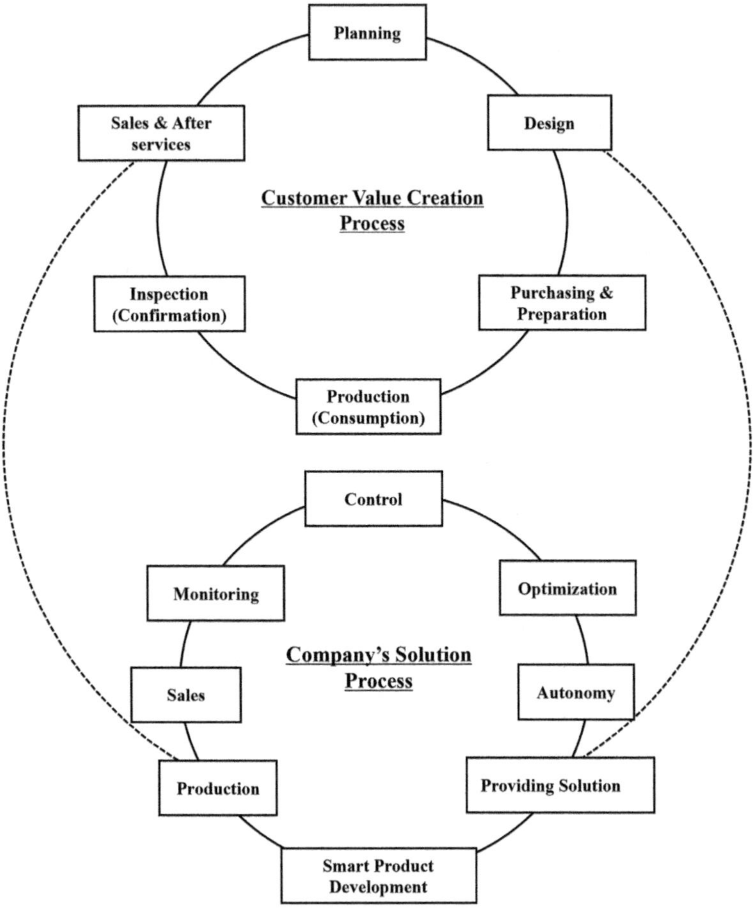

Fig. 5.1 Digital servitization process from the perspective of customer value creation (adopted from Porter & Heppelmann, 2014)

5.4.2 Customer Value Creation Process in B2C Business

In the context of value creation in a B2C setting, consumers follow a structured process that involves planning the purchase of a product or service, determining its intended use, procuring necessary items, producing (or consuming) the product, evaluating the outcome post-consumption, and, if needed, receiving after-sales service.

For example, consider a four-person family preparing and enjoying dinner together. The process begins with menu planning, where the family decides what dishes to prepare. Next, they determine the necessary ingredients, seasonings, cooking utensils, and recipes, ensuring they have everything required for meal preparation. Once the menu is finalized, they visit a supermarket to purchase the necessary ingredients, completing the procurement and preparation phase. Following this, they cook and consume the meal, engaging in the production and consumption phase. If the food is well-prepared and the dining experience is enjoyable, it enhances overall satisfaction, contributing to a positive and fulfilling experience.

Through this structured value creation process, the family transforms raw materials into a meaningful and enjoyable dining experience. This example illustrates how consumer decision-making and resource utilization contribute to value generation, aligning with the broader principles of B2C business models.

5.4.3 Company's Servitization Process

The digital servitization process of the company, which supports customer value creation, follows a structured sequence. After developing, manufacturing, and selling smart, connected products, the company continuously monitors their usage, collects relevant data, and uploads it to the cloud. This data is then analyzed and optimized using AI and other analytical tools. Ultimately, this process enables autonomous collaboration between smart, connected products and other systems, delivering optimal solutions that enhance efficiency and performance.

A key aspect of this framework is the intersection—represented by the dotted line—where the customer value creation process and the company's servitization process overlap. This represents critical touchpoints between the company and its customers. The company interacts with customers through its sales activities and the provision of smart, connected products. Therefore, it is essential to leverage these touchpoints effectively to provide solutions that align with and enhance the customer value creation process.

In the B2B context, consider the example of a security sensor company. This company develops and manufactures smart, connected products, such as security sensors with integrated cameras, and sells or leases them to security firms. Once acquired, these sensors enable real-time monitoring, data collection, and cloud-based storage, where AI-driven analytics enhance operational efficiency. Beyond basic functionality, the security sensor company aims to provide more sophisticated solutions. For instance, by integrating outdoor sensors with camera systems, the company can offer advanced image verification capabilities, reducing false alarms and minimizing unnecessary police dispatches.

In the B2C business context, the servitization process of a supermarket can be designed to enhance the shopping experience for consumers. This process begins with the development of a smart app, which consumers install on their devices, establishing a key touchpoint between the supermarket and its customers. Through the app, supermarkets can distribute promotional information, encouraging consumers to visit the store and engage with personalized offers.

Once in the store, in-store monitoring is facilitated by smart shopping carts equipped with tablets, allowing consumers to make purchases while generating valuable data on shopping behavior. This data is then collected, stored in the cloud, and analyzed using AI-driven optimization techniques. The insights obtained enable supermarkets to enhance their shopping experience, offering personalized recommendations tailored to individual purchasing habits.

For instance, when a consumer purchases vegetables or meat, the tablet interface may suggest recipes, seasonings, and complementary ingredients, helping them make informed purchasing decisions. By integrating such digital servitization strategies, supermarkets can facilitate customer value creation, transforming the shopping process into a personalized and interactive experience.

As demonstrated in this example, digital servitization strategies enhance customer engagement by visualizing and optimizing key aspects of the value creation process. Therefore, the digital servitization process can be understood as a framework that actively supports and enhances customer value creation by leveraging data-driven insights and personalized services.

5.5 A Case Study of Komatsu[1]

5.5.1 Overview of Construction Machinery Business

Founded in 1921, Komatsu celebrated its 100th anniversary on May 13, 2021, marking a century of innovation and growth. Over the years, the company has achieved significant sales growth, with revenue increasing from 1432 billion yen in FY2009 to 3865 billion yen in FY2023. As of FY2023, Komatsu reported an operating margin of 15.7% and employed 65,738 people (Fig. 5.2).

Approximately 90% of Komatsu's revenue is derived from its construction machinery and vehicles division, with sales distributed across multiple global regions. The revenue composition includes Japan (9%), China and Asia (14%), Oceania (10%), North America (28%), South America (18%), Europe (9%), and other regions. According to Statista, Komatsu held a 10.4% global market share in the construction machinery industry in 2023, ranking second only to Caterpillar (16.8%) in the United States.

One of Komatsu's most technologically advanced innovations is its ICT construction machinery, which integrates information and communication technology for semi-automated control. These machines are equipped with on-board sensors and GNSS antennas from satellite positioning systems, enabling them to autonomously measure, track, and recognize work progress at construction sites. By precisely determining the machine's position and the cutting edge of the work machinery, these semi-automated systems significantly enhance efficiency, accuracy, and productivity in construction operations.

[1]This case study is written with reference to Kurashige et al. (2012), Shike (2021), Takanashi (2017), Tomita (2022), and Komatsu IR Report 2009–2024.

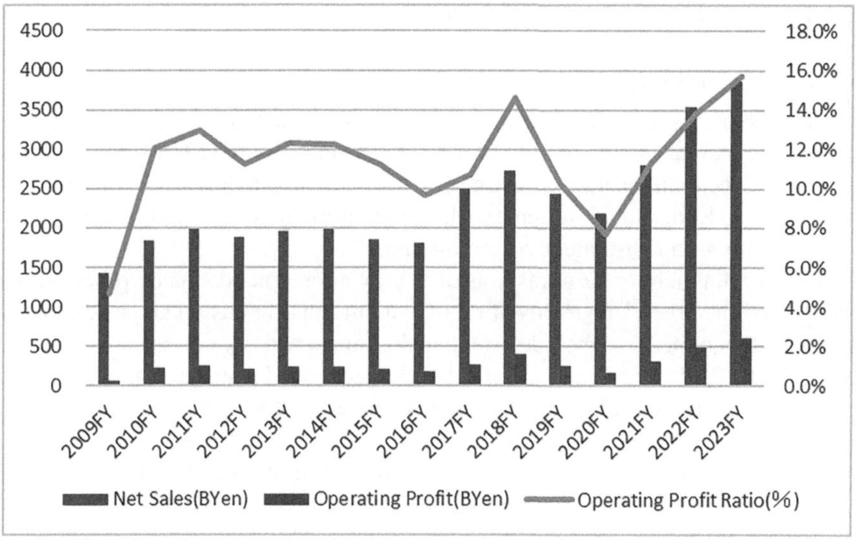

Fig. 5.2 Financial Data of Komatsu (Komatsu IR Report 2009–2023)

The construction machinery industry is relatively small compared to the automobile industry, with a production volume of approximately 1/300 and a market value of 1/14 that of the automotive sector. However, in terms of product life cycle, automobiles typically have an operational lifespan of 2500 h, whereas construction machinery operates for 12,000 h, making its lifespan five times longer. Consequently, the life cycle cost per unit is substantially higher, with a new automobile costing around 200,000 yen, while new construction machinery costs approximately 20 million yen—a 100-fold difference.

Given these characteristics, the construction machinery industry operates on a "recurring business" model, in which companies manufacture equipment in-house, distribute it through sales agents or rental services, and generate continuous revenue through repairs, maintenance, inspections, spare parts sales, and the resale of used machinery.

To further strengthen this recurring revenue model, Komatsu has long emphasized the development of "DANTOTSU products", which focus on high quality and added value. Additionally, the company offers "DANTOTSU services", leveraging IoT and AI technologies to enhance the visibility and operational efficiency of construction machinery. Furthermore, Komatsu provides "DANTOTSU solutions", designed to support value creation at construction sites by optimizing operational processes and enhancing productivity.

5.5.2 The Development of Komtrax

As part of its DANTOTSU Service, Komatsu successfully developed Komtrax in the late 1990s, a system that utilizes communication satellites to monitor and manage the operational status of construction machinery. Komtrax transmits real-time data on the location, operating conditions, and remaining fuel levels of construction equipment to Komatsu's data center, where the information is shared with Komatsu, its distributors, and customers via the Internet.

The development of Komtrax was driven by an internal company proposal that recognized the potential benefits of remote monitoring for customers. Additionally, Komtrax functioned as a theft deterrent, addressing a widespread issue in the industry at the time. If a machine moved beyond a predefined distance from a construction site, its location could be tracked, and if necessary, the engine could be remotely disabled via a server command.

Initially introduced as an optional feature, Komtrax later became a standard offering, positioning Komatsu ahead of its competitors and expanding the company's range of services. With real-time access to the operating status and fuel levels of construction machinery, distributors were able to provide proactive maintenance solutions. By leveraging Komtrax, distributors could optimize repair and replacement schedules, enhance operational efficiency, and reduce the risk of unexpected breakdowns, ultimately improving customer satisfaction and increasing equipment uptime.

5.5.3 Smart Construction

Komatsu has achieved steady growth through the development of Komtrax and ICT construction machinery. However, the company recognized that while these innovations had improved the utilization rates of construction machinery in specific processes, such as embankments and slopes, they had not yet led to the comprehensive optimization of the entire construction process at job sites.

Additionally, the construction industry, which serves as the primary market for construction machinery, has been experiencing a chronic labor shortage. It is projected that within a few years, the industry will face a shortfall of approximately 1.2 million skilled construction workers, further exacerbating operational challenges and increasing demand for automation and efficiency improvements.

In response to these challenges, Komatsu launched the "Smart Construction" initiative in 2015 with the objective of enhancing safety and productivity at construction sites while mitigating labor shortages. This initiative leverages digital technologies, including IoT, to integrate data from workers, construction sites, and machinery, thereby visualizing and optimizing the entire construction process (Fig. 5.3). By implementing Smart Construction, Komatsu aims to provide safe,

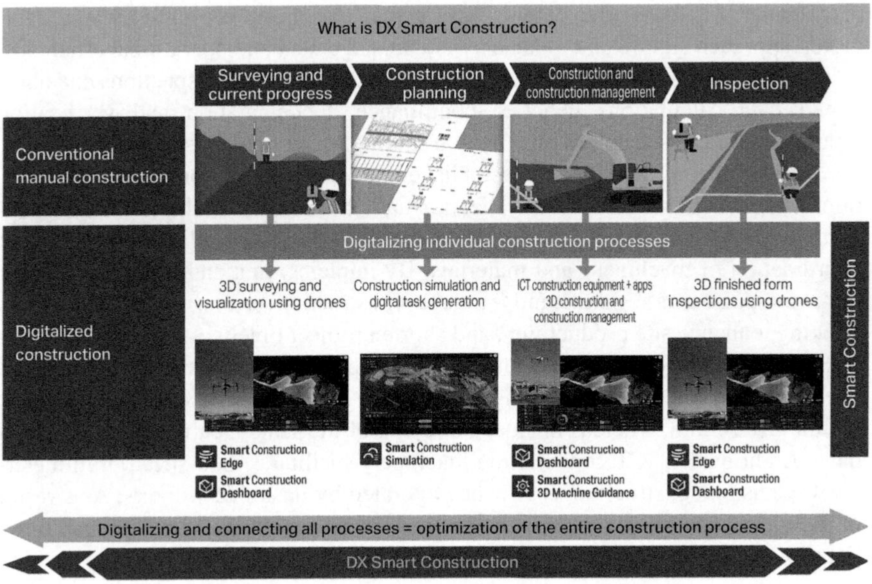

Fig. 5.3 Approaches through Smart Construction and other solutions (Komatsu IR Report, 2024)

efficient, and intelligent solutions that contribute to industry-wide DX and enhance operational efficiency.

Figure 5.3 illustrates the overall flow of the construction process, which consists of four key stages: surveying and progress assessment, construction planning, construction and construction management, and inspection.

In the conventional manual construction process, these stages rely on manual surveying, 2D construction planning, non-ICT-based construction, and manual inspection, all of which require significant manpower and time. To address these inefficiencies, digitalization is first applied to each stage, a process referred to as vertical digitalization.

In the surveying and visualization phase, a topographical drone is deployed to survey the site, generating high-resolution 3D topographical data of the current conditions. In the construction planning phase, 3D topographical data and 3D as-built drawing data are superimposed using Smart Construction applications. By analyzing discrepancies between these datasets, the precise excavation area and soil volume can be calculated, allowing for the creation of an optimal construction plan through simulation.

During the construction phase, semi-automated ICT construction machinery executes tasks using machine control based on 3D design data, eliminating the need for "hinging", the traditional process of manually marking precise construction areas before initiating work. In the construction management phase, daily drone surveys capture landform data, while Smart Construction applications monitor the operating status of construction machinery, track completed work and analyze discrepancies between the planned and actual construction progress. This data-driven approach significantly enhances site management and operational efficiency.

Finally, in the inspection phase, 3D drone-based formwork inspections and automated data calculations generate compliance documentation and verification reports, facilitating seamless information sharing across project stakeholders.

Once vertical digitalization is established, the next step is horizontal digitalization, which enables real-time visualization and optimization of the entire construction workflow. This process interconnects digitized processes, allowing for efficient coordination of machinery and materials. By implementing end-to-end digitalization, companies can significantly reduce construction errors, improve progress tracking, enhance site productivity, and shorten project timelines.

Several digital solutions have already been deployed to improve construction site operations. On-site drone surveying now enables high-precision 3D data acquisition within just 20 min, whereas manual surveying of the same area would take several days. Additionally, ICT construction machinery facilitates high-speed, millimeter-level precision construction, even when operated by unskilled workers. As a result of these innovations, some construction sites have been able to reduce project timelines by one-third and triple productivity.

5.5.4 The Roadmap for DX Smart Construction

Figure 5.4 illustrates the future direction of Komatsu's roadmap for DX Smart Construction in 2021 (Shike, 2021). The company aims to establish safe, highly productive, smart, and environmentally sustainable workplaces in the future. Achieving this vision requires not only Komatsu's own DX but also the optimization and automation of the entire construction process for its customers. This involves both autonomous construction machinery and end-to-end process optimization to enhance efficiency and sustainability.

The horizontal axis of the roadmap represents products level of construction machinery, progressing from Level 1 (limited operational support) to Level 5 (Advance Decision-making Autonomous Operation). To enhance safety and productivity at construction sites, construction machinery must operate collaboratively and autonomously. This is being facilitated through the adoption of 5G and high-precision GNSS (Global Navigation Satellite System) technology.

The vertical axis represents process optimization level, ranging from Level 1 (3-Dimensional Design Data) to Level 5 (optimization of construction). Before reaching Level 5, Level 4 focuses on automation of construction planning, where the goal is to integrate digitalization across all processes. For example, real-time drone surveys capture current site conditions, which are then processed into 3D digital models and deployed throughout the entire construction workflow to enhance decision-making and planning.

As of now, Komatsu has achieved Level 3, which signifies advanced solo automation on the horizontal axis and 3-Dimentonal construction planning capabilities

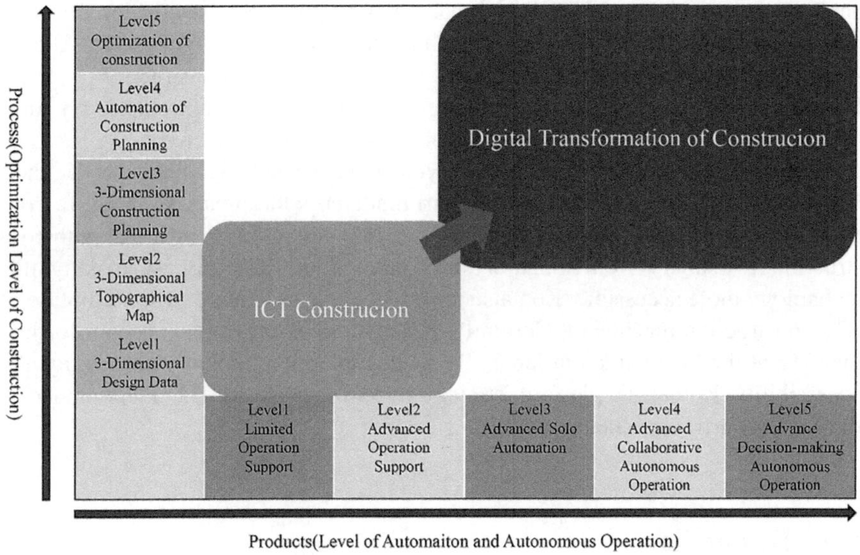

Fig. 5.4 Komatsu's roadmap for DX Smart Construction (adapted from Shike, 2021)

on the vertical axis. The company continues to progress toward higher levels of automation, integration, and efficiency, further accelerating DX in the construction industry.

To advance DX reforms, Komatsu has adopted an open platform strategy that integrates surveying, design, construction, and maintenance/management within the entire construction process. Komatsu aggregates data collected from drones, construction machinery cameras, and sensors onto a centralized digital platform. This data is then analyzed using AI and other advanced analytical tools and made available to external stakeholders. The disclosed information includes 3D topographical data, machine costs, labor costs, and material costs, all of which are critical for developing data-driven construction solutions.

Komatsu actively shares its extensive data with external entities because acquiring comprehensive application and device development technologies for the entire construction process within a short timeframe is impractical. Instead of relying solely on in-house development, Komatsu fosters open innovation by collaborating with various hardware and software manufacturers. To facilitate this, the company has opened its APIs to third-party developers. For example, the 3D topographical data captured and processed by drones is provided by a partner company based in Silicon Valley, USA. By embracing open innovation, Komatsu enables continuous platform evolution, expands user adoption, and strengthens industry-wide DX.

One notable example of Komatsu's DX-driven platform strategy is the "Smart Construction Retrofit Kit," launched in 2020. This kit integrates ICT functions into conventional construction machinery that lacks digital capabilities. Specifically, it includes an IMU (Inertial Measurement Unit) sensor, which detects the attitude angle of the machinery, a GNSS antenna, which acquires precise position information, and a controller, which enables machine guidance for highly accurate 3D construction based on 3D design data and real-time monitoring of construction progress. By eliminating the need for manual hinging work and reducing reliance on additional personnel, the retrofit kit enhances both safety and productivity.

Although similar retrofit kits have previously been available on the market, their high cost (approximately ten million yen) hindered widespread adoption. In contrast, Komatsu's solution is priced at just 700,000 yen and is compatible with construction machinery from manufacturers other than Komatsu. By making this technology more accessible, Komatsu aims to significantly increase the adoption of ICT construction machinery. Currently, ICT hydraulic excavators account for less than 2% of the total market in Japan. By scaling adoption through affordability and accessibility, Komatsu's platform strategy seeks to accelerate DX adoption across the entire construction industry.

5.6 Discussion

Applying smart, connected product concept to Komatsu, Komtrax, and ICT construction machinery effectively transforms construction equipment into smart, connected products. Specifically, Komtrax, which is installed as a standard feature across all Komatsu models, enables real-time monitoring of machinery location and operational status. This data facilitates optimal maintenance planning and predictive parts replacement based on operating hours.

For ICT construction machinery, Komtrax leverages on-board cameras, sensors, and antennas to detect the attitude angle of the vehicle body, capture operational data, and store this information in the cloud. This data visualization enhances system-wide construction optimization, enabling proactive decision-making and efficiency improvements throughout the construction process.

A construction company's value creation process encompasses several key stages. The planning stage involves formulating a project plan, including scope, objectives, and timeline. The design stage includes conducting site surveys and measurements, performing structural calculations, and estimating project costs. The procurement and preparation stage consist of acquiring construction machinery, assessing existing site conditions, refining detailed designs, and preparing for construction execution (Fig. 5.1).

The production stage includes executing core construction activities such as excavation, loading, earthmoving, filling, compaction, embankment, and pavement. The inspection stage involves conducting completion assessments, verifying construction results, and preparing delivery documentation. The sales and after-sales service stage include the handover of constructed facilities, followed by repairs, maintenance, and renewal services. By successfully completing this end-to-end process, construction companies generate customer value and drive project success.

As the solution company, Komatsu develops and manufactures smart, connected products such as ICT construction machinery, distributes them through sales agents and rental services. When customers purchase or rent ICT construction machinery, real-time monitoring becomes possible. Data is collected, stored in the cloud, analyzed, and optimized using AI-driven analytics (Fig. 5.1).

Looking ahead, Komatsu aims to enhance its solutions through greater coordination and autonomy among construction machinery. The company currently supports nearly all value creation processes within the construction industry and seeks to further expand its capabilities.

Komatsu has reached Level 3 in its DX Smart Construction roadmap and aspires to progress toward the upper-right quadrant of Fig. 5.3. The company's goal is to develop advanced autonomous machinery capable of task creation and work order optimization, thereby maximizing customer value through AI-driven decision-making.

However, achieving true industry-wide optimization requires collaboration beyond Komatsu's own ecosystem. To fully optimize the construction process, Komatsu must establish data-sharing partnerships not only with suppliers and complementary partners but also with competitors. Overcoming competitive barriers and fostering industry-wide cooperation will be the most significant challenge in Komatsu's pursuit of a fully digitalized and automated construction sector.

5.7 Conclusions

This chapter presented a case study of Komatsu from the perspective of the digital servitization process, clarifying the underlying logic that drives business model innovation and customer value maximization.

Komatsu has established itself as a leader in DX. In addition to developing Komtrax and ICT construction machinery, the company has recently been promoting Smart Construction, an initiative that visualizes and optimizes the entire construction process through digital technologies such as IoT. Both Komtrax and ICT construction machinery fall under the category of smart, connected products. By leveraging these technologies, big data is continuously collected and stored in the cloud, enabling data-driven decision-making and the delivery of optimized solutions. These efforts are strategically designed to support the customer value creation process.

From a customer value creation perspective, companies must prioritize customer touchpoints by leveraging smart, connected products. This requires collecting relevant data, optimizing solutions, and enhancing customer satisfaction. As digital servitization strategies and processes become more widely adopted, they are expected to expand across various industries, fostering greater innovation and efficiency.

However, a key limitation of this research is that it is based on a single case study. To develop a more comprehensive understanding of digital servitization typologies and generalizations, future research should incorporate multiple case studies and survey-based methodologies. Expanding this research will provide deeper insights into the evolving role of digital servitization across different industries and business models.

References

Allmendinger, G., & Lombreglia, R. (2005). Four strategies for the age of smart services. *Harvard Business Review, 83*(10), 131–138.

Baines, T., & Lightfoot, H. W. (2013). Servitization of the manufacturing firm: Exploring the operations practices and technologies that deliver advanced services. *International Journal of Operations & Production Management, 34*(1), 2–35.

Coreynen, W., Matthyssens, P., & Van Bockhaven, W. (2017). Boosting servitization through digitization: Pathways and dynamic resource configurations for manufacturers. *Industrial Marketing Management, 60*, 42–53.

Fujimoto, T. (2017). *The theory of corporate strategy from the field [Genba kara miageru kigyo senryakuron]*, Kadokawa Shinsho. (in Japanese).

Gebauer, H., Saul, C. J., Haldimann, M., & Gustafsson, A. (2017). Organizational capabilities for pay-per-use services in product-oriented companies. *International Journal of Production Economics, 192*, 157–168.

Holmström, J., & Partanen, J. (2014). Digital manufacturing-driven transformations of service supply chains for complex products. *Supply Chain Management: An International Journal, 19*(4), 421–430.

Komatsu IR Report 2009–2024 (https://www.komatsu.jp/en/ir/library/annual).

Kurashige, M., Hirano, M., Osanai, A., & Sakakibara, K. (Eds.). (2012). *Aftermarket Strategy: Komatsu's Solution Business to Prevent Commoditization* [Aftermarket Senryaku: Komoditika wo fusegu Komatsu no solution business], Hakuto Shobo. (in Japanese).

Neely, A. (2008). Exploring the financial consequences of the servitization of manufacturing. *Operations Management Research, 1*, 103–118.

Paschou, T., Rapaccini, M., Adrodegari, F., & Saccani, N. (2020). Digital servitization in manufacturing: A systematic literature review and research agenda. *Industrial Marketing Management, 89*, 278–292.

Porter, M. E., & Heppelmann, J. E. (2014). How smart, connected products are transforming competition. *Harvard Business Review, 92*(11), 64–88.

Porter, M. E., & Heppelmann, J. E. (2015). How smart, connected products are transforming companies. *Harvard Business Review, 93*(10), 96–114.

Shen, L., Sun, W., & Parida, V. (2023). Consolidating digital servitization research: A systematic review, integrative framework, and future research directions. *Technological Forecasting and Social Change, 191*, 1–24.

Shike, C. (2021). DX in the construction industry realized through smart construction [Smart Construction de Jitsugen suru Kensetsu Sangyo no DX]. In Ohmae, K. (Ed.), *DX Revolution [DX Kakumei]*, President, 167–194. (in Japanese).

Takanashi, C. (2017). Platform construction of manufacturing companies and its requirements: From the perspective of CPS and servitization [Monodukuri Kigyo no Platform Koutsiku to sono Youken]. *The Journal of Science Policy and Research Management [Kenkyu Gijutsu Keikaku], 32*(3), 316–333. (in Japanese).

Tomita, J. (2022). Digital business strategy and process [Digital Keiei no Senryaku to Process]. In Ito, M., Matsuo, H., & Tomita, J. (2022). *Digital management from 1 [Ichi karano Digital Keiei]*, Sekigakusha, 80–95. (in Japanese).

Vandermerwe, S., & Rada, J. (1988). Servitization of business: Adding value by adding services. *European Management Journal, 6*(4), 314–324.

Chapter 6
Environmental Sustainability Data Management in Supply Chains

Aleksi Harju, Mika Immonen, and Jukka Hallikas

Abstract This chapter explores the critical role of data management and supply chain analytics in enhancing environmental sustainability during the digital transformation of supply chains. It introduces a framework for leveraging sustainability-related data to address environmental challenges while detailing the barriers and drivers influencing its adoption. Through qualitative interviews with thirteen firms and a case study on logistics emissions, the research highlights the importance of data quality, integration, and utilization in achieving supply chain transparency and sustainability. The findings emphasize how advanced analytics processes can inform decision-making and improve operational efficiency. Practical implications for integrating sustainability into supply chain strategies are discussed, alongside recommendations for overcoming data-related challenges. The chapter discusses the necessity of robust data management practices and analytics capabilities for organizations striving to align with emerging environmental objectives and regulatory demands.

Keywords Digital transformation · Sustainable supply chain management · Data management · Supply chain analytics · Climate change

6.1 Introduction

The digital transformation of supply chains and its relationship to environmental sustainability continues to elicit interest from scholars and practitioners alike. Digital transformation, defined as "a process that aims to improve an entity by triggering

---

A. Harju (✉) · M. Immonen · J. Hallikas
LUT University, Business School, Lappeenranta, Finland
e-mail: aleksi.harju@lut.fi

Y. Khojasteh, S.M. M. Jahan Abadi (eds.), *Digital Transformation of Supply Chains*, International Series in Operations Research & Management Science 372, https://doi.org/10.1007/978-3-032-01206-7_6

significant changes to its properties through combinations of information, computing, communication, and connectivity technologies" (Vial, 2019) has arguably a great role in the development of more sustainable supply chains. For example, research has broadly established that many novel digital technologies may be important in achieving sustainability in supply chains (Yang et al., 2021). Similarly, Schilling and Seuring (2024), investigated the literature on the linkage between digital transformation and environmental sustainability, finding that it is enabling environmentally sustainable practices for managing supply chains. Recent studies have also often extended the discussion on beneficial connection of supply chain digitalization and environmental sustainability to the context of circular economy (Sadaf Khan et al., 2021; Yu et al., 2022).

Although prior studies have contributed mostly by focusing on digital technology adoption and sustainability improvements derived from the use of technologies, current research is lacking a more granular approach that considers the environmental sustainability-related data and its key role in the digital transformation of supply chains. In fact, environmental sustainability data is arguably the de facto basis for the use of digital technologies that improve sustainability, as technologies that support managerial decision-making are mostly data-driven, and thus, dependent on the quantity and quality of data (Melville & Whisnant, 2014; Shi et al., 2024). Despite evidence from research showing that analytics capabilities may increase performance related to environmental and social sustainability (Dubey et al., 2019; Kamble et al., 2020), current research lacks a sufficient understanding of contingent factors that drive the development of sustainability-oriented supply chain analytics forward or act as barriers for its implementation (Agrawal et al., 2025). This is particularly important, as many organizations may currently have a low degree of sustainability data utilization and suffer from transparency issues in their supply chains. Similarly, studies have lacked an empirically established understanding on how digitally transforming supply chains will have to adapt their data management to comply with the future's new environmental sustainability objectives. It is important to study this issue further, as a holistic understanding of factors contributing to the development of a data-driven approach to sustainable supply chain management is required for actionable solutions to increase sustainability, and, for example, to more effectively mitigate climate change (Dahlmann & Roehrich, 2019). It is also evident that there is an increasing complexity to how global supply chains operate, therefore, producing an understanding on how rigorous data management and reporting processes can implemented is essential in achieving transparency and environmental sustainability in global supply chains.

This chapter argues that the leveraging of data management and supply chain analytics capabilities is essential for addressing sustainability issues in digitally transforming supply chains. The purpose of this chapter is to empirically investigate factors contributing to or preventing change towards the utilization of data and data-related practices for environmental sustainability. We also complement this discussion by presenting a simulation case, where logistics data derived from an industrial supply chain is used to evaluate different kinds of outcomes for environmental sustainability in comparison to competitors. Based on the collected data, we discuss the future implications for sustainability data management and performance in supply chains.

6.2 Literature Review

The integration of data-driven analytics for sustainable supply chain management is generally considered to be a complex, yet valuable task. By leveraging multiple data sources and advanced analytical techniques, organizations can enhance decision-making processes and implement proactive sustainability strategies (Agrawal et al., 2025). Effective data management can support collaborative decision-making processes for sustainability (Beske & Seuring, 2014), improve transparency (Fosso Wamba et al., 2015), and facilitate compliance with environmental regulations (Darnall et al., 2008). The role of sustainability data in supply chains extends beyond mere information storage; it enables predictive, prescriptive, and descriptive analytics to drive environmental and social improvements (Holsapple et al., 2014; Wang et al., 2016). Prior research has also examined how data-driven approaches enable sustainability in supply chains by enhancing collaboration, improving decision-making efficiency, optimizing supply-demand matching, and reducing inventory, carbon emissions, and overall supply chain costs through integrated data platforms (Hu et al., 2019). Research has argued that the proper utilization of data sources and analytics techniques for sustainability may improve environmental and social performance in supply chains (Dubey et al., 2019; Kamble et al., 2020). Bag et al. (2020) emphasize that human resources may significantly impact the adoption of analytics for improved sustainability in the supply chain.

6.2.1 *Enablers of Sustainability-Driven Analytics*

Research indicates that multiple factors may influence the successful implementation of data management and analytics for enhancing sustainability in supply chains. For example, top management commitment can play a pivotal role in driving the implementation of more efficient data management in supply chains, which has implications for managing sustainability (Lamba & Singh, 2018). Similarly, government and policy support may serve as institutional enablers by incentivizing more transparency for data and the adoption of analytics and data-driven sustainability practices (Janssen et al., 2012; Vinuesa et al., 2020). Moreover, studies argue that the digitization of supply chain processes and the alignment of IT infrastructure with data governance improves data collection and analysis, which allows firms to better identify sustainability challenges and implement corrective measures through analytics (Grover et al., 2018; Gupta et al., 2021). Collaborative practices with outsourcing partners have been suggested to support the implementation of analytics practices for sustainability (Akbari & Hopkins, 2022). Moreover, the identification of novel business opportunities for sustainable products and services could further promote data-driven initiatives within organizations (Côrte-Real et al., 2019). Finally, institutional pressures which affect the setting of sustainability objectives has been found to be linked with improved data management and analytics efforts to support achieving environmental goals (Bag et al., 2021). In a similar vein, research has argued that human resource policies, employee education, and personnel incentive systems that guide sustainability-oriented decision-making may better encourage employees to engage with data-driven sustainability initiatives (Mani et al., 2018; Yadav et al., 2020).

6.2.2 Barriers to Sustainability-Driven Analytics

Mesquita et al. (2025) posited that effective data utilization enhances sustainability but found no direct link between sustainability data practices and environmental performance, highlighting the necessity for more extensive research. Studies have identified organizational barriers—such as a lack of commitment from top management and insufficient policy support—as significant obstacles to the implementation of sustainability analytics (Suprayoga et al., 2020; Perçin, 2023). Additionally, the inadequate integration of sustainability into corporate culture has been highlighted as a major impediment, with many organizations struggling to foster stakeholder awareness and engagement (Le et al., 2024). Furthermore, technological barriers such as data privacy and security, lack of data scalability, and complexity of data integration and management have been noted as significant challenges (Agrawal et al., 2025). Other data management challenges, including fragmented data sources, poor data quality, and the lack of standardized collection guidelines, continue to complicate data-driven sustainability efforts (Le et al., 2024). The complexity of developing and adapting advanced analytics models further challenges these issues, particularly if such models require highly specialized human skills and technological resources (Movahedipour et al., 2017). Human barriers are also significant, as many organizations face a shortage of skilled personnel and insufficient training, which undermines successful sustainability initiatives (Suprayoga et al., 2020; Naheed et al., 2024). On the technological front, high implementation costs and inadequate IT infrastructures significantly hinder the adoption of big data techniques intended to support sustainability strategies (Jain et al., 2023). Furthermore, the integration of emerging technologies—particularly artificial intelligence and machine learning—poses considerable technological complexity that requires ongoing innovation and resource investment (Menon & Ravi, 2022; Harfouche et al., 2024).

6.3 A Process Framework for Sustainable Supply Chain Analytics

Based on prior research, we conceptualize the supply chain analytics process framework for improving environmental sustainability in Fig. 6.1. The process framework employs a systematic and stepwise approach, integrating elements from the CRISP-DM methodology (Schröer et al., 2021). Initially, Step 1 is dedicated to

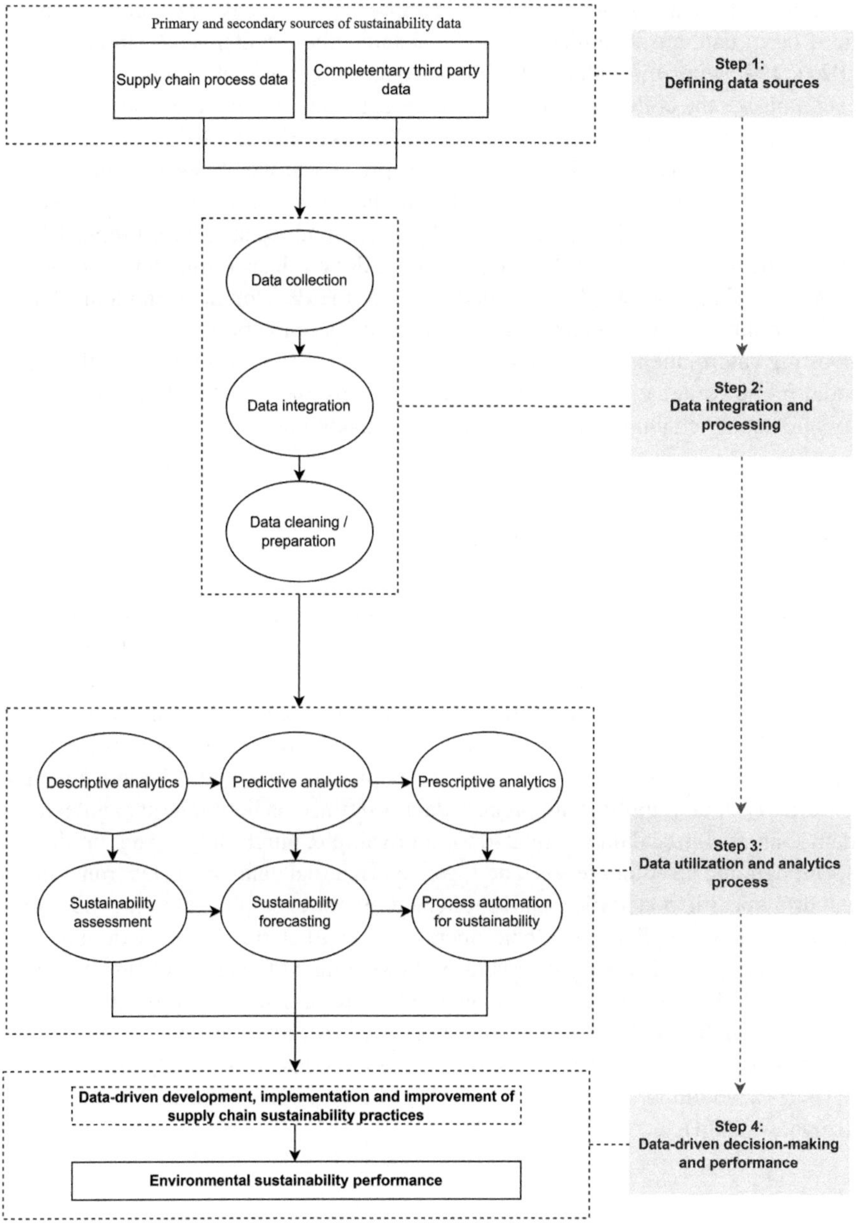

Fig. 6.1 Supply chain analytics process framework for improving environmental sustainability

defining data sources, where primary supply chain process data and additional third-party data are identified to tackle sustainability challenges (Hamdani et al., 2024). Following this, Step 2 highlights data integration and processing, which encompasses the collection, cleaning, and preparation of data to guarantee its quality and readiness for subsequent analysis (Plotnikova et al., 2022). In Step 3, the focus shifts to data utilization and analytics processes, which leverage descriptive, predictive, and prescriptive analytics techniques to perform sustainability assessments, forecasting, and the automation of processes aimed at sustainability (Schröer et al., 2021). Finally, Step 4 centers on data-driven decision-making and performance, utilizing the insights obtained to guide the development, implementation, and continuous enhancement of supply chain sustainability practices, thereby boosting environmental performance (Kamble et al., 2020). By following this structured framework, a comprehensive approach is ensured for embedding sustainability into supply chain operations and decision-making processes.

6.3.1 Defining Data Sources

The first step of the sustainable supply chain analytics process is the identification of relevant data sources within the supply chain. Accurate and comprehensive definition of sustainability data sources is essential, because the quality of the data affects the usefulness of the following analytics process (Schoenherr & Speier-Pero, 2015). We identify two primary data categories related to environmental sustainability in supply chains: (1) supply chain process data, and (2) complementary third-party data. Supply chain process data is defined as both internal and external data sources. Internal data is related to, for example, purchasing spend, production specifications, inventory levels, and logistics. External data originates from outside the firm-specific boundaries, and is reflected by, for example, environmental information about suppliers and their operations, market trends and external risks. Complementary third-party data is information related to supply chain operations, which is collected from public or licensed sources. These are typically different types of databases that can contain supportive information about, for example, environmental or reputational risks or emissions-related data components used in carbon accounting (Bloomberg, 2024; Carbon Disclosure Project, 2024; Ecovadis, 2024).

6.3.2 Data Integration and Processing

The second step is data integration and processing, which includes the collection, integration and preparation of sustainability data from the defined sources. This step typically involves rigorous data science practices, which are often aimed at reducing noise or identifying outliers in the data and creating integrated data sets (Rehman et al., 2016). Data collection involves gathering information from the specified sources to ensure comprehensive coverage of the supply chain's essential sustainability indicators. As this may often generate large volumes of data, it is important to ensure sufficient architecture for storing data, as well as controls for data protection and information security. For example, this could include extracting data on carbon emissions from transportation logs, sourcing information from supplier sustainability reports, and energy consumption data from manufacturing facilities. Data integration refers to merging this data into a unified framework, which increases data quality and consistency between the selected indicators (Hazen et al., 2014). An example is combining carbon emissions data with supplier sustainability ratings and energy consumption figures to create an estimated view of the supply chain's environmental impact. Data preparation includes cleaning, transforming, and organizing the data, making it suitable for subsequent analysis. This often involves removing duplicate records, correcting errors, standardizing data formats, and filling missing values for the collected data.

6.3.3 Data Utilization and Analytics Process

The data analytics process enables firms to derive insights for decision-making to enhance operational efficiency and sustainability of supply chains. A typical taxonomy for understanding a business analytics process in the supply chain context can be categorized into three main types of analytics (Holsapple et al., 2014; Souza, 2014): (1) descriptive, (2) predictive, and (3) prescriptive.

Descriptive analytics is used to summarize and interpret historical data to understand what has happened in the past. It is a form of data-driven sensemaking of what has or is happening (Holsapple et al., 2014), and often comprises of using of various tools and techniques to visualize data and identify trends, patterns, and anomalies. With a proper operationalization of sustainability information, descriptive analytics can be used to produce sustainability assessments, which the most basic form of reporting that supports the management of environmental sustainability. In this context, descriptive analytics driven assessments include analyzing past performance data on, for example, carbon emissions, energy usage, and waste production (Zhu et al., 2007; Schaltegger & Csutora, 2012). Firms can use dashboards and reports to track key sustainability metrics over time, identifying periods of high emissions or inefficient resource use. These visualizations help identify areas where the supply chain is performing well and where improvements are needed, providing a clear, and potentially real-time picture of environmental performance.

Predictive analytics extends on the descriptive analytics by using historical data to forecast future events and trends (Waller & Fawcett, 2013). This type of analytics may be used to employ, for example, statistical models, machine learning algorithms, and other advanced techniques to predict potential outcomes based on historical performance and market data (Choi et al., 2018). Due to this extended capacity, predictive analytics can serve as basis for sustainability forecasting, and thus, is a more comprehensive approach when compared to the static assessment of sustainability. Analytics-based forecasting can be used to more accurately assess future environmental performance, such as carbon emissions under different production scenarios and supply chain configurations or anticipate resource shortages based on consumption trends. Moreover, by analyzing historical data on energy consumption and integrating it with market forecasts, firms can predict periods of high energy demand and take preemptive measures to mitigate their environmental impact. Thus, predictive analytics enables proactive decision-making, allowing firms to address sustainability challenges before they arise.

Prescriptive analytics goes a step further by not only predicting future outcomes but also recommending specific actions to achieve desired sustainability goals. It involves using optimization algorithms, simulation models, and decision analysis to determine the best course of action (Bertsimas & Kallus, 2020). Prescriptive analytics is the most advanced form for supply chain analytics, which allows firms to design and implement process automation for improving sustainability. This prescriptive approach is based on providing automatic recommendations to optimize operations and minimize environmental impact. For example, if supply chain analytics models indicate a potential increase in carbon emissions, prescriptive analytics techniques can suggest interventions into production schedules, routing strategies, or sourcing decisions to mitigate this impact.

6.3.4 Data-Driven Decision-Making and Environmental Performance

The last step of the sustainability analytics process is to leverage the information refined into actionable insights and solutions that improve environmental sustainability in the supply chain. Therefore, the analytics process is used for developing, implementing and continuously improving sustainability-driven strategy, practices, governance internally and collaboratively with supply chain partners (Beske et al., 2014; Raut et al., 2019). This requires that all of the preceding steps in the process are well-structured and efficient, so that the uncertainty related to data-driven insights is minimized and that the underlying analytics process is actually purposeful in achieving organizational sustainability goals. The managerial orientation should be reinforcing towards the use of supply chain data for environmental performance monitoring, so that the analytics process remains robust in its implementation and aligns with organization-specific priorities for both environmental sustainability and financial considerations. Without managerial commitment and support, a transition towards a data-driven sustainable supply chain management is unlikely to succeed.

6.4 Methodology

The purpose of this research is to elaborate on how environmental sustainability data can be integrated to manage sustainability in supply chains. The research process consisted of two stages: (1) a qualitative interview assessment of critical change drivers and implementation barriers, and (2) an empirical application of the supply chain analytics method to showcase how sustainability data management can be utilized for benchmarking and strategic decision-making concerning logistics operations.

The first stage of this research involved a qualitative interview study, utilizing data collection conducted with representatives from 13 firms. To ensure consistency across all participants, a semi-structured interview form was employed (Bryman & Bell, 2015), designed to collect information about sustainability data management issues, particularly focusing on identifying change drivers and implementation barriers. Questions explored the reliability and accuracy of data collection methods, the obstacles to effective information sharing and collaboration across supply chain partners, and the organizational and technical challenges in integrating sustainability data into decision-making processes. Additionally, the interviews addressed factors that hinder engagement and motivation among stakeholders, risks associated with failing to meet sustainability targets, and criteria for identifying and prioritizing areas for improvement. These targeted inquiries provided a comprehensive understanding of the difficulties organizations face in adopting and utilizing sustainability data practices.

The sample consisted of firms from various industrial sectors, including manufacturing, logistics, and retail, all operating in Finland. Purposive sampling (Patton, 2002) was used to select participants who were actively involved in assessing and developing sustainability data management within their supply chains, ensuring the inclusion of individuals with relevant expertise and direct engagement in sustainability initiatives. The interviews were conducted remotely, with each session lasting approximately 60 min. Two interviewers participated in every interview to ensure that critical aspects of the data collection were adequately addressed. This approach also minimized potential biases and allowed for a more thorough capture of the informants' responses. To ensure the reliability and validity of the collected data, several measures were implemented. The interview form was pilot-tested to refine the questions for clarity and relevance. Triangulation was applied by cross-validating insights from the interviews with secondary data sources, such as sustainability reports and industry benchmarks. The qualitative data analysis involved iterative coding by multiple researchers to ensure consistency in interpretation. Additionally, detailed notes and recordings were maintained for all interviews, and verbatim responses were transcribed, enabling subsequent review and verification. This systematic approach to data collection ensured the generation of insights into the challenges and enablers of sustainability data management in supply chains.

The primary data collected during the qualitative interviews was analyzed using a qualitative content analysis approach (Miles et al., 2014) to identify change drivers and implementation barriers in the transition towards improved sustainability data management in supply chains (Fig. 6.2). The analysis process involved reducing, displaying, and drawing conclusions from the data in alignment with the research problem (Miles & Huberman, 1994). A structured codebook was developed to organize the observed challenges and implementation barriers systematically, ensuring consistency throughout the analysis. The coding process followed an inductive, data-driven approach, allowing for the emergence of key themes directly from the data. This enabled the identification of both change drivers and implementation barriers without imposing preconceived categories or theoretical constructs. Nvivo 13 software was used to facilitate the coding and analysis process. To ensure the reliability of the results, triangulation was employed, with multiple researchers independently validating the codes and ensuring alignment across their interpretations. Data source triangulation was also applied, drawing on insights from participants across diverse industries, including manufacturing, logistics, and retail. This diversity of perspectives provided a richer understanding of the challenges and barriers and ensured that findings were not specific to a single sector. A workshop was

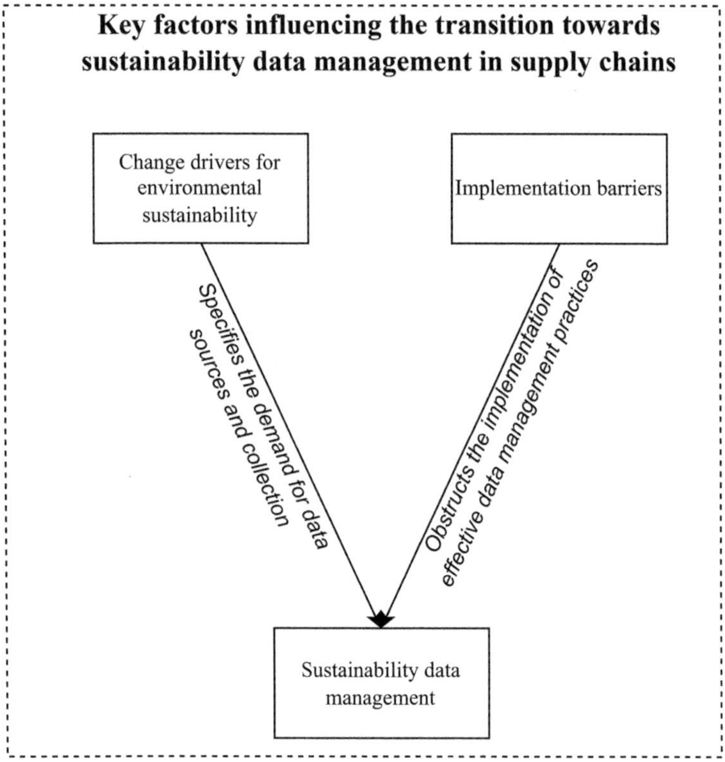

Fig. 6.2 Research approach for understanding the factors influencing the transition towards sustainability data management in supply chains

also held, where preliminary findings were presented to selected study participants, enabling their feedback to further validate and refine the results.

In the second stage of the research, we present an application of the analytics process in a case study conducted through an advanced data processing and analysis methodology, operationalized through a KNIME 5 workflow. The analysis investigates the case company's sales and delivery data over a single-year period. The goal of the analysis is to create a carbon emission assessment applying post-hoc routing data extraction (GHG Protocol, 2024). Using the selected approach, the case also provides an example for comparative performance assessment against competitors. In this process, the case firm's delivery locations from the customer base have been replicated against the main competitors' locations to enable performance comparisons and spatial advantage assessment.

The research applies a structured analytics framework to address supply chain and sustainability challenges through data preparation, integration, and advanced analysis. The process begins with data import and cleaning, where raw sales data is standardized, filtered, and grouped to ensure consistency. This is followed by data enrichment using Azure Maps geocoding, which incorporates geospatial information to provide location-based insights. Supplier-specific routing analysis calculates distances and durations for supply chain routes using OSRM-based systems, with workflows preparing and filtering geocoded data for accurate computations.

Emissions per transaction are then calculated by combining sales and routing data, factoring in delivery tonnage and route characteristics. Aggregation and manipulation techniques refine the data, and advanced formulae adjust emissions calculations based on maximum delivery weight and proportional contributions. The results are prepared for actionable decision-making, providing insights to optimize supply chain operations and improve environmental sustainability. This framework demonstrates an efficient and scalable method for addressing complex sustainability challenges in supply chains.

6.5 Empirical Study

6.5.1 Change Drivers

6.5.1.1 Climate Change Disruption Risks

Climate change poses significant risks to supply chains, affecting everything from raw material availability to transportation and distribution networks. This puts pressure for firms to adopt digital systems and data management practices to improve the measurement, monitoring and reporting of climate change risks in supply chains. Many firms expect that an increasing use of sustainability data for analytics purposes will enable them to better assess climate risks, optimize routes, predict disruptions, and implement mitigation strategies, thereby enhancing their resilience against the adverse effects of climate change.

6.5.1.2 Increasing Regulatory Pressure and Compliance Risks

Many informants were greatly concerned about the uncertainty surrounding regulatory frameworks, which are expected to become increasingly stringent. For example, key regulatory developments in Europe include carbon taxation, carbon pricing, and the carbon border adjustment mechanism, all designed to incentivize the reduction of greenhouse gas emissions. International agreements such as the Paris Agreement, and regional directives like the Corporate Sustainability Reporting Directive (CSRD) and the Corporate Sustainability Due Diligence (CSDD) initiative, mandate firms to disclose their environmental impact and take responsibility for their supply chains' sustainability. Many firms stated that compliance with such regulations requires accurate and comprehensive data, which can be effectively managed through the development of sustainability data management.

6.5.1.3 Public Perception

Some firms noted that public perception plays a crucial role in driving their sustainability data management efforts. Consumers are increasingly aware of environmental issues and are demanding greater transparency and responsibility from firms. Moreover, brands perceived as environmentally irresponsible were considered to face heightened reputational risks and potential loss of market share. Conversely, firms that demonstrate a commitment to sustainability were considered to potentially enhance their brand image and foster customer loyalty. Therefore, the interviewed firms expected the utilization of sustainability data to be crucial in responding transparently to consumer expectations and increase the validity of their sustainability reporting.

6.5.1.4 Employer Branding and Competitive Recruitment

Employer branding was observed as another significant driver of change. For example, many firms considered that the competition for top talent is continuously intensifying, and there is an increased demand for personnel that possess sustainability and analytics-related expertise. Some also considered that with strong sustainability credentials they can better attract and retain skilled employees that support the realization of the firm's goals for environmental sustainability and digital transformation. Therefore, having in-place and functioning sustainability strategy and corresponding data management practices may support branding and recruitment-related aspects, because firms are better able to factually communicate their commitment to reducing environmental impact and creating a sustainability-oriented culture in their organization.

6.5.1.5 Investor Expectations

There is a broad expectation that investor demand for environmental sustainability is rapidly increasing. Investors are recognizing that sustainable firms are better positioned for long-term success and are less exposed to risks associated, for example, with climate change and regulatory changes. As a result, we observed a growing trend towards incorporating Environmental, Social, and Governance (ESG) criteria into firms' investment decisions. This incentivizes the collection of sustainability data and the development of supply chain analytics capabilities, because firms can better provide the detailed data required by investors related to them and their business partners' sustainability performance.

6.5.2 Implementation Barriers

6.5.2.1 Data Quality Issues

A primary barrier to transitioning towards data-driven management of environmental sustainability in supply chains is considered to be the quality of sustainability-related data. For example, many firms were found to still rely on revenue-based emission allocations, which is an inaccurate methodology for assessing greenhouse gas emissions. Such an approach may lead to significant discrepancies in sustainability data. Furthermore, the informants considered to have a general lack of data validation throughout their supply chains, which undermines the reliability of reported sustainability issues. Firms also reported a lack of expertise in developing robust information systems for measuring and reporting about environmental sustainability.

6.5.2.2 Lack of Explicit Environmental Sustainability Reporting Requirements

We found that the absence of explicit environmental sustainability reporting requirements had lead to inconsistent data production across the interviewed firms. As an example, one of the informants noted that organizations in their supply chains may produce emission data of varying granularities, creating a "wild west" scenario for emission factors, which can lead to inaccuracies. This makes it difficult for firms to commit to implementing new data management processes, as information-systems and employee competence-based investments for data management may be very high and may not fit into the changing regulatory standards of the future.

6.5.2.3 Problems in Emission Allocation Practices

Data-driven management of environmental sustainability is challenged by issues related to the allocation practices for emissions, such as misallocations, forgetting allocations, or double-counting emission-generating activities. This was observed to be caused by firms often having low visibility into processes that indirectly generate emissions in their supply chains, leading to incomplete accounting of data sources. Additionally, we found that there may not be clear decision criteria for precisely allocating emissions categories internally, complicating a data-driven and integrated management of performance across different units. For example, in multi-stakeholder projects, the studied firms found it often unclear who is responsible for specific value chain emissions, creating ambiguity and inefficiencies during data collection, allocation and reporting processes.

6.5.2.4 Supply Chain Complexity

Large and complex industrial supply networks were seen as one of the key barriers to using sustainability data for decision-making purposes in the studied firms. Such complexity presents significant barriers, primarily because it encompasses various interconnected factors that affect the collection and verification of data, and thus, the quality of data components from the supply chain, which are utilized in the analytics process. Supply chain complexity was perceived to be related to many things; for example, the increased complexity of products, region-spanning supply chain operations and varying regulatory requirements within distinct market areas of the supply chain. Essentially, firms implied that an increased supply chain complexity affected the resource demands particularly for data collection, preparation and integration, severely affecting the validity of conclusions that can be made about environmental sustainability in the supply chain.

6.5.2.5 Confidentiality of Sustainability Data

Interestingly, we found that transparent sustainability reporting may be linked to sensitive revenue streams and operation-specific data of firms. Many publicly traded firms noted that they must carefully consider what data they release to avoid compromising commercially sensitive information. This can reduce the accuracy of public disclosures, but we also found instances where a non-disclosure or more conservative information sharing about environmental sustainability issues in the supply chain is incentivized. When this problem exists in the supply chain, it directly affects the effectiveness of sustainability data management and analytics, as accurate data may be intentionally made unavailable for the supply chain partners.

6.5.2.6 Manual Labor and the Lack of Integrated Information Systems

Most studied firms reported that sustainability data management and reporting processes are still heavily reliant on manual entry and fragmented data sources, such as Excel spreadsheets and PDFs, resulting in slow and resource-intensive workflows. We also found that integrating databases and management information systems for environmental sustainability poses significant challenges for firms. Moreover, a crucial implication was put forward by one the informants: the ICT sector may not have sufficient resources to meet the high demand for tailored environmental sustainability monitoring and reporting solutions across different supply chains. Also, many firms spontaneously brought up the issue of third-party sustainability platforms; they were rated low by study participants, particularly due to their mismatch to firm-specific requirements, lack of integration with other systems, and contrary to expectations, their continued reliance on manual processes post-implementation.

6.5.3 Example Case: Comparative Modelling of Emissions in Company Logistics

The example case study focuses on calculating the emissions of a company's logistics operations using a data enrichment approach that completes sales data with data fetched from a routing API ('application programmer interface'). The goal is to highlight the importance of data-driven approaches to managing and reducing logistics-related emissions. The case also provides insights into the methodologies and tools necessary for carbon dioxide—emissions ('CO2') assessment when the current information systems and data have remarkable gaps regarding content. The analysis begins with geocoding locations of customers from the case company's one-year sales data, which includes delivery information for each sales activity. Subsequently, the routing API is utilized to create detailed route data for unique deliveries in the logistics operations. Using this route data, the emissions calculation is generated for over one year of deliveries. The computations are conducted in (1) relation to customer locations and deliveries, and (2) categorized by customer categories (ABC). To provide a comparative perspective, the case analysis replicates the calculation procedure for competitors' locations. This comparison aims to highlight differences between location strategies based on logistics performance and environmental impacts.

6.5.3.1 Analysis Process

The applied structured analysis approach directs the emission assessment of logistics and addresses gaps in the current information system. Through the steps presented, the case study develops a method to compute logistics-related emissions for enhancing overall sustainability of companies. This streamlined data processing highlights the importance of a data-driven approach in managing logistics emissions. The structured workflow for assessing emissions in the case company logistics is as follows:

1. Data import,
2. Geolocation enrichment,
3. Route calculation,
4. Emissions computation.

The process starts with importing sales data. Unique identifiers are created, country names are standardized to ISO codes, and data cleaning and consolidation is performed. Next, the sales data is enriched with geolocation information using geocoding services, preparing the address field for geocoding and adding geolocation data. For route calculation, the workflow uses a routing API (e.g. OpenStreetMap Routing) to determine distances and durations of routes. Emissions are computed on a per-transaction basis. In this process, a unique identifier for sales deliveries is created, and sales data is aggregated by customer ID and transaction date. In the last phase, CO2 emissions per route are calculated and adjusted based on delivery tons and the maximum weight of vehicles. Finally, the sales dataset is readjusted to include competitors' site locations to which calculations are replicated.

A two-step approach is applied to assess total emission per customer based on sales data. In the first step, the computation determines the initial CO2 emissions from truck transport by route (Formula 1a). The formula considers the distance traveled by the truck (m), the specific CO2 emission factor for the truck, and the gross weight of the truck (kg) to calculate the initial emissions. In the second step, the process applies an adjustment to the initial CO2 emission values based on the actual tons transported (Formula 1b). The adjusted CO2 emissions describe the efficiency of the transport, which depends on the actual load carried by the truck. Using the selected procedure, sales-delivery emissions can be roughly estimated without complete transportation data from logistics operators.

Formula 1a and 1b: A two-step approach for logistics sustainability assessment

(a) *Initial Route Emissions (full payload emissions)*

$$\text{CO2Emission}(\text{initial}) = \left(\frac{\text{Distance} \times \text{CO2Emission Factor per Truck}}{1000}\right) \times \left(\frac{\text{Gross Weight of Truck}}{1000}\right)$$

(b) *Adjusted Emissions by Share of Delivery Total Payload*

$$\text{CO2Emission}(\text{corrected}) = \text{CO2Emission}(\text{initial}) \times \left(\frac{\text{Tons}}{\text{Gross Weight of Truck} / 1000} \right)$$

6.5.3.2 Empirical Data-Source and Generated Data

The dataset used in the case contains detailed sales data from the focal firm, including 23,716 rows, each representing a unique sales ID (Table 6.1). These rows encompass various attributes such as customer addresses, deliveries, and customer home countries. The sales data includes 5283 unique deliveries identified by a dedicated identifier. Customer addresses are recorded in a standardized format including total of 751 unique addresses. Customer names comprise 229 unique entries from 34 distinct countries.

The base data provides grounds for conducting a detailed CO2-emissions analysis after geocoding locations and calculating route features for logistics. The structured data allows for an in-depth comparison of emissions across different factory locations, highlighting the importance of data-driven approaches in environmental performance assessment. The routing approach involved geocoding customer locations to obtain precise GPS coordinates, transforming addresses into geographical data. An OpenStreetMap-based route engine was used to generate routes for the focal company and competitors, targeting the specified customer base. Routing parameters were set to find the shortest road-based routes for cars, ensuring practical and efficient logistics routes. The analysis was focused on Europe, ensuring relevant and consistent routing data. This method provided accurate distance and duration calculations, essential for emissions analysis and environmental performance assessment.

Table 6.1 Case data description

Attribute	Unique Values
Customer IDs	801
Unique Address	751
Customer Countries	34
Customer Name (Company names)	229
Sales Delivery ID (Unique deliveries)	5283
Sales Rows (Source data events, Sales ID)	23,716

6.5.3.3 Empirical Findings

The analysis compares the logistics emissions across four factory locations representing the case company and four competitors. The factories analyzed were located as follows; one in Northern Europe ('N-EUR'), two in Southern Europe ('S-EUR'), and one in Western Europe W-EUR. Each factory location's performance is evaluated in relation to the unique routes and deliveries to provide a comprehensive assessment of their environmental impact. The empirical analysis presents three separate comparisons: (1) naïve factory-to-customer route assessment, (2) customer category weighted route assessment, and (3) sustainability performance evaluation. The primary objective was to assess regional logistics providers' performance metrics focusing specifically on average route lengths within distinct geographic zones encompassing Northern-European Region, Southern regions, and Western European Markets (Table 6.2). In the first phase, the naïve factory-to-customer route assessment showed slight differences impacting operations. The results are reported as "all cases" in the table. The customer base was relatively evenly distributed across Europe, where competitive advantage or disadvantage by location can be seen only

Table 6.2 Summary of calculations based on sales data

Customer class	Customer locations[a], n	Deliveries count[a], n (%)	Location[b] (factory)	Distance [km], Mean (SD)	CO_2 emission[c] [metric-tn] (Sum)	CO_2 emission[c] [metric-tn] (Mean)
All cases	792 (100)	5176 (100)	S-EUR	1667 (822)	5073.0	1.0
			N-EUR	1693 (1061)	5802.5	1.1
			W-EUR	1654 (875)	4943.6	1.0
			S-EUR	1621 (829)	4889.0	0.9
A	29 (3.6)	1758 (33.9)	S-EUR	1504 (772)	3012.5	1.7
			N-EUR	1778 (900)	3423.2	1.9
			W-EUR	1463 (786)	2924.3	1.7
			S-EUR	1438 (776)	2883.3	1.6
B	203 (25.7)	2325 (44.9)	S-EUR	1649 (802)	1771.8	0.8
			N-EUR	1793 (1111)	2159.1	0.9
			W-EUR	1636 (885)	1726.3	0.7
			S-EUR	1610 (812)	1721.1	0.7
C	560 (70.7)	1093 (21.2)	S-EUR	1970 (860)	288.7	0.3
			N-EUR	1344 (1118)	220.2	0.2
			W-EUR	1998 (892)	293.0	0.3
			S-EUR	1940 (852)	284.7	0.3

[a]Non-missing values after geocoding, some addresses fail in OpenStreetMap geocode API
[b]Regions: S-EUR = South Europe, N-EUR = North Europe, W-EUR = West Europe
[c]After payload share correction of each delivery

in the standard deviation level. In the other words, locations near to central Europe achieve a more stable performance basis for operations.

In the second analysis phase, including customer category in performance evaluation provides more precise insights into the effects of location strategy and competitive positioning between regions. The category-adjusted analysis considers delivery frequencies and the importance of certain operations to value creation. Using the ABC classification based on net sales cumulative percentage, the study showed how different customers contribute to overall sales. Categories were defined using cut points, where category "A" includes customer contributing up to 60% of total sales, category "B" includes customers contributing more than 60% but up to 95%, and Category "C" includes customers contributing over 95%, reflecting high, moderate, and low contributions to overall sales, respectively. The customer class-based assessment shows differences in the customer related strategic positioning. Class A includes the lowest count of customer locations, which strongly biases overall competitive positioning of the factory location regions. In this case, S-EUR is clearly the most advantageous position for location distribution in the sales data. On the other, N-EUR has clearly the greatest challenges form logistics perspective to perform in competition. Class B customers have a similar distribution to class A, indicating that locations share the same description with category A. Class C customer locations have a distinct locational distribution, which is biased more towards the northern parts of central Europe and Northern Europe. In this case, the N-EUR location has performance advantages in operations.

In the third phase, the computation of CO2 emissions was included to measure the impact of location strategies on sustainability performance. Results from the emission calculation broadly follow the results from the category-based distance analysis, showing that S-EUR has an advantage in customer categories A and B, whereas category C customers are more favorable for the N-EUR region. The emission results can be explained by customer demand and order sizes, which are higher for customers in A and B classes. According to the data, mean values for delivery capacity ratio were multiple times higher for A class than for the others. The findings also show the logistics sustainability management challenges that companies face for different customers. Main customer deliveries are relatively easy to optimize because logistics run at maximum capacity. However, small-scale customers create management challenges and issues for emission computations. In these cases, payloads dedicated for single customers are only a fragment of the total delivery, making true emissions hard to evaluate and manage.

6.6 Discussion and Conclusions

We discussed the critical role that data management and supply chain analytics play in driving environmental sustainability within digitally transforming supply chains. This chapter highlights that the granularity and quality of environmental sustainability data are paramount for leveraging digital technologies effectively. This aligns with prior literature, which posits that data-driven technologies are essential for informed managerial decision-making and achieving sustainability goals (Melville & Whisnant, 2014; Dubey et al., 2019; Kamble et al., 2020; Agrawal et al., 2025). By establishing a detailed understanding of the data requirements and the contingent factors that enable and hinder their collection and utilization, this chapter contributes to research on how organizations can enhance their sustainability performance through robust data management practices (Raut et al., 2019; Schilling & Seuring, 2024; Agrawal et al., 2025). Moreover, this chapter reveals that while digital transformation holds significant potential for improving sustainability outcomes, it is contingent on the successful integration of comprehensive data management strategies. The barriers to effective data utilization identified in our analysis, such as low data transparency and the complexity of global supply chains, echo concerns raised in existing research (Movahedipour et al., 2017; Suprayoga et al., 2020; Le et al., 2024). Addressing these barriers requires technological advancements and also organizational changes that prioritize sustainability and data governance. Therefore, the findings presented in this chapter serve as guides for organizations seeking to navigate the challenges of data-driven supply chain management while prioritizing environmental sustainability.

Through the example case, the examination of logistics efficiency across diverse regions in Europe reveals that geographical placement significantly influences operations. Our research highlights the importance of geographic location in crafting efficient logistic systems for different clientele segments. While major client shipments are manageable due to optimized capacities, smaller clients pose significant managerial challenges for emissions calculations. Individualized cargo allocations contribute only fractions to overall shipment volumes, making precise environmental impact assessments challenging to quantify and manage. The simulation case study presented in this chapter illustrates the practical implications of our findings. Using logistics data from an industrial supply chain, we demonstrated how different data management practices can lead to varying outcomes in terms of environmental sustainability. This comparative analysis provides valuable insights for practitioners, showing that proactive data management and analytics can significantly enhance sustainability performance. The case study also shows the need for continuous adaptation and improvement of data practices to meet evolving environmental objectives. Thus, this chapter contributes to a deeper understanding of the interplay between digital transformation, data management, and environmental sustainability, offering a comprehensive framework and practical solutions for advancing environmental sustainability in supply chains.

Overall, the empirical findings of this study highlight the growing importance of sustainability-focused data management in supply chains. As companies navigate an increasingly complex regulatory and market landscape, integrating sustainability considerations into data strategies is essential for enhancing transparency and environmental performance. Effective data management enables organizations to leverage digital technologies for more accurate sustainability reporting, informed decision-making, and improved supply chain efficiency. A key practical implication is the need for robust data systems that support the strategic use of sustainability data. Our analysis of logistics efficiency across European regions highlights how data-driven decision-making can optimize operational performance and improve emissions tracking. The case study further demonstrates that incorporating geospatial and routing information enhances carbon footprint assessments, allowing companies to implement targeted measures that drive sustainability outcomes. From a managerial perspective, continuous adaptation of data management practices is essential to staying ahead of regulatory changes and shifting market expectations. Companies that proactively invest in advanced analytics, ensure seamless interoperability across data systems, and embed sustainability into their decision-making frameworks will be better positioned to achieve long-term environmental and economic benefits.

References

Agrawal, R., Islam, N., Samadhiya, A., Shukla, V., Kumar, A., & Upadhyay, A. (2025). Paving the way to environmental sustainability: A systematic review to integrate big data analytics into high-stake decision forecasting. *Technological Forecasting and Social Change, 214*(February), 124060. https://doi.org/10.1016/j.techfore.2025.124060

Akbari, M., & Hopkins, J. L. (2022). Digital technologies as enablers of supply chain sustainability in an emerging economy. *Operations Management Research, 15*(3–4), 689–710. https://doi.org/10.1007/s12063-021-00226-8

Bag, S., Pretorius, J. H. C., Gupta, S., & Dwivedi, Y. K. (2021). Role of institutional pressures and resources in the adoption of big data analytics powered artificial intelligence, sustainable manufacturing practices and circular economy capabilities. *Technological Forecasting and Social Change, 163*, 120420. https://doi.org/10.1016/j.techfore.2020.120420

Bag, S., Wood, L. C., Xu, L., Dhamija, P., & Kayikci, Y. (2020). Big data analytics as an operational excellence approach to enhance sustainable supply chain performance. *Resources, Conservation and Recycling, 153*, 104559. https://doi.org/10.1016/j.resconrec.2019.104559

Bertsimas, D., & Kallus, N. (2020). From predictive to prescriptive analytics. *Management Science, 66*(3), 1025–1044. https://doi.org/10.1287/mnsc.2018.3253

Beske, P., Land, A., & Seuring, S. (2014). Sustainable supply chain management practices and dynamic capabilities in the food industry: A critical analysis of the literature. *International Journal of Production Economics, 152*, 131–143. https://doi.org/10.1016/j.ijpe.2013.12.026

Beske, P., & Seuring, S. (2014). Putting sustainability into supply chain management. *Supply Chain Management, 19*(3), 322–331. https://doi.org/10.1108/SCM-12-2013-0432

Bloomberg, L. P. (2024). *ESG Data*. Available at: https://www.bloomberg.com/professional/products/data/enterprise-catalog/esg/

Bryman, A., & Bell, E. (2015). *Business research methods* (4th ed.). Oxford University Press.

Carbon Disclosure Project. (2024). *Data packages*. Available at: https://www.cdp.net/en/data

Choi, T.-M. M., Wallace, S. W., & Wang, Y. (2018). Big data analytics in operations management. *Production and Operations Management, 27*(10), 1868–1883. https://doi.org/10.1111/poms.12838

Côrte-Real, N., Ruivo, P., Oliveira, T., & Popovič, A. (2019). Unlocking the drivers of big data analytics value in firms. *Journal of Business Research, 97*(December 2018), 160–173. https://doi.org/10.1016/j.jbusres.2018.12.072

Dahlmann, F., & Roehrich, J. K. (2019). Sustainable supply chain management and partner engagement to manage climate change information. *Business Strategy and the Environment, 28*(8), 1632–1647. https://doi.org/10.1002/bse.2392

Darnall, N., Jolley, G. J., & Handfield, R. (2008). Environmental management systems and green supply chain management: Complements for sustainability? *Business Strategy and the Environment, 17*(1), 30–45. https://doi.org/10.1002/bse.557

Dubey, R., Gunasekaran, A., Childe, S. J., Papadopoulos, T., Luo, Z., Wamba, S. F., & Roubaud, D. (2019). Can big data and predictive analytics improve social and environmental sustainability? *Technological Forecasting and Social Change, 144*, 534–545. https://doi.org/10.1016/j.techfore.2017.06.020

Ecovadis. (2024). *Scope 3 Carbon Management*. Available at: https://ecovadis.com/solutions/carbon/

Fosso Wamba, S., Akter, S., Edwards, A., Chopin, G., & Gnanzou, D. (2015). How "big data" can make big impact: Findings from a systematic review and a longitudinal case study. *International Journal of Production Economics, 165*, 234–246. https://doi.org/10.1016/j.ijpe.2014.12.031

GHG Protocol. (2024). *Corporate Value Chain (Scope 3) Standard*. Available at: https://ghgprotocol.org/corporate-value-chain-scope-3-standard

Grover, V., Chiang, R. H. L., Liang, T.-P., & Zhang, D. (2018). Creating strategic business value from big data analytics: A research framework. *Journal of Management Information Systems, 35*(2), 388–423. https://doi.org/10.1080/07421222.2018.1451951

Gupta, H., Kumar, S., Kusi-Sarpong, S., Jabbour, C. J. C., & Agyemang, M. (2021). Enablers to supply chain performance on the basis of digitization technologies. *Industrial Management & Data Systems, 121*(9), 1915–1938. https://doi.org/10.1108/IMDS-07-2020-0421

Hamdani, F.-E., Quintero, I. A. Q., Enjolras, M., Camargo, M., Monticolo, D., & Lelong, C. (2024). Agile supply chain analytic approach: A case study combining agile and crisp-dm in an end-to-end supply chain. *Supply Chain Forum: An International Journal, 25*(1), 96–110. https://doi.org/10.1080/16258312.2022.2064721

Harfouche, A., Merhi, M. I., Albizri, A., Dennehy, D., & Thatcher, J. B. (2024). Sustainable development through technological innovations and data analytics. *Information Systems Frontiers, 26*(6), 1989–1996. https://doi.org/10.1007/s10796-024-10570-2

Hazen, B. T., Boone, C. A., Ezell, J. D., & Jones-Farmer, L. A. (2014). Data quality for data science, predictive analytics, and big data in supply chain management: An introduction to the problem and suggestions for research and applications. *International Journal of Production Economics, 154*, 72–80. https://doi.org/10.1016/j.ijpe.2014.04.018

Holsapple, C., Lee-Post, A., & Pakath, R. (2014). A unified foundation for business analytics. *Decision Support Systems, 64*, 130–141. https://doi.org/10.1016/j.dss.2014.05.013

Hu, J., Lim, M. K., Tan, K. H., Tseng, M.-L., & Sarkis, J. (2019). Big data application in sustainable supply chains: A transportation industry case. In *Handbook on the sustainable supply chain* (pp. 429–443). Edward Elgar.

Jain, P., Tambuskar, D. P., & Narwane, V. S. (2023). Is the implementation of big data analytics in sustainable supply chain really a challenge? The context of the indian manufacturing sector. *International Journal of Innovation and Technology Management, 20*(05), 2350033. https://doi.org/10.1142/S0219877023500335

Janssen, M., Charalabidis, Y., & Zuiderwijk, A. (2012). Benefits, adoption barriers and myths of open data and open government. *Information Systems Management, 29*(4), 258–268. https://doi.org/10.1080/10580530.2012.716740

Kamble, S. S., Gunasekaran, A., & Gawankar, S. A. (2020). Achieving sustainable performance in a data-driven agriculture supply chain: A review for research and applications. *International Journal of Production Economics, 219*, 179–194. https://doi.org/10.1016/j.ijpe.2019.05.022

Lamba, K., & Singh, S. P. (2018). Modeling big data enablers for operations and supply chain management. *The International Journal of Logistics Management, 29*(2), 629–658. https://doi.org/10.1108/IJLM-07-2017-0183

Le, T. T., Priya, J. C., Le, H. C., Le, N. V. L., Duong, M. T., & Cao, D. N. (2024). Harnessing artificial intelligence for data-driven energy predictive analytics: A systematic survey towards enhancing sustainability. *International Journal of Renewable Energy Development, 13*(2), 270–293. https://doi.org/10.61435/ijred.2024.60119

Mani, V., Gunasekaran, A., & Delgado, C. (2018). Supply chain social sustainability: Standard adoption practices in Portuguese manufacturing firms. *International Journal of Production Economics, 198*, 149–164. https://doi.org/10.1016/j.ijpe.2018.01.032

Melville, N. P., & Whisnant, R. (2014). Energy and carbon management systems – Organizational implementation and application. *Journal of Industrial Ecology, 18*(6), 920–930. https://doi.org/10.1111/jiec.12135

Menon, R. R., & Ravi, V. (2022). An analysis of barriers affecting implementation of sustainable supply chain management in electronics industry: A grey-dematel approach. *Journal of Modelling in Management, 17*(4), 1319–1350. https://doi.org/10.1108/JM2-02-2021-0042

Mesquita, L. L., Lizarelli, F. L., & Duarte, S. (2025). Big data analytics and lean practices: Impact on sustainability performance. *Production Planning & Control, 36*, 333–356. https://doi.org/10.1080/09537287.2023.2267512

Miles, M. B., & Huberman, A. M. (1994). *Qualitative data analysis* (2nd ed.). Sage.

Miles, M. B., Huberman, A. M., & Saldaña, J. (2014). *Qualitative data analysis: A methods sourcebook* (3rd ed.).

Movahedipour, M., Zeng, J., Yang, M., & Wu, X. (2017). An ism approach for the barrier analysis in implementing sustainable supply chain management. *Management Decision, 55*(8), 1824–1850. https://doi.org/10.1108/MD-12-2016-0898

Naheed, R., Waqas, M., Ahmad, N., Iqbal, M., & Ismail, M. (2024). Fostering sustainability and green innovation reporting in manufacturing firms: An investigation of barriers through ism-micmac approach. *Environment, Development and Sustainability* [Preprint] No. 0123456789. https://doi.org/10.1007/s10668-024-04771-7

Patton, M. Q. (2002). *Qualitative research & evaluation methods* (3rd ed.). Sage.

Perçin, S. (2023). Identifying barriers to big data analytics adoption in circular agri-food supply chains: A case study in turkey. *Environmental Science and Pollution Research, 30*(18), 52304–52320. https://doi.org/10.1007/s11356-023-26091-5

Plotnikova, V., Dumas, M., & Milani, F. P. (2022). Applying the CRISP-DM data mining process in the financial services industry: Elicitation of adaptation requirements. *Data & Knowledge Engineering, 139*, 102013. https://doi.org/10.1016/j.datak.2022.102013

Raut, R. D., Mangla, S. K., Narwane, V. S., Gardas, B. B., Priyadarshinee, P., & Narkhede, B. E. (2019). Linking big data analytics and operational sustainability practices for sustainable business management. *Journal of Cleaner Production, 224*, 10–24. https://doi.org/10.1016/j.jclepro.2019.03.181

Rehman, M. H. U., Chang, V., Batool, A., & Wah, T. Y. (2016). Big data reduction framework for value creation in sustainable enterprises. *International Journal of Information Management, 36*(6), 917–928. https://doi.org/10.1016/j.ijinfomgt.2016.05.013

Sadaf Khan, I., Ovais Ahmad, M., & Majava, J. (2021). Industry 4.0 and sustainable development: A systematic mapping of triple bottom line, circular economy and sustainable business models perspectives sustainable development triple bottom line circular economy sustainable business models. https://doi.org/10.1016/j.jclepro.2021.126655

Schaltegger, S., & Csutora, M. (2012). Carbon accounting for sustainability and management. Status quo and challenges. *Journal of Cleaner Production, 36*, 1–16. https://doi.org/10.1016/j.jclepro.2012.06.024

Schilling, L., & Seuring, S. (2024). Linking the digital and sustainable transformation with supply chain practices. *International Journal of Production Research, 62*(3), 949–973. https://doi.org/10.1080/00207543.2023.2173502

Schoenherr, T., & Speier-Pero, C. (2015). Data science, predictive analytics, and big data in supply chain management: Current state and future potential. *Journal of Business Logistics, 36*(1), 120–132. https://doi.org/10.1111/jbl.12082

Schröer, C., Kruse, F., & Gómez, J. M. (2021). A systematic literature review on applying CRISP-DM process model. *Procedia Computer Science, 181*(2019), 526–534. https://doi.org/10.1016/j.procs.2021.01.199

Shi, Y., Van Toorn, C., & McEwan, M. (2024). Exploration–exploitation: How business analytics powers organisational ambidexterity for environmental sustainability. *Information Systems Journal, 34*, 894–930. https://doi.org/10.1111/isj.12456

Souza, G. C. (2014). Supply chain analytics. *Business Horizons, 57*(5), 595–605. https://doi.org/10.1016/j.bushor.2014.06.004

Suprayoga, G. B., Witte, P., & Spit, T. (2020). Identifying barriers to implementing a sustainability assessment tool for road project planning: An institutional perspective from practitioners in indonesia. *Journal of Environmental Planning and Management, 63*(13), 2380–2401. https://doi.org/10.1080/09640568.2020.1724083

Vial, G. (2019). Understanding digital transformation: A review and a research agenda. *The Journal of Strategic Information Systems, 28*(2), 118–144. https://doi.org/10.1016/j.jsis.2019.01.003

Vinuesa, R., et al. (2020). The role of artificial intelligence in achieving the sustainable development goals. *Nature Communications, 11*(1), 233. https://doi.org/10.1038/s41467-019-14108-y

Waller, M. A., & Fawcett, S. E. (2013). Data science, predictive analytics, and big data: A revolution that will transform supply chain design and management. *Journal of Business Logistics, 34*(2), 77–84. https://doi.org/10.1111/jbl.12010

Wang, G., Gunasekaran, A., Ngai, E. W. T., & Papadopoulos, T. (2016). Big data analytics in logistics and supply chain management: Certain investigations for research and applications. *International Journal of Production Economics, 176*, 98–110. https://doi.org/10.1016/j.ijpe.2016.03.014

Yadav, G., Kumar, A., Luthra, S., Garza-Reyes, J. A., Kumar, V., & Batista, L. (2020). A framework to achieve sustainability in manufacturing organisations of developing economies using industry 4.0 technologies' enablers. *Computers in Industry, 122*, 103280. https://doi.org/10.1016/j.compind.2020.103280

Yang, M., Fu, M., & Zhang, Z. (2021). The adoption of digital technologies in supply chains: Drivers, process and impact. *Technological Forecasting and Social Change, 169*, 120795. https://doi.org/10.1016/j.techfore.2021.120795

Yu, Z., Khan, S. A. R., & Umar, M. (2022). Circular economy practices and industry 4.0 technologies: A strategic move of automobile industry. *Business Strategy and the Environment, 31*(3), 796–809. https://doi.org/10.1002/bse.2918

Zhu, Q., Sarkis, J., & Lai, K. (2007). Green supply chain management: Pressures, practices and performance within the chinese automobile industry. *Journal of Cleaner Production, 15*(11–12), 1041–1052. https://doi.org/10.1016/j.jclepro.2006.05.021

Part II
Applications of Digital Transformation in Retails and Logistics

Chapter 7
The Future of Brick-and-Mortar in an E-Commerce-Driven Supply Chain

Alan D. Smith

Abstract This chapter explores the evolving retail landscape, highlighting the impact of the pandemic on consumer preferences and the resulting gap between e-commerce companies. Amazon and more traditional brick-and-mortar stores like Walmart particularly highlight the strengths of each model, with e-commerce excelling in convenience and broad product availability, while brick-and-mortar stores leverage in-person experiences and immediate customer interaction. The analysis uncovers the need for brick-and-mortar stores to adapt and innovate to remain competitive. To close this gap, solutions are proposed. These include enhancing point-of-sale technology, prioritizing engaging in-store experiences, optimizing convenience models, focusing on customer service, and strengthening supplier relationships. Ultimately, closing this gap is crucial for the sustained success of both retail models, but especially brick-and-mortar.

Keywords Automatic identification and data capture · Brick-and-click · Brick-and mortar · Digital technology · Digital twins · e-Commerce · Supply chain management

7.1 Introduction

7.1.1 Factors Affecting Brick-and-Mortar Stores

When the global COVID-19 pandemic hit the U.S in early 2020, many brick-and-mortar stores closed their locations for the time being. While the option of online shopping was available pre-pandemic, customers were pivoting to e-commerce to fulfill their needs during this time when businesses had to close their doors opening

A. D. Smith (✉)
Department of Marketing, Robert Morris University, Pittsburgh, PA, USA
e-mail: smitha@rmu.edu

Y. Khojasteh, S.M. M. Jahan Abadi (eds.), *Digital Transformation of Supply Chains*, International Series in Operations Research & Management Science 372, https://doi.org/10.1007/978-3-032-01206-7_7

new consumer shopping behaviors (Nair & Shams, 2021). In fact, it sparked the e-commerce industry and continues to be a threat to brick-and-mortar retailers today. For example, Shopify's relatively recent trend roadmap expects the e-commerce market to grow approximately US$11 trillion between 2021 and 2025, with global sales potentially reaching US$6 trillion by 2024 (Marose, 2022). As noted by Sheth (2021), in a post-COVID-19 global economy, brick-and-mortar stores thrive by repositioning from merely selling merchandise to a more value-added marketing approach by moving to a more high-tech customer experience in store shopping.

During the height of COVID-19, consumers now value convenience more than ever, as Amazon allows customers to browse a firm's large inventory via any mobile device and makes it easier for customers to shop all while doing it from their own safe and familiar surroundings. Unfortunately, there was much supply chain disruptions, resulting in significantly longer delays and higher prices/inflation (Mohammed et al., 2023). The rise of e-commerce has made products and services available to customers at their fingertips and in some cases, allowing potential customers not to leave the convenience of their homes and not shopping at brick-and-mortar stores as desired items are sent directly to their door. Another advantage of e-commerce is the idea that the business is not limited to shelf space, so the product selection is wider for online sales platforms. The reach of e-commerce is ubiquitous and available to everyone with sufficient technology and connectivity. Whereas a brick-and-mortar store might not be as readily available for the same customer base, especially rural populations that have longer distances to reach physical stores (Toy, 2001). Time management is very valuable for customers as a cornerstone for the exponential growth of logistics, and they definitely do not want to spend it traveling just to get to the store only to find out that they did not have the item they were looking for.

On-the-other hand, there is some value behind brick-and-mortar stores that e-commerce might not be able to compete with. As suggested by Sheth (2021), the in-store experience advantage still has its appeals. With the development of omnichannel strategies, physical stores can take advantage of placing orders online, using home delivery or arranging physical pickup at the store location. This is a common phenomenon today. This is also called the omnichannel. Based on a recent Raydiant's second-annual State of Consumer Behavior Report, it was reported that 46% of surveys completed preferred in-store shopping over online and 33% wanting the ability to touch and interact with the product (Marose, 2022). To start, think about a clothing item such as a t-shirt. Going into a store gives you the opportunity to try the shirt on, feel the fabric, amongst other things that a consumer might want to know about that they will not be able to do behind a screen. The idea of a face-to-face interaction with a sales associate is also valued for asking questions about a particular product or service.

Similarly, both types of commerce must deal with customer returns. Brick-and-mortar stores have the upper hand here where they might experience less returns as you can try on or interact with the product before purchasing. Alternatively, when purchasing online, you have to wait for the product to get to you before you can truly know if the item is going to work out for you, leading to an increased number of returns. The return process in store is quite simple. Take your item or items being returned up to a store associate and they can give you an instantaneous refund back in the method you chose to pay with at that time or a store credit. The process of returning items ordered online can be more complicated. Most times you have to take the product back to a logistics provider yourself to be returned and then you have to wait until that product makes its way back to the online retailer before one can receive their refund. In fact, a significant portion of retail consumers (32%) stated that taking returns to the original logistics provider was the most inconvenient experience in the returns process (Buxbaum, 2023). This can create problems for online retailers and their supply chain processes which will be touched on later in this assessment. This poses the question of, will e-commerce be the future and will brick-and-mortar stores cease to exist, or are there pros and cons associated with each that allow both to co-exist?

7.1.2 Purpose of This Chapter

With the exponential expansion of inexpensive data store, automatic identification, and data capture (AIDC) technologies, the possibilities for data mining and analyzing for the betterment of firms' stakeholders are equally great. This new array of data and the statistical techniques used to gain useful information that provide opportunities for growth and proper supply chain management (SCM) generates several difficulties that management perceives in harnessing these data-driven applications and technologies. Undoubted, it is expected that companies will strategically leverage such advances that ultimately influences behavior and positively affects decision-making. The examination of two large global retailing firms will be examined to emphasize the impact of the recent COVID-19 pandemic on consumer preferences. The gap between e-commerce focused Amazon and more traditional brick-and-mortar Walmart will be further developed. This case study follows a classic best-business practices qualitative case study (Bogdan & Biklen, 1992). The sections in the case study include company description and supply chain operations and problems, with discussion on digital technologies in the respective companies' SCM-related operations and some of their future challenges. Undoubtedly, these firms are so large and economically powerful that antitrust concerns are an apparent concern for management (Haucap & Heimeshoff, 2014). Table 7.1 lists some of the major terms that will be used in the present study.

Specifically, the major goals of this chapter are to inspect some recent trends that traditional global brick-and-mortar retail establishments are undertaking to provide a rewarding customer experience that can only be experienced in a face-to-face environment. By showcasing such point-of-sale technologies, strategic recommendations can be generated and explained in some level of detail that may be of importance to managers. Hence, the research propositions (RPs) are as follow:

Table 7.1 Common terms and definitions found in dealing with online and brick-and-mortar industries

Terms	Definitions
Two-sided markets	A business model in which two individuals interact through a platform. Both sides must participate in order for the model to be a success.
Online markets	A market in which all participants conduct business through the Internet.
Digital economy	An economy based on digital computing technologies.
Antitrust	Legislation preventing or regulating business monopolies to promote competition.
Storefront Technologies	Storefront windows can differentiate stores from each other and create visual impacts to those who walk by. As technology is becoming more prominent, changes are being made to the layouts of the stores as delivery is becoming favorable for consumers. Advanced technologies are being integrated into the storefront to increase the shopping experience

RP1: After the relative end of the COVID-19 pandemic, management at many global brick-and-mortar retail establishments are realizing that to explain the customer experience, face-to-face locations are well equipped and logical locations to perform these enhancements.

RP2: Many of these customer enhancements can be accomplished by behind the customer interfaces trough digitalization of their supply chains, especially logistical-related services.

To accomplish these goals, a discussion of digital innovations in SCM and logistics, IoTs (Internet-of-Things)-based network applications, and legal considerations are highlighted. With the ever-changing nature of Internet markets, more regulation will surely follow and brick-and-mortar stores may directly benefit. Case studies and successful digital solutions are designed to promote further discussion of this chapter's goals.

7.2 Background

7.2.1 Digital Innovations in SCM and Logistics

In recent years, many digital innovations involving the strategic management of transportation and warehousing including supply chain methods, techniques, and technologies have evolved, such as blockchains, digital twins, and generative AI-tracking/prediction techniques (Park et al., 2021). In general, blockchains, in the case of SCM, are elaborate digital tracking-ledgers that record vital shipping, logistics, documentary, and item descriptions real-time that aids in properly protecting information from cybercriminals and illicit manipulation (Liu et al., 2023). By digitally encrypting a distributed ledger across several computers and related platforms, added layers of protection are created. Digital twins, for example, are designed to reproduce an object or process via simulations and, more recently, machine learning and generative AI applications identifying appropriate levels of inventory, cash replenishment policies, supply chain disruptions, and gaps by helping decision-makers create effective rules (Badakhshan & Ball, 2023). In some cases, many of these applications have been in practice for a number of years (Llaguno et al., 2021). However, the supply chain environment is changing rapidly, as are the major technological innovations in pure click and brick-and-mortar retail industries. Hence, new supply chain techniques are emerging with it. As noted by Kache and Seuring (2017), several corporate level opportunities would include predicting customer purchasing trends, supply chain transparency, operational efficiency and maintenance, information management, responsiveness, and innovative strategies models. At the SCM level, these digital innovations and opportunities include real-time logistics decision-making, SCM visibility and transparency, operations efficiency and maintenance, integration and collaboration, innovative product/service designs, and the optimizing of inventory stock levels.

For example, a typical IoT application in SCM systems classifies what Ben-Daya et al. (2017) refers to as 4Rs (e.g., responsiveness, reliability, resilience, and relationships. An IoT system is made up of four parts: A sensing layer that integrates physical access points (i.e., RFID tags, sensors, actuators); networking layer that promotes information transfers via network configurations; service layer for allows for services and applications through a middleware technology; and an interface layer to display user information and interaction with the system. An IoT-based network application for a firm must have technology that is synchronized. Manuel data entry slows time down and can be sloppy. Systems at a corporation must be able to effectively communicate with each other. This will allow the data to be interpreted more smoothly. Inventory ordering can be automated, and the travel of the product can be monitored from production to deliver to the end consumer. The RFID tags can monitor the temperature and evaluate the conditions of freight. This provides the supply chain and inventory management departments clear details and insights to their cargo. Perhaps if a third-party logistics (3PL) provider, for example, may not handle the freight correctly, this information would help determine any negligence on the packing and handling of the products.

These innovative methods can be observed in either Amazon's and/or Walmart's supply chains. For the purpose of citing this information, as well as keeping it organized, sections have been created. For Amazon, some techniques were listed by Banker (2022). Likewise, the terminology pertaining to Walmart was derived from applications reported by Rickerby (2023). Digitalization in the retail industry has a number of interesting customer-centric challenges (Pantano et al., 2019). For example, Jiang and Zou (2020) took a closer look at how consumer behavior and online retail platforms intertwine when it comes to the search and filtering process. Specifically, Jiang and Zou were looking to see how the consumer's search cost and filtering on a retail platform affects the platform, the third-party sellers, and the consumers. The authors researched this phenomenon of search cost and platform optimization of fees and profits. Essentially, depending on the platform's percentage referral fee, the lower search cost of a consumer can depict an increase or decrease in the platform's profit; however, with this in mind, if the platform adjusts its referral fee, the lower search cost will ultimately increase the profits of the platform. Furthering the research, the authors studied the effects of filtering a platform. With this custom feature added to retail sites, filtering will push consumers to search for fewer products but purchase products with higher match values. Thus, when filtering a product to a custom search result in only a small amount of match-value, there will be benefits all around.

As suggested by Sheth (2021), there are a number of strategies that increase the value-added proposition of brick-and-mortar stores. A practical approach is to utilize ordering online but delivering elsewhere (home of store pickup) via an omnichannel strategy. Amazon is the leader in an omnichannel through its Prime™ shipping options. Another service strategy is financing and having payments over time. One of the largest industries that survives on financing is the automobile dealers. Very few consumers can afford such an expensive durable product. And financing does not have to be limited to durable or expensive products, such as home mortgages or car financing. One can finance grocery products as they do in emerging markets where the local merchant (often a very small owner-managed grocery store) has a lifelong relationship with consumers in the neighborhood. A store employee would come to the home of the customer and deliver the products, take the order, and the store owner also finances in case the customer does not have income that particular month. Financing is a very key ingredient in the retail setting. There are companies such as large furniture retailers who will offer multiple years of financing given such low-interest rates.

An important strategy is using intelligently designed and operated loyalty programs (Kamm et al., 2021) via big-data analytics and data-mining activities. Properly using historical records, machine learning, and predictive analytics, anticipating customers' needs and preferences would provide retailers with powerful tools. Hence, the knowledge about operational performance from developing such business intelligence would allow management to adapt to external factors and disruptions, such as was experienced with supply problems and high inflation in the post-COVID era. Hence, a business intelligence (BI) system for small shops becomes an essential strategic initiative. "A loyalty program could offer additional services to analyze and strengthen the position of small brick-and-mortar stores" (p. 283). It is important to provide value-added opportunities that include more than financing and home delivery, but installation and maintenance as well as subscription-based apps (Sheth, 2021). In terms of successful methods, techniques, and technologies for the two companies showcased in this case study, Amazon and Walmart, Tables 7.2 and 7.3 illustrate basic terms/trends and highlight some of the major processes that characterize Amazon and Walmart, respectively.

Table 7.2 Selected successful methods, techniques, and technologies for Amazon

Methods/Processes	Description
Long-haul delivery	Long-haul delivery refers to the transportation of goods over an extended distance, typically involving the movement of products from manufacturing or distribution centers to final destinations such as retail stores. Long haul delivery ensures timely and cost-effective delivery.
Last-mile delivery	Last-mile delivery refers to the final aspects of a product's journey from the distribution center to the customer's doorstep. It's the crucial step that ensures timely and direct delivery to the end consumer, often involving local transportation and logistical solutions to efficiently complete the supply chain process.
Driver safety	Refers to such practices and technologies as generative AI backed cameras - Driver safety with AI-backed cameras refers to the use of smart cameras in vehicles to monitor and enhance safe driving practices. These cameras use AI technology to analyze drivers' behavior, detect potential risks, and promote overall safety on the road. This rather new implementation ensures continued delivery success and customer satisfaction.
Last-mile fulfillment centers	Last mile fulfillment centers are facilities located close to the final destination where products are stored and processed for quick and efficient delivery to customers. As the name suggests, these fulfillment centers form a delivery system with last-mile delivery vehicles.
Robotics within the warehouse and warehouse management systems (WMS)	Robotics within the warehouse refers to the use of automated machines and robots to perform various tasks in a storage facility. These robots can efficiently handle tasks like picking, packing, and transporting goods, enhancing the speed and accuracy of warehouse operations.

7.2.2 *Antitrust Concerns*

Haucap and Heimeshoff (2014) discussed these three Internet giants (e.g., Amazon, Google, eBay) and their ability to monopolize their respective markets. The authors begin by questioning whether current antitrust laws are sufficient to protect online markets from one company taking over the majority of the market share. The fact that these websites do not generate their own content, but rather provide access to goods and services provided by other companies make them unique from typical brick-and-mortar businesses. They simply ask the question: "If government intervention is necessary, how can an online monopoly be dealt with?" The authors conclude that intervention beyond the current competition laws is not necessary for these three Internet market giants. The fact that Google's costs for consumers are so low forces it to constantly defend itself against innovation and other search engines entering the market. This makes Google less of a threat in terms of monopolization. In terms of multi-homing, Facebook's costs are relatively low and it operates on a relatively simple model. This leaves room for many other competitors to attempt to duplicate what Facebook has achieved. eBay is undoubtedly the closest to monopolizing its market. Haucap and Heimeshoff noted that there have been arguments made for stronger regulation of eBay, but there are just as many counterarguments

Table 7.3 Selected successful methods, techniques, and technologies for Walmart

Methods/ Processes	Description
Vendor-managed inventory	Vendor Managed Inventory (VMI) is a supply chain practice where the supplier takes responsibility for maintaining optimal stock levels at the retailer's location. In this setup, the supplier monitors inventory and ensures timely replenishment, helping the retailer minimize stockouts and manage inventory more efficiently.
Safety stock/ buffer Inventory	Safety stock, also known as buffer inventory, refers to the extra stock that a company holds to mitigate the risk of running out of products due to uncertainties in demand or supply chain disruptions.
Cross docking	Cross docking is a supply chain strategy where products from suppliers are directly transferred to outbound transportation. It involves quickly moving items through a distribution center, minimizing the need for storage.
Barcoding and scanning	Barcoding in the supply chain involves assigning unique codes (barcodes) to products and using electronic devices (scanners) to read and track these codes throughout the logistics process. Ultimately, it enhances overall efficiency and reduces errors.
RFID technology	RFID technology is a system that uses electronic tagging to track and manage inventory. This electronic tag allows for easy and efficient monitoring of products from production to delivery.

deeming intervention unnecessary. With the ever-changing nature of Internet markets, more regulation could become necessary in the future. However, the authors believe that current law is sufficient.

7.3 Methodology

The basic methodology is the qualitative case study method (Yin, 2003), which uses best business practices. A classical qualitative business-case study approach has the advantage of inspecting the various technical and strategic initiatives within the context of the organizations' involved. Instead of an isolated piecemeal approach, a more holistic analysis and description can make the discussion more relevant to the readership interested in the procedures, policies, and processes involved (More & Babu, 2012). The historical aspects of digitization of supply chains and related logical concepts will be considered in addition to the impacts of the COVID-19 pandemic on global supply chains (Smith, 2022, 2024) as advantages and disadvantages of repositioning warehouses and tactically applying software warehouse management systems (WMS) and transportation management systems (TMS) at the present time are presented. Amazon, a retail and logistical giant, has a global imprint that will be profiled to show how these concepts can prove advantageous to an industry leading retailer to best optimize their network and provide expediency to their consumers. Walmart, another traditional retail giant, which has recently lost market share, is trying to improve its supply chain and the customer experience as well.

To accomplish these tasks and the stated goals of the chapter, each company will begin with a company description, followed by supply chain operations (i.e., fulfillment centers, delivery fleet), supply chain problems, possible solutions, and ending with suggestions for enhancing the customer experiences. The last section is a discussion of advantages and disadvantages associated with these enhanced strategies/tactics.

7.3.1 CASE 1: Amazon

7.3.1.1 Company Description

Amazon began as an online bookselling company that was created by Jeff Bezos in his garage in 1994. Bezos had high hopes for his company, as just a year later, Amazon was an online bookseller delivering to all 50 US states and 45 countries. Though experiencing some ups and downs throughout the beginning stages, Amazon has quickly taken the world by storm as the world's largest online retailer and prominent cloud service provider. With such a large assortment of inventory, including but not limited to, clothing, beauty products, food, jewelry, books, movies, electronics, pet supplies, furniture, toys, household goods etc., customers can find just about anything they might want or need (Yasar, 2023). Not only can customers find anything they may need, but they can also get it at next-day delivery through Amazon's subscription service titled Amazon Prime, where members can access exclusive shopping and entertainment services, discounts, and more. A company built on continuous innovation, Amazon has expanded beyond just the e-commerce space, with other subscription services such as Amazon Prime Video, several consumer technology devices like the Kindle, Alexa, Echo, and Fire TV, and even purchasing a famous grocery store chain, Whole Foods in 2017. For the purpose of this present chapter, and the digital transformation of the supply chain, the focus will be on Amazon's e-commerce offerings, specifically how their supply chain has been created to maintain its top position in the e-commerce industry and the steps taken to be successful. Although headquartered in Seattle, WA, Amazon has expanded globally with fulfillment centers across the world and when the pandemic hit, the rise of in-home shopping grew, making Amazon perhaps consumers' saviors. This change in customer shopping experience is just the beginning and Amazon has a tremendous head start over any potential competitors in the e-commerce space with their highly advanced supply chain operations.

7.3.1.2 Supply Chain Operations

Amazon has created state-of-the-art supply chain operations that are giving them a promising competitive edge over its competitors. Much of the same tactics that were used during the launch of Amazon in 1995 are still being used today, but on a much larger scale. Amazon has third party sellers among their websites, which make up about 55% of all sales and has created two different fulfillment options for its sellers, fulfillment by Amazon (FBA) or fulfillment by merchant (FBM). With the FBA option, the company can carefully manage logistics- and customer support-related activities. 3PLs are able to transport bulk products to many fulfillment centers for them to pick, pack, and eventually ship to customers. On-the-other-hand, FBM allows these providers/sellers to be in charge of listing their products and effectively handle storage and fulfillment aspects (Bauer, 2021).

The key parts of their supply chain operation that are attributed to their success are their fulfillment centers and the Amazon delivery fleet. The promise of next-day delivery is accomplished by their 175 fulfillment centers globally in more than 150 million square feet of space and an estimated 500 warehouses located near major metropolitan areas. The company basically divides into sortable and non-sortable fulfilment centers. Sortable fulfillment centers generally are about 800,000 square feet in size and employ more than 1500 full-time workers. Within these structures, employees order pick, pack, and ship to customers a myriad of products. These centers are very famous and serve as best-in-class examples of separate autonomous robotic machines that often work alongside human workers in the same environment. Non-sortable fulfillment centers typically range from 600,000 to one million square feet and can employ more than 1000 full-time workers. These fulfilment centers, workers order pick, pack, and ship significantly heavier and bulky items. Below is a figure outlining where the fulfillment centers are located in the U.S., with California having 21.

7.3.1.2.1 Fulfillment Centers

When products make their way to their fulfillment centers, they are unloaded using forklifts or manual labor is used to formulate them into pallets. Once unloaded, Amazon stores their items randomly, rather than by type and then they are placed in what they call the "stow" area. Employees then remove the items and place them in coded yellow bins. The use of computers helps track these items when they are placed in yellow bins. Once a consumer has selected the buy button online, it comes into the system where a robot that is guided by 2D barcodes on the floor will bring pods full of items to the Amazon associates working at the numerous pick stations and get the item requested and move it into a tote, where it is then pushed onto a conveyor belt to move it to the next part of the process.

From there it is then organized and scanned into different shipments where it is then packed by Amazon associates. Through the use of barcodes, this tells the system what is located inside the package as well as recommendations for box sizes and measures out the appropriate amount of tape needed in order to reduce waste. Next another barcode is put on the box containing the information to create a shipping label where it then heads onto what Amazon refers to as the SLAM machine, which stands for scan, label, apply, and manifest. For quality assurance, the box travels across a scale where it calculates the weight to detect any discrepancies between expected and actual weight of the item. The next step is where the boxes are sorted based on shipping location. Again, a scanner is used to read the barcodes and assigns each package a chute to go down to leading to the correct trailer for shipment.

7.3.1.2.2 Delivery Fleet

Once the items are loaded onto the trailers at the fulfillment center, Amazon's trucks transport them to their sort centers where the packages are again sorted based on location and delivery speed required. By using intermodal transportation, Amazon has an airplane fleet that flies out from more than 20 airports in the U.S. On ground transportation vehicles include trucks, small vans we see traveling in local neighborhoods, bikes, and even sometimes robots. In the future we could potentially see drones delivering packages as Amazon is looking to invest in this new shipment method.

This is a small overview of the process that makes Amazon the company they are today. With the use of robots, material handling devices, human manpower, barcodes, and delivery fleets, Amazon has a supply chain operation that cannot be challenged by any other competitor and they will continue to make advancements to ensure the quickest most efficient methods of delivery to its customers.

7.3.1.3 Supply Chain Problems

As Amazon has had much success in their supply chain operations in terms of receiving orders, picking them, packing them, and delivering them, we should evaluate the reverse logistics and how Amazon has dealt with it. Should a customer be unsatisfied with their purchase for any reason Amazon has a return policy that allows for full refund within 30 days of the receipt for most of their products – software and groceries have a shorter window and sometimes are not eligible for return at all (Robinson, 2023).

By simply clicking on the button to return or replace items, Amazon provides them with a return label to print and attach to their package. Once a customer printed the return label, they will need to have an appropriate box to return the product, once it is sealed with the label attached. Amazon has given their customers numerous ways to then ship the item out using a drop off service at locations that are convenient for them, using pickup options, and using lockers and Amazon Hub Counter locations for smaller packages ("Return shipping-Amazon customer service," 2024). Using Amazon's platform, they allow you to track your return and your refund status. After this process is complete it can take two weeks to process the return.

As previously mentioned, returns are quite frequent for an online retailer. While Amazon has streamlined the process of getting the item via the last mile logistical process relatively hassle free for the customer, the return process is bothersome for a customer. For example, in Amazon's case, customers simply need access to a printer for the shipping label and a cardboard box to send the items back into. Not everyone has access to a printer, nor do they have cardboard boxes lying around their house, which is why returns to a brick-and-mortar store might be more attractive to customers. When returning to a brick-and-mortar store, customers can bring the item and the receipt and the associate working at the store can easily take care of your return and even give you an instantaneous refund. Amazon, however, might not be able to refund their customers in fewer than two weeks. It was found that 60% of retail shoppers prefer to make in-store returns, even if they bought the items online (Buxbaum, 2023).

Returns are expensive for any retailer (Kumar et al., 2011; Mathirajan et al., 2011; More & Babu, 2012). This situation is especially true for Amazon if it is offering to pick up the product from customers' homes or at specified locations. On top of that, some products could come back damaged due to the fact that consumers were the ones packaging the items up. Now Amazon can't resell the item and ultimately takes a loss on that specific item. They now have to take the time to dispose of the item properly and in an environmentally friendly way. Another issue that e-commerce deals with is fraudulent returns. Recently, Amazon filed a lawsuit against alleged scammers that had more than a dozen transactions from June 2022 to May 2023. They had tricked Amazon into processing refunds for products that are never returned. In a brick-and-mortar store this could not happen as the item has to be present at the time of return including the receipt in most cases ("Amazon sues alleged scammers …," 2023). Obviously, management at Amazon needs to find a way to better process their returns that makes the process easier for their customers, but also ensures actual returns are being made and the company is not losing money. This is a potential downfall in the e-commerce way of shopping, both for the consumer and the online retailer.

7.3.2 CASE 1: Walmart

7.3.2.1 Company Description

Retrospectively, it now seems Sam Walton knew there was an opportunity to gain a competitive edge in the retail business. With this growth Walton took interest in his employees' opinions, where the conclusion was that Walmart needed to expand their operations by opening up their first distribution center in Bentonville, Arkansas. Sam Walton was quoted saying "the key to success is to get out into the store and listen to what the associates have to say" (McKinnon, 2023, p. 1), which is exactly why Walmart saw an immediate positive impact from opening up their first distribution center. While already achieving successful retail practices for discounted items, Walton believed there was another opportunity in order to sustain brand loyalty. Thus, in 1983, the company decided to open up their first Sam's Club store in Midwest City, OK.

Sam's Club are all owned and operated by Walmart, but they offer their own membership for customers. In competition with Costco, Sam's Club is known for their competitively low prices and bulk products. They offer fresh food, electronics, clothing, just like Walmart but all under warehouse pricing. Sam's Club currently is the biggest retail warehouse chain in the United States with 600 locations, 13 more than Costco (Koopman, 2023). In 2023, Walmart operated around 10,500 stores and clubs in 19 different countries and e-commerce websites. They employ 2.1 million associates worldwide, with 1.6 million of them coming from the U.S.

7.3.2.2 Supply Chain Operations

A major reason Walmart has grown to where they are today is because of their strong supply chain operational practices. In the 1980s, the company decided to cut out the middleman, who was the distributor, to deal only with manufacturers in order to cut costs and have better direct oversight of their supply chain. Via disintermediation, management developed relationships among different vendors (manufacturers) and offered guaranteed plus consistent business in exchange for low, wholesale prices. With this guarantee and Walmart's desire for bulk merchandise, it was an easy decision for the vendors. Traveling to the current day, with numerous technological advancements, Walmart decided to implement a vendor-managed inventory (VMI) system, which puts most of the inventory responsibility on the vendors. With this system, vendors are responsible for managing their own inventory in Walmart's warehouses.

Through sending data from centralized databases of stock inventory levels, vendors can know right away when their products are getting low and need more to be sent to that specific location. While dealing with so many vendors this saved Walmart significant financial resources and time since this technology almost guarantees that items are in stock at all times. Another supply chain practice Walmart successfully implemented is cross-docking. Cross-docking is defined as the practice of unloading goods from inbound vehicles and loading them directly into outbound vehicles (Essex, 2023). With the amount of business Walmart does, they try to limit their inventory storage amount as much as possible in order to eliminate the threat of being overstocked, which is quite costly. One way they use cross docking is by accepting the products at the Walmart warehouse, where they are then repackaged and instantly distributed to its assigned Walmart location without ever sitting in storage. Another way Walmart uses cross docking, for example with Proctor and Gamble, is at Walmart distribution centers, Proctor and Gamble unload their products directly onto trucks that are headed to that assigned Walmart location.

7.3.2.3 Supply Chain Problems

Even though Walmart has a very strong grasp of their supply chain, there are many problems that can arise from the number of products and vendors they deal with, especially from a size and scale of their operations perspective. Walmart has millions of products moving across long distances, that they are challenged to coordinate and manage the goods, as well as guarantee that those products are delivered to the store and customers in a timely and cost-effective manner. The next issue Walmart deals with is maintaining relationships with suppliers and vendors. With thousands of suppliers and vendors Walmart has to ensure that both parties agree on a price and quality that must be upheld throughout the entirety of the relationship. This often requires a procurement process to best identify if this relationship is beneficial for both parties. Another supply chain issue is the emergence of eCommerce. With this recent surge, Walmart has been forced to adapt and evolve by investing money into new technology and processes to improve the supply chain as a whole.

7.4 Managerial Implications

7.4.1 Overall Concerns

Referring to the work by Davis (2021), in which there was a detailed comparison of brick-and-mortar and pure click of online stores in terms of sales in 2021, many factors have reshaped the evolving landscape of the retail industry. While consumer preferences have been trending towards e-commerce shopping for quite some time, it defines the COVID-19 pandemic as the catalyst for its recent growth. From a managerial perspective, this rapid change in consumer preference can be alarming. As a result, many managerial implications must be understood. Recognizing that the future of commerce is moving online, brick-and-mortar stores should embrace innovation and invest in necessary changes to remain competitive. Strategies like creating engaging in-store experiences, mentioned in the article, and will later be referenced in this report, are crucial for the success of brick-and-mortar stores.

Another consideration the retail manager may employ is the reduction of workforce in brick-and-mortar stores. Worker shortage results in longer wait times, reduced store operating hours, productivity, and efficiency. Because companies expect customer service levels to be met, regardless of today's labor landscape, a large amount of strain is placed on the available workers. Naturally, this creates a poor-work environment for those employed, leading to worker shortages, as illustrated in Fig. 7.1, the basic model illustrating the major assumptions of the present chapter that are essential to develop intrinsic and extrinsic drives for producer performance and customer service. The model described in Fig. 7.1 illustrates the factors of proper management, quality assurance, professional development, human-capital development, and digital technologies that are very important for producer performance and customer service.

Ultimately, this scenario does not bode well for brick-and-mortar stores, since it reduces the in-store experience required to compete with e-commerce while simultaneously increasing the cost, especially during the recent COVID-19 pandemic. Well-trained and loyal workers will require higher wages, which will then be reflected in item pricing. To mitigate these risks and managerial implications, there are a few solutions that can be implemented by brick-and = mortar stores to bridge the gap and cut into the e-commerce growth.

7.4.2 Possible Solutions

So, what can be done by Walmart, as well as many other brick-and-mortar stores alike, in order to stay competitive with the growing consumer trends? For this to happen, the following improvements are suggested priorities.

7.4.2.1 Point-of-sale Technology

Physical stores should continue to deploy technology like self-checkout, contactless payments, and cashier-free setups to make the buying process smoother (Davis, 2021; Smith, 2005). This not only reduces potential wait times at checkout, but also

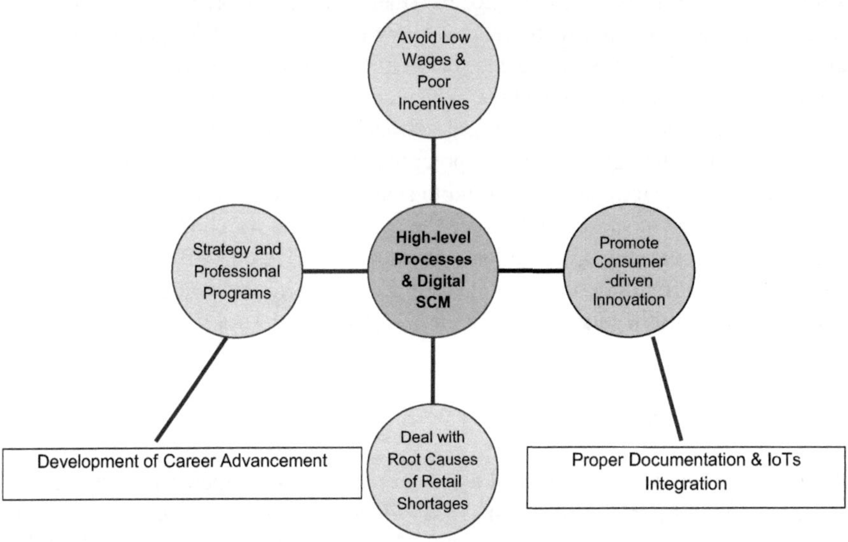

Fig. 7.1 The research model

takes care of many health- and safety-related issues, as well as issues with employment (Farzandipour et al., 2010; López-Roig et al., 2010; Rooney, 2009).

7.4.2.2 In-store Experiences

To compete with online shopping, physical stores should prioritize a customer-friendly in-store experience (Davis, 2021; Smith et al., 2019). This means appealing displays, themed showcases, and digital solutions to make shopping more engaging. Add entertainment to make it enjoyable, going beyond just a transaction. Because viewing the product in stores provides an edge over e-commerce product viewing capabilities, it is important to fully exploit this feature. To further stress the importance of in-store experiences, an empirical study was conducted to determine the influence (if any) of store attributes on consumers. An empirical study examining the perceived impact of store attributes on grocery shopping behavior in emerging markets by Nair and Shams (2021) resulted in positive feedback for eight attributes. These attributes include atmosphere, convenience, facilities, institutional factors, merchandise, promotion, store personnel interaction, and service.

7.4.2.3 Convenience and Delivery Models

Undoubtedly, one of the driving factors of e-commerce shopping is convenience and many SCM systems try to enhance these characteristics in their manufacturing and logistical functions, especially delivery services (Kamalul et al., 2018; Kim et al., 2004, 2009; Kim, 2017; Nandkeolyar et al., 2011; Park & Nahm, 2011; Tomita et al., 2011; Yang & Park, 2011). That being said, stores should continue to invest in online shopping methods that allow consumers to take part in curb-side pickup. Running a cost-effective and successful warehouse industry requires several key factors to maintain a competitive advantage. Being able to understand the complex roles of planning effectiveness, decision-rule complexity, control sophistication, and information systems are all potential problem areas that need to be solved in order to create strategies that promote sustainable competitive advantages. It is equally important to strategically place the business of a warehouse in a more competitively advantageous scenario that can naturally drive cost-effective approaches down the line. Providing better automation systems with technology are issues that warehousing is still trying to solve in the post-COVID-19 era.

7.4.2.4 Customer Service

An important area where physical stores excel over e-commerce is customer service. Instant and personalized assistance will result in increased customer loyalty and satisfaction (Davis, 2021; Smith et al., 2019). According to Griffis et al. (2012), online retailers walk a fine line between building customer loyalty and allowing for a more liberal returns policy. Although it is typical of today's online retailers to opt for an ease and open-return policy, the effectiveness of said approach had not been proved so far, and the link to returns processing and customer loyalty had not been yet established. In fact, several retailers carry the preconceived idea that they can either make returns difficult and maximize their sales, by encouraging customers to avoid returns, or offering a simple and easy return policy but facing the risks of losing money due to "retail borrowing" and other types of return frauds. Griffis et al. (2012) suggested that operations managers begin looking at customer return rates as an opportunity to delight customers and develop more loyal customers, rather than challenging numbers that need to be overcome. Retailers can do that by investing in training and other elements to increase the speed with which returns are processed. Another recommendation is that the benefits of easy-to-make returns, retailers invest on distinguishing legitimate from illegitimate returns, so return policies can benefit the loyal and profitable customers and simultaneously identify those customers who are abusing the system.

7.5 Conclusions

7.5.1 Managerial Implications

The COVID-19 pandemic significantly accelerated the shift towards e-commerce, presenting both challenges and opportunities for traditional brick-and-mortar retailers. The convenience offered by online shopping, coupled with a broader product selection and timely deliveries, has fueled the growth of the e-commerce industry. At the same time, brick-and-mortar stores still hold value for consumers who appreciate the ability to physically interact with products, receive immediate assistance, and make hassle-free returns. As the retail landscape evolves, managerial implications include the need for brick-and-mortar stores to innovate, enhance in-store experiences, and adapt technology to remain competitive. To address these challenges, solutions such as implementing advanced point-of-sale technology, prioritizing engaging in-store experiences, optimizing convenience and delivery models, investing in customer service, and strengthening supplier relationships can help bridge the gap and ensure the success of both retail models in the changing market.

Potential downsides and/or challenges of implementing the proposed solutions include both customer and employee resistance to change. The attractiveness associated with the status quo is a difficult factor to counter. Probably few would argue that investing in customer service and optimization convenience and delivery models are not good investments, but they require a mindset to accept more advanced point-of-sale technology and prioritizing engaging in-store experiences. Online shopping is extremely popular and will continue to grow in its attractiveness to customers. A major challenge will be to promote brand identity (i.e., specific design and message elements that are distinctive to a particular product line of company and/or image a company can be associated with) (Ordabayeva et al., 2022; Pluntz & Pras, 2020). Such elements can be both intangible and tangible (i.e., package, promotions, loyalty member benefits and discounts) in their appeal to customers.

7.5.2 Future Research Directions

The goal of being able to accurately forecast demand trends should be at the top of the priority list for any company involved in the supply chain process. In today's competitive supply chain environment, companies that do not embrace the emerging analytical tools, such as digital technologies, IoTs, and big-data analytics, will be left behind. It is not enough to simply be able to capture large quantities of data; supply chain companies must be able to efficiently identify and extract the data that will prove useful to them and their processes. Management at many companies, especially brick-and-mortar stores, may hesitate at the initial upfront cost of installing and learning how to operate systems that can capture and analyze big data, or fail to see the benefits in partnering with firms that can perform these analytical functions for them, but companies must be forward thinking about the benefits digital innovations could bring for them. Much research opportunities exist to study how digital transformations of SCM will close the gap between retailing configurations.

References

Amazon sues alleged scammers who stole millions by bribing and hacking their way to refunds for products they never returned. (2023). [Online]. Retrieved April 16, 2024 from *Fortune*, fortune.com/2023/12/07/amazon-fraud-thieves-refunds-millions-of-dollars/

Badakhshan, E., & Ball, P. (2023). Applying digital twins for inventory and cash management in supply chains under physical and financial disruptions. *International Journal of Production Research, 61*(15), 5094–5116.

Banker, S. (2022). Amazon supply chain innovation continues. *Forbes*. [Online]. Retrieved April 16, 2024 from https://www.forbes.com/sites/stevebanker/2021/04/01/amazon-supply-chain-innovation-continues/?sh=1bceb46677e6

Bauer, M. (2021). How Amazon supply chain works. *Teikametrics* [Online]. Retrieved April 16, 2024 from www.teikametrics.com/blog/how-amazon-supply-chain-works/

Ben-Daya, M., Hassini, E., & Bahrouna, Z. (2017). Internet of Things and supply chain management: A literature review. *International Journal of Production Research, 57*(3), 1–12.

Bogdan, R. C., & Biklen, S. K. (1992). *Qualitative research for education: An introduction to theory and methods*. Boston: Allyn and Bacon.

Buxbaum, P. (2023). With e-commerce on the rise, reverse logistics takes center stage: Returns processes emerge as important element of customer experience, but there are some pitfalls to avoid. *American Journal of Transportation, 750*, 20–24. [Online]. Retrieved December 3, 2023 from https://www.ajot.com/premium/ajot-with-e-commerce-on-the-rise-reverse-logistics-takes-center-stage

Davis, T. (2021). The 60 second marketer: Brick and mortar vs. e-commerce: The state of sales in 2021. *Newstex*. [Online]. Retrieved April 16, 2024 from https://reddog.rmu.edu/login?url=https://www.proquest.com/blogs-podcasts-websites/60-second-marketer-brick-mortar-vs-e-commerce/docview/2541847434/se-2

Essex, D. (2023). What is cross-docking? Definition from TechTarget. *ERP*. [Online]. Retrieved April 16, 2024 from https://www.techtarget.com/searcherp/definition/cross-docking

Farzandipour, M., Sadoughi, F., Ahmadi, M., & Karimi, I. (2010). Security requirements and solutions in electronic health records: Lessons learned from a comparative study. *Journal of Medical Systems, 34*(4), 629–642.

Griffis, S., Rao, S., Goldsby, T. J., & Niranjan. (2012). The customer consequences of returns in online retailing: An empirical analysis. *Journal of Operations Management, 30*(4), 282–294.

Haucap, J., & Heimeshoff, U. (2014). Google, Facebook, Amazon, eBay: Is the Internet driving competition or market monopolization? *International Economics and International Policy, 11*(1–2), 49–61.

Jiang, B., & Zou, T. (2020). Consumer search and filtering on online retail platforms. *Journal of Marketing Research (JMR), 57*(5), 900–916.

Kache, F., & Seuring, S. (2017). Challenges and opportunities of digital information at the intersection of Big Data Analytics and supply chain management. *International Journal of Operations and Production Management, 37*(1), 10–36.

Kamalul, A. S., Mohan, T., & Goh, Y. N. (2018). Influence of consumers' perceived risk on consumers' online purchase intention. *Journal of Research in Interactive Marketing, 12*(3), 309–327.

Kamm, M. R., Kucklick, J.-P., Schneider, J., & vom Brocke, J. (2021). Data mining for small shops: Empowering brick-and-mortar stores through BI functionalities of a loyalty program. *Information Systems Management, 38*(4), 270–286.

Kim, J. (2017). The influence of graphical versus numerical information representation modes on the compromise effect. *Marketing Letter, 28*(3), 397–409.

Kim, J., Jin, B., & Swinney, J. L. (2009). The role of retail quality, e-satisfaction and e-trust in online loyalty development process. *Journal of Retailing and Consumer Services, 16*(4), 239–247.

Kim, H. W., Xu, Y., & Koh, J. (2004). A comparison of online trust building factors between potential customers and repeat customers. *Journal of the Association for Information Systems, 5*(10), 392–420.

Koopman, A. (2023). *Is a Sam's Club membership worth it?* SFGATE. [Online]. Retrieved April 16, 2024 from https://www.sfgate.com/shopping/article/sams-club-membership-cost-deals-18206732.php

Kumar, P., Shankar, R., & Yadav, S. S. (2011). Global supplier selection and order allocation using FQFD and MOLP. *International Journal of Logistics Systems and Management, 9*(1), 43–68.

Liu, J., Zhang, H., & Zhen, L. (2023). Blockchain technology in maritime supply chains: Applications, architecture and challenges. *International Journal of Production Research, 61*(11), 3547–3563.

Llaguno, A., Mula, J., & Campuzano-Bolarin, F. (2021). State of the art, conceptual framework and simulation analysis of the ripple effect on supply chains. *International Journal of Production Research, 60*(6), 2044–2066.

López-Roig, S., Pastor, M. Á., & Rodríguez, C. (2010). The reputation and professional identity of family medicine practice according to medical students: A Spanish case study. *Atencion Primaria, 42*(12), 591–601.

Marose, J. (2022). *Why e-commerce needs brick & mortar*. [Online]. Retrieved April 4, 2024 from www.mapbox.com/blog/ecommerce-or-brick-mortar

Mathirajan, M., Manoj, K., & Ramachandran, V. (2011). A design of distribution network and development of efficient distribution policy. *International Journal of Logistics Systems and Supply Management, 9*(1), 108–137.

McKinnon, T. (2023). *39 of the best Sam Walton quotes about success, innovation & more.* [Online]. Retrieved April 4, 2024 from https://www.indigo9digital.com/blog/samwaltonquotes

Mohammed, A., Lopes de Sousa Jabbour, A. B., & Diabat, A. (2023). COVID-19 pandemic disruption: A matter of building companies' internal and external resilience. *International Journal of Production Research, 61*(8), 1–22.

More, D., & Babu, A. S. (2012). Benchmarking supply chain flexibility using data envelopment analysis. *International Journal of Logistics Systems and Management, 12*(3), 288–317.

Nair, S. R., & Shams, R. (2021). Impact of store-attributes on food and grocery shopping behavior: Insights from an emerging market context. *EuroMed Journal of Business, 16*(3), 324–343.

Nandkeolyar, U., Tran, O. T., & Hwang, W. (2011). High-speed surface transportation corridor: A conceptual framework. *International Journal of Services and Operations Management, 8*(3), 347–364.

Ordabayeva, N., Cavanaugh, L. A., & Dahl, D. W. (2022). The upside of negative: Social distance in online reviews of identity-relevant brands. *Journal of Marketing, 86*(6), 70–92.

Pantano, E., Constantinos, V. P., & Foroudi, P. (2019). Innovation starts at the storefront. *International Journal of Retail and Distribution Management, 47*(2), 202–219.

Park, Y., & Nahm, A. Y. (2011). Classification of mass customisation: A Socio-Technical System perspective. *International Journal of Services and Operations Management, 8*(3), 322–334.

Park, K. T., Son, Y. H., & Noh, S. D. (2021). The architectural framework of a cyber physical logistics system for digital-twin-based supply chain control. *International Journal of Production Research, 59*(19), 5721–5742.

Pluntz, C., & Pras, B. (2020). Exploring professional human brand identity through cultural and social capital: A typology of film director identities. *Journal of Marketing Management, 36*(9/10), 830–866.

Return shipping-Amazon customer service. (2024). [Online]. Retrieved April 17, 2024 from www.amazon.com/gp/help/customer/display.html?nodeId=GFLBEJCLHMVFEPA8

Rickerby, M. (2023). *The Walmart supply chain: Its success explained.* [Online]. Retrieved April 4, 2024 from https://www.extensiv.com/blog/supply-chain-management/walmart#:~:text=The%20foundation%20of%20Walmart's%20successful,and%20streamlined%20supply%20chain%20operations

Robinson, A. (2023). *How does Amazon process returns? A comprehensive guide.* [Online]. Retrieved April 4, 2024 from shipscience.com/how-does-amazon-process-returns-a-comprehensive-guide/

Rooney, K. (2009). Consumer-driven healthcare marketing: Using the web to get up close and personal. *Journal of Healthcare Management, 54*(4), 241–251.

Sheth, J. N. (2021). Future of brick and mortar retailing: How will it survive and thrive? *Journal of Strategic Marketing, 29*(7), 598–607.

Smith, A. D. (2005). Exploring the inherent benefits of RFID and automated self-serve Checkouts in a B2C environment. *International Journal of Business Information Systems, 1*(1/2), 149–183.

Smith, A. D. (2022). Updating an empirical investigating risk perceptions associated with national ID cards in the wake of the global Covid-19 pandemic. *Health Marketing Quarterly, 38*(2–3), 70–90.

Smith, A. D. (2024). Comparison of selected supply chain management practices of three major regional/metropolitan hospitals within the context of the dual COVID-19 and opioid crises. *International Journal of Services and Operations Management, 48*(2), 155–175.

Smith, A. D., Shock, J. R., & Beaves, R. G. (2019). Customer relationship management and the impact of e-coupons on B2C retail markets. *International Journal of Business Information Systems, 30*(2), 203–231.

Tomita, J., Park, Y. W., & Hong, P. (2011). Process architecture impact: A comparative study of two glass industries. *International Journal of Services and Operations Management, 8*(3), 390–403.

Toy, B. (2001). E-Business from the front line of a 140-year-old brick and mortar company. *Transportation Journal (American Society of Transportation and Logistics Inc), 40*(4), 27.

Yang, M., & Park, K.-H. (2011). Self-Service Technologies (SSTs): Determinants of adoption and its post-usage outcomes from a focal company perspective. *International Journal of Services and Operations Management, 8*(3), 305–321.
Yasar, K. (2023). *What is Amazon?* [Online]. Retrieved April 4, 2024 from www.techtarget.com/whatis/definition/Amazon
Yin, R. K. (2003). *Case study research: Design and method* (3rd ed.). Sage.

Chapter 8
Identifying Critical Factors in Cold Chain Logistics System: A Review

Goutam Sutar, Krantiraditya Dhalmahapatra, and Sayan Chakraborty

Abstract Cold chain logistics involves transporting perishable goods in temperature-controlled environments, such as cold storage reefers and containers, from farm to consumer. Despite large-scale global food production, significant losses occur due to insufficient cold storage, particularly in emerging economies. This chapter explores recent trends, challenges, and the role of digital transformation in optimizing cold chain logistics. Research mapping categorizes studies based on key issues, including distribution challenges, monitoring methods, efficiency improvements, sustainability, and associated risks. It also classifies research by perishable goods, such as fruits, vegetables, dairy, meat, vaccines, grains, and flowers. The study identifies seven critical factors influencing cold chain efficiency: shelf life, timely delivery, distribution route optimization, inventory planning, wireless technology, temperature prediction, and infrastructure. Addressing these factors is essential for improving operations and minimizing losses. The chapter provides insights for new entrants and existing practitioners in the cold chain sector, helping them enhance their strategies. Furthermore, it highlights the transformative impact of digital technologies in increasing efficiency, visibility, and ensuring the integrity of perishable goods throughout the supply chain.

Keywords Cold chain logistics · Perishable products · Critical factors · Digital transformation

---

G. Sutar
Thiagarajar School of Management, Madurai, Tamil Nadu, India

K. Dhalmahapatra
Indian Institute of Management Shillong, Shillong, Meghalaya, India

S. Chakraborty (✉)
ICFAI Business School Hyderabad, ICFAI Foundation for Higher Education, Hyderabad, India
e-mail: sayan@ibsindia.org

Y. Khojasteh, S.M. M. Jahan Abadi (eds.), *Digital Transformation of Supply Chains*, International Series in Operations Research & Management Science 372, https://doi.org/10.1007/978-3-032-01206-7_8

8.1 Introduction

The digital transformation of the cold supply chain represents a groundbreaking evolution in how perishable goods are sourced, stored, and distributed. By integrating advanced technologies like Internet of Things (IoT) sensors, blockchain, and data analytics, companies operating in cold storage and distribution can revolutionize their processes, ensuring the preservation of product quality and safety while optimizing efficiency and minimizing waste (Tang et al., 2024). Real-time monitoring and tracking of temperature-sensitive goods, coupled with predictive analytics, enable proactive decision-making to prevent spoilage (Mayr et al., 2003) and ensure compliance with regulatory standards. Furthermore, digital technologies (Gupta et al., 2019) enhances transparency and traceability throughout the cold chain, fostering trust among stakeholders and empowering consumers with access to valuable information about the origin and handling of the products they purchase (Jiang et al., 2024). Again, a cold chain system that is robust in all aspects can be profitable to the company and ensures prompt delivery of high-quality products to the customer with the intact quality (Chermala et al., 2025). Thus, embracing digital transformation in the cold supply chain not only drives operational excellence but also facilitates innovation and resilience in an increasingly dynamic and competitive marketplace.

The cold chain is a widely used terminology applied to the handling of perishable goods and its distribution maintained under a controlled or suitable temperature conditions (Bogataj, 2005; Dai et al., 2020; Bishara, 2006) from farmer level through cooling or freezing process to the end-user level (Vrat et al., 2018). In this supply chain, factors such as transportation of perishable products require reefer vehicles and cold storages in which goods are stored and transported in closed refrigerated space with temperature-controlled surroundings, be it on a truck, a container, or any other vehicle (Li et al., 2019; Yavari & Zaker, 2020). If any perishable goods stay longer in the supply chain, it will lead to high operating costs and deterioration in the quality of the perishable goods (Bogataj et al., 2017). Therefore, transportation of perishable goods imposes critical and serious challenges if the supply chain network and its logistics operation is not adequately considered factors like transhipment time, environment factors, warehouse management, vehicle routing and many more (Al Theeb et al., 2020; Rahbari et al., 2019).Again, real-time monitoring of distribution condition which includes temperature, oxygen, humidity, etc. (du Plessis et al., 2024) poses another important challenge that can cause a loss in the quality of perishable goods, loss for the business, and consequences even put human lives in danger in the case of pharmaceutical transportation (Mejjaouli & Babiceanu, 2018).

There are many arguments found in the literature discussing food wastages such as demand-supply mismatch (Babagolzadeh et al., 2020; FAO, 2014; Taylor & Fearne, 2006), government regulations, awareness, transportation facility, lack of cold storage facility, exports (Sharma et al., 2020; van Donselaar et al., 2006), inadequate processing facilities (Lin et al., 2020), marketing, and service innovation as well as innovative practices (Dai et al., 2020) and many others. Among these reasons, lack of proper cold chain logistic system with adequate cold storage facility is a major area of study where rapidly growing economies like India (Joshi et al., 2009) and China (Dai et al., 2020) need to focus to check the wastages of food not only for food security but also for distribution during crucial natural calamities and situations like epidemic (Dasaklis et al., 2012) and pandemic such as Covid-19 pandemic. To further depict this problem Fig. 8.1 illustrates the gap between cold storage and fruits and vegetable production state wise in India.

Apart from above issues, other key factors that impacts the cold chain logistics system is recently emerging situations in global economies. For example, many news media reported how events like Russia-Ukraine conflicts, Covid-19 Pandemic, and natural disasters were key deciders in crude-oil price hike and that directly impacts the transportation cost in cold supply chain (Chen et al., 2025). Similarly, global climate change has adverse contributions in cold supply chain affecting its logistics significantly (du Plessis et al., 2024). The natural disasters such as cyclones, hurricanes, are becoming very frequent across the globe and researchers attributed it mostly to climate change (Sebatjane, 2024). The same author also advocates that greenhouse gas emissions are key contributors in global warming and highlights how green technology under the regulation carbon tax to reduce the cost in cold chain logistics. Not only this, the lack of technology adoption in temperature control, mismanagement, but non-adherence to food safety protocols also poses various level of risks to cold chain logistics as shown by (Mustafa et al., 2024) in their literature review paper based on food cold chain logistics. In addition to this, United Nation's Sustainable Development Goals (SDGs) also prioritising sustainable food value chain (Jabbour Al Maalouf et al., 2024) that adds more challenges to cold chain logistics in terms of implementation. Keeping all the above facts in mind it is very clear that managing cold chain logistics in present scenario is quite challenging. On a similar line MAERSK website report in December 2024 (Silvia La Face, 2024) identified five upcoming trends in cold chain logistics. They are namely-market change due to recent geo-political events, better visibility of cold chain, alternative products such as protein, innovative and upgraded storage facilities, and smoother distribution through the help of various technological and digital transformation. Therefore, integration of innovation and technology is inevitable in the today's cold chain logistic systems for which need to identify various critical factors to drive it.

In this study, an attempt is made to identify some of the salient critical factors from the existing body of literature through a systematic literature review that can be referred as sustainable criteria for practitioners and managers while developing an uninterrupted cold chain system. Not only this, but the criterions also identified in this study will be helpful to various government bodies particularly in the countries like India where the cold chain sectors are raising in current years. Again, as a vast literature shows importance of the sustainable management of a cold chain (Singh & Shabani, 2016; Deng et al., 2019) that has capability to ensure the improvements in food security and reduce the wastages of agricultural produce, this study also tried to explore various trends and challenges in implementing cold chain logistics. The categorization of literature included in this chapter has been incorporated

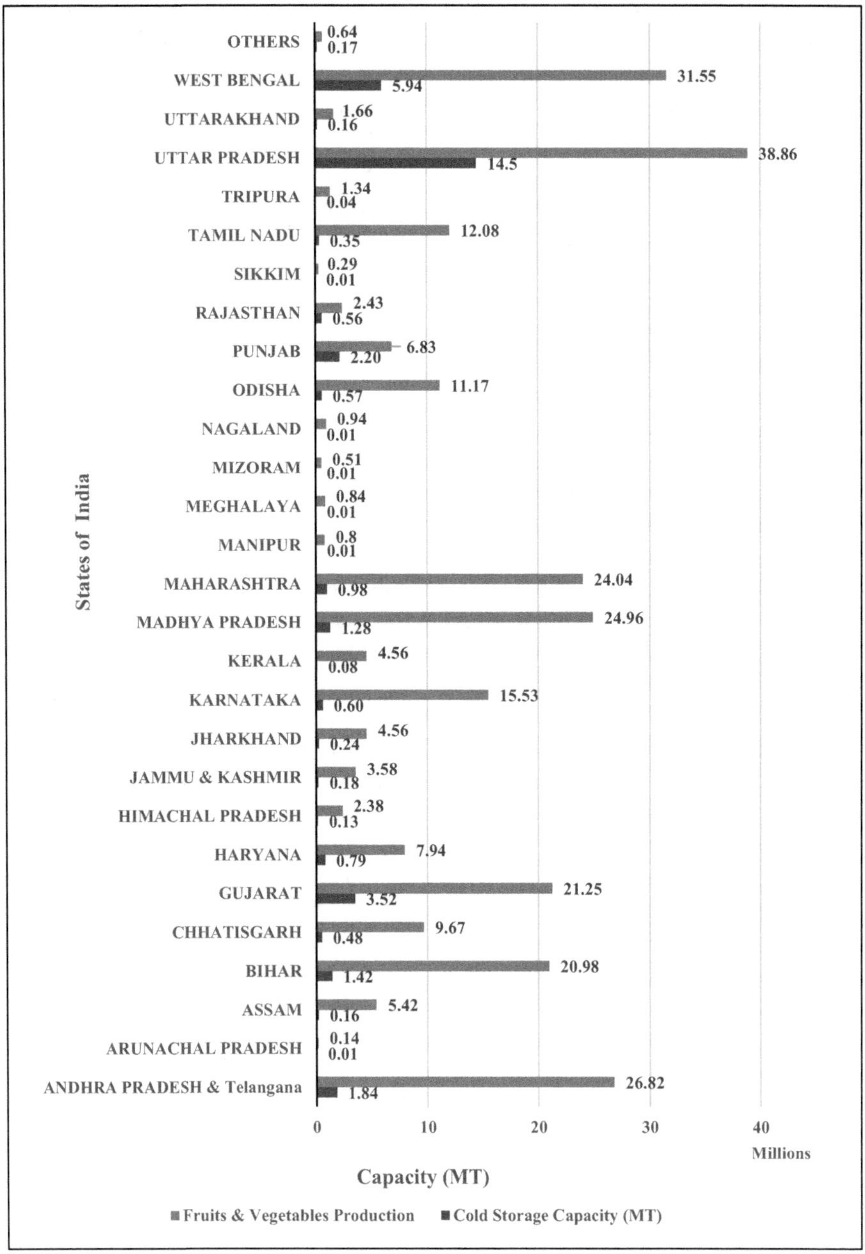

Fig. 8.1 State-wise fruits and vegetables production and cold storage capacity in India

based on their key research questions with required journal analysis from the pre-pandemic era to maintain consistency and avoid confounding variables. Therefore, we limit the study to the research that is published till 2021 and primarily deals with pre-covid data. From this categorization, seven critical factors are identified that are key parameters to help the policy makers, industries, government bodies or similar institutions while implementing or establishing sustainable cold chain system. We have organized this chapter as follows; Section 8.2 represents the methodology and material collection; Section 8.3 discusses the results based on research statements and article analysis. Sections 8.4 and 8.5 represents the identification of key critical factors and conclusions respectively.

8.2 Research Methodology

In this section, the proposed methodology in the current study is discussed. The primary objective of this review process is to explore the literature on cold chain logistics and related areas. The literature collection process in the current study includes only articles having high scientific value. The three-phase analysis considered for this study are briefly discussed below. A schematic chart of screening and selection process is given in Fig. 8.2. Initially, 156 articles retrieved through databases like Science Direct, Emerald, Wiley, Springer, and Google Scholar. Around 28 articles are removed as their title and abstract are not relevant to above key research themes described earlier in this sub-section and duplicity. To further scrutiny, 128 number of articles are identified at screening phase. Around 19 articles are excluded again as their research statements and scopes are beyond the purpose of this study. For example, we have excluded articles addressed the research questions regarding management of cold chain system during exceptional situations like natural disasters and articles on humanitarian logistics. With this elimination we have focused on 109 articles out of which 13 number of articles are removed due to its discontinuation from Scopus list (as per the database on (Scimago journal and country rank). Finally, a total of 96 articles and 7 institutional reports included out of which 83 number of articles are considered to identify the objective of this study.

8.2.1 Selection Criteria

This study incorporated a systematic literature review process by analyzing existing empirical, theoretical, and conceptual literature on cold chain logistics and similar areas. All the retrieved literature included in this study are from peer reviewed journals within the time span of 2004 to 2020, where majority of articles (approximately 80%) are from the year 2015 to 2020. This is due to the importance of research statement reflected in the included articles during the later stated time horizon. The selection criteria are based on the themes that deal with the handling of perishable

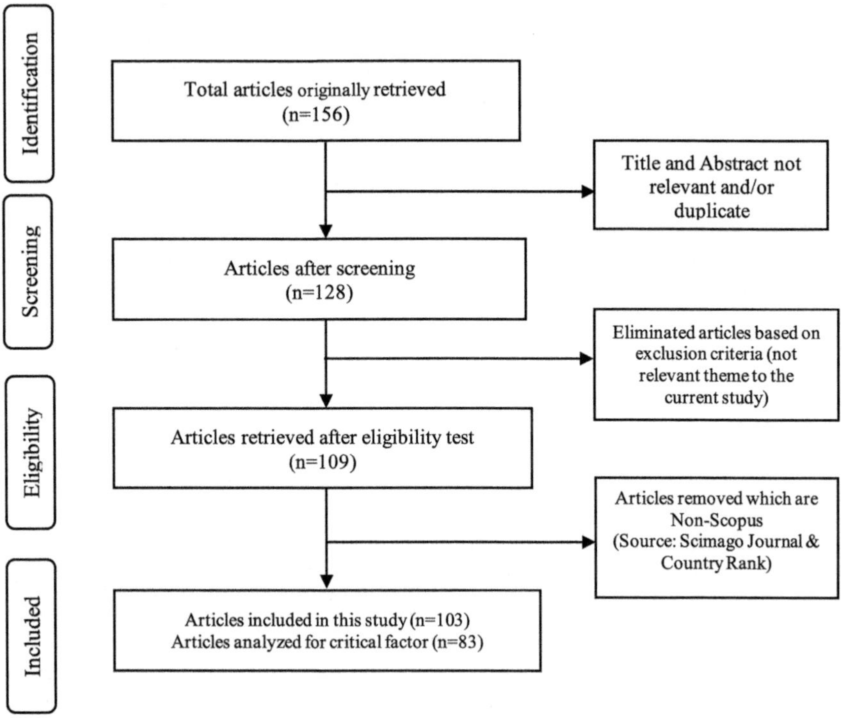

Fig. 8.2 Screening process of articles

products such as fruits, vegetables, dairy products, meat, vaccines, crops/grains, food, and flowers. We have considered the products mentioned earlier due to its high volume of human usage, proneness to the loss and wastages during the supply chain process (Mercier et al., 2017; Rahbari et al., 2019) as well as at consumption side (Principato et al., 2019). The studies such as commentaries, book chapters, conference proceedings, and non-abstracted documents are excluded in this work while identifying the critical factors. However, we have acknowledged various reports published by government agencies like Ministries of Food Processing Industries, Government of India, and credible institutional body such as abbreviation (FAO) of United Nations, Food and Sustainability Index of BCFN foundations and some others.

8.2.2 Key Word Search and Inclusion Criteria

The inclusion of articles for this wholly related to various important research themes to cold chain logistics and its problems. The aim of this study is to identify critical factors crucial for implementing cold chain system and therefore, the study is carried out on existing literature based on important themes like shelf life of perishables, distribution system, timely delivery, infrastructure of cold chain, temperature prediction, re-routing decisions, real-time monitoring, risks/challenges while implementing etc. The major keywords used to collect the articles for research are "Perishable products", "Perishable supply chain", "cold chain logistics", "Perishable logistics", "Sustainability", "Perishable food products" "food supply chain" etc. Articles which mention the way of cold chain logistics design for a natural disaster or also called as humanitarian logistics (like flood, earthquake, tsunami, hurricane, etc.) are categorically excluded.

8.3 Results and Discussions

In this section, results from the descriptive analysis, scient metric analysis, and analysis related to digital transformation are discussed.

8.3.1 Descriptive Analysis

In this section, findings related to articles focusing on various product category, research theme wise publication of articles, year wise publication of articles, and journal wise publication of articles are presented. This analysis shows an overview of the current research status of cold chain system.

8.3.1.1 Distribution of the Number of Articles as per Product Category

The collected articles are first grouped based on the perishable food product type. The various perishable product type are fruits, vegetables, dairy products, meat, vaccine, crops/grains, food and flower shown in Fig. 8.3. Common to all represents that it can be applicable to every type of perishable products (i.e. fruits, vegetable, meat, diary, vaccines, etc.).

8.3.1.2 Distribution of the Number of Articles as per Selected Research Themes

The collected articles are grouped based on the research themes. Figure 8.4 shows how percentage has been contributed by the collected articles based on research problem.

Table 8.1 shows number of articles addressing the five research themes taking each perishable product category into consideration.

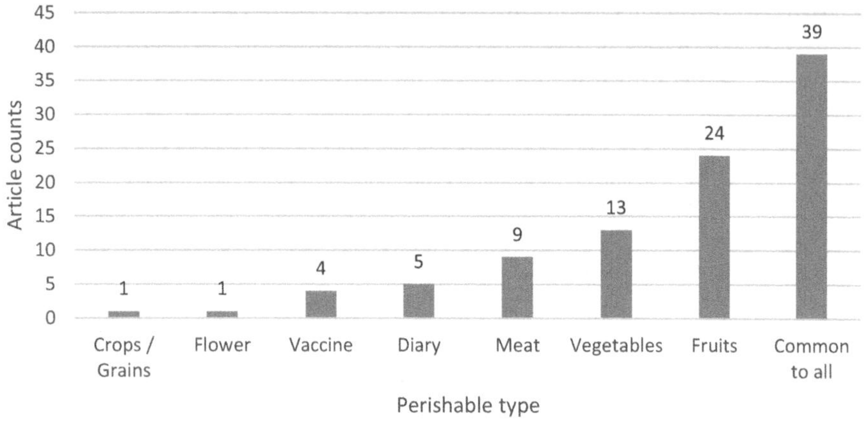

Fig. 8.3 No of articles addressed in each perishable product category

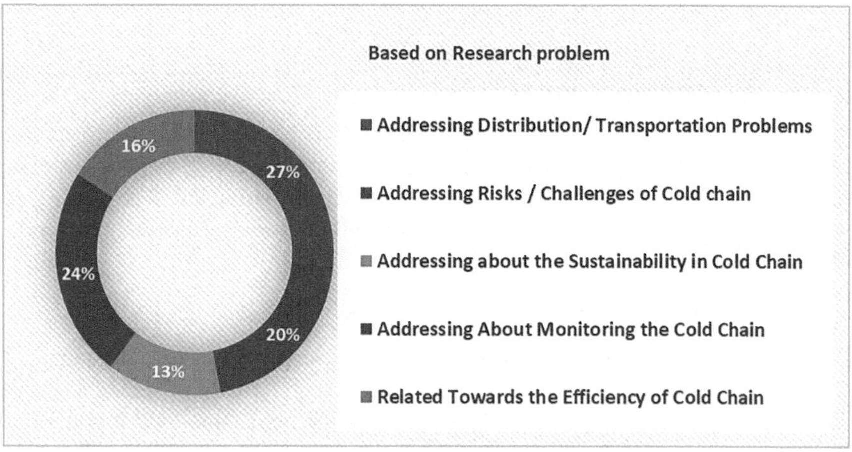

Fig. 8.4 Number of articles based on classification of reserch themes

8.3.1.3 Distribution of Number of Articles by Year of Publication

Figure 8.5 shows the year wise distribution of collected articles (out of 96 numbers). Most of the research conducted were from 2015–2024. This interval were the research and development of cold chain were taken serious by the researchers.

Table 8.1 Number of articles based on reserch themes considering product categories

Perishable Products\ Research Problems	Addressing Distribution/ Transportation Problems	Addressing about the Sustainability in Cold Chain	Addressing About Monitoring the Cold Chain	Related Towards the Efficiency of Cold Chain	Addressing Risks/ Challenges of Cold chain
Fruits	5	4	7	3	2
Vegetables	5	1	1	3	2
Dairy Products	–	3	–	1	1
Crop/Grains	1	–	–	–	–
Meat	2	1	4	1	1
Vaccine	–	1	–	–	2
Flowers	–		1	–	–
Common to all Perishables	11	5	7	5	10
Food	–	–	1	–	–

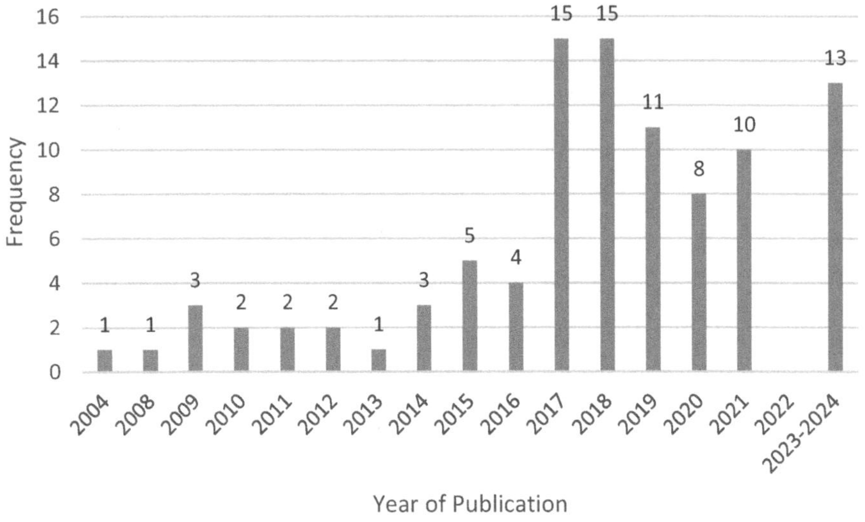

Fig. 8.5 All published articles from 2004 to 2024

8.3.1.4 Distribution of the Number of Articles Published by Journals

Table 8.2 gives us the count of journal distribution (out of 96 numbers) contributed to our study. The highest contribution is from "Industrial Management & Data Systems" journal followed by "International Journal of Production Economics", and "Food Control".

8.3.2 Scientometric Analysis

Scientometric analysis will provide network visualization between nodes based on distance where the nodes are the keywords. The two-dimensional visualization scientific map was created using the software VOSviewer. The size of each keyword and the font of its label indicates the frequency of occurrence. The bigger the keyword size or the label form, the higher the frequency of occurrence. If the keywords connected through links, they indicate the relationship between other keywords. For instance, co-occurrence links explains links between keywords of the cited articles. Each link has its certain strength, a positive value that might vary depending on the degree to strength in the relationship between keywords. The greater the strength of the link, the stronger the co-occurrence. However, keywords are strongly related when they are located closely, and vice versa. The items/keywords constructed of the network are grouped into clusters which shows their relatedness and are similarity in characteristics.

Table 8.2 Frequency distribution over various journals

Journal Names	Count(s)	Journal Names	Count(s)
2020 IEEE 17th Annual Consumer Communications & Networking Conference (CCNC)	1	International Journal of Physical Distribution and Logistics Management	1
IEEE Access	1	International Journal of Production Economics	7
American Pharmaceutical Review	1	Internet Research	1
Applied and Environmental Microbiology	1	Journal of Agriculture and Food Research	1
Applied and Environmental Microbiology	1	Journal of Business & Industrial Marketing	1
Applied Mathematical Modelling	1	Journal of Business & Industrial Marketing	1
Benchmarking: An International Journal	1	Journal of cleaner Production	2
British Food Journal	2	Journal of Food Engineering	1
Business Strategy and the Environment	1	Journal of Food Process Engineering	2
Cleaner Engineering and Technology	1	Journal of Food, Agriculture & Environment	1
Cleaner Logistics and Supply Chain	2	Journal of Operations and Supply Chain Management	1
Comprehensive Reviews in Food Science and Food Safety	1	Measuring Business Excellence	1
Computers & Industrial Engineering	3	Operations Research Perspectives	1
Computers and Chemical Engineering	1	Opsearch	1
Computers and Electronics in Agriculture	4	Physica A: Statistical Mechanics and its Applications	2
Computers in Industry	1	Production & Operations Management	1
Eurasip Journal on Wireless Communications and Networking	1	Production Planning & Control	1
European Journal of Operational Research	2	Research in Transportation Economics	1
European Journal of Pharmaceutics and Biopharmaceutics	1	Resources, Conservation and Recycling	2
Food additives and containments	1	Sensors	1
Food Control	6	Soft Computing	1
Food packaging and shelf life	1	Supply Chain Management	2
Heliyon	2	Sustainability	1
IEEE Internet Computing	1	Transportation Research Interdisciplinary Perspectives	1
IFAC-PapersOnLine	1	Transportation Research Part D	1
Industrial Management & Data Systems	8	Trends in Food Science & Technology	2
Industrial Marketing Management	2	Vaccine	2

(continued)

Table 8.2 (continued)

Journal Names	Count(s)	Journal Names	Count(s)
International Journal of Logistics Management	3	Other	3
International Journal of Online and Biomedical Engineering	1		
Total			96

Due to constraints in accessing Scopus database, we have used keyword co-occurrence type of bibliometric network to observe the importance of the keywords that have been included in this study. This technique employed with three visualization approaches like distance-based, timeline-based and graph-based. VOSviewer software is tool used to visualize the bibliometric networks to find out the most occurred keywords from the collected research articles. It uses distance-based approach to visualize the bibliometric networks (van Eck & Waltman, 2014). Keywords are acquired from the title and abstract of the collected research articles. The no of occurrence a keyword is the no of publications in which that is used in either in the title or in the abstract part of the research articles. Once the RIS file is loaded into the software we could find the total no of terms (2817) present from our collected articles. The minimum occurrence of a keyword term is set to 4. Where 254 terms met this minimum threshold out of 2817 terms. With these 154 terms there will be some unwanted terms which won't give relevant meaning when presented alone are de-selected. Finally, 59 terms were used for visualization. Table 8.3 represents the cluster of keywords visualized from the collected articles.

Keywords are acquired from the title and abstract of the collected research articles. The no of occurrence a keyword is the no of publications in which that particular is used in either in the title or in the abstract part of the research articles. Once the RIS file is loaded into the software we could find the total no of terms (2817) present from our collected articles. The minimum occurrence of a keyword term is set to 4. Where 254 terms met this minimum threshold out of 2817 terms. With these 154 terms there will be some unwanted terms which won't give relevant meaning when presented alone are de-selected. Finally 59 terms were used for visualization. From Fig. 8.6 shows the keywords are mapped based on their occurrence link strength. Each circle represents a keyword term. Larger the size of the term occurrence of that particular term will be huge. The colour of the term indicates the cluster they belong to. The shorter the distance between two terms shows the higher relatedness between those terms based on their number of occurrences. Figure 8.7 gives the occurrence of a keyword corresponding to a particular year. In overlay visualization colour represents the year here we have considered years from 2014 to 2020. It represents a keyword how much times cited in other research articles.

Table 8.3 Most Occurred keywords cited in the collected article

Terms	No of occurrences	Terms	No of occurrences
Cluster 1		Cluster 3	
Cold Supply chain	14	Cold chain management	6
Distribution	19	Fresh food	4
Distribution Cost	7	Fruit	18
Efficiency	12	Information technology	4
Food safety risk	4	Logistic	25
Food waste	7	Quality level	9
Freshness	11	Shipment	8
Inventory routing problem	4	Vegetable	10
Perishability	5	Cluster 4	
Route	14	Agricultural products cold chain logistic	8
Supply chain management	21	Cold chain logistic	20
Time	39	Delivery	10
Time Window	5	Food cold chain	7
Vehicle routing problem	7	Food quality	9
Cluster 2		Originality value	17
Challenge	27	Risk	33
Humidity	7	Cluster 5	
Infrastructure	7	Cold chain	73
IoT	17	Distributor	6
Monitoring	22	Performance	51
Perishable food supply chain	9	Quality	47
Product quality	9	Temperature Monitoring	5
RFID	4	Transportation	27
Sensor	19	Vaccine	12
Shell life	19	Cluster 6	
Technology	32	Distribution Centre	13
Temperature	36	Location	24
Wireless sensor network (WSN)	20	Perishable food product	5
Cluster 3		Sustainability	31
Cold chain management	6	Sustainable supply chain	6
Fresh food	4	Transportation link	4
Fruit	18	Cluster 7	
		Intelligent Container	5
		Transport	9

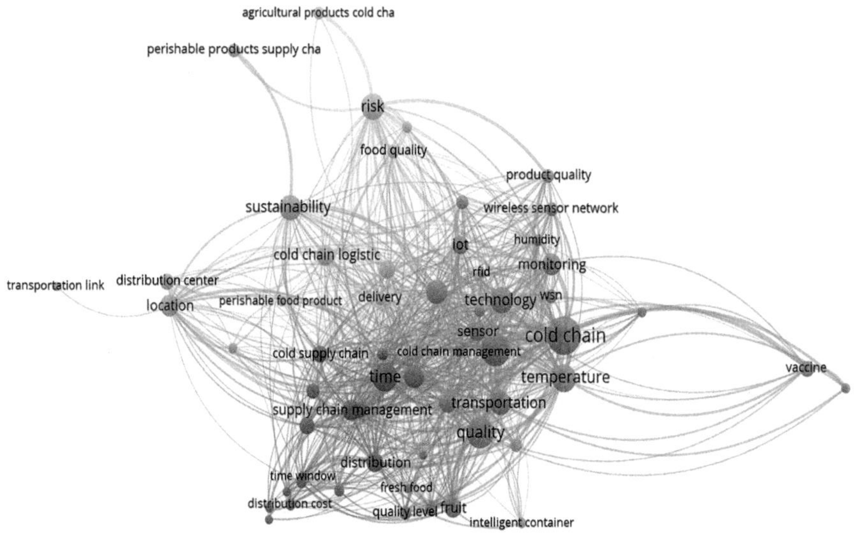

Fig. 8.6 Network Visualization of keywords used in the collected articles

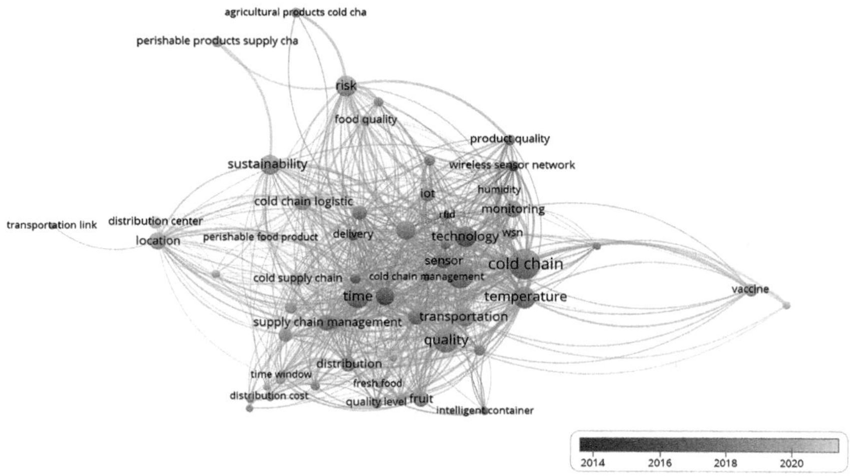

Fig. 8.7 Overlay Visualization of keywords used in the collected articles

8.3.3 *Addressing Digital Transformation and Monitoring of the Cold Chain*

The discussions in Sect. 8.1 and various research on food waste (FAO, 2014, 2019; Heising et al., 2017; Principato et al., 2015, 2019, 2021; Schanes et al., 2018; Sharma et al., 2020; Xiao, 2021) are evidential that a huge amount of food commodities is being wasted from source to the consumption points. Factors such as the lack of maintaining required temperature (Badia-Melis et al., 2018), improper handling and inadequate cold-chain logistic infrastructure (Raut et al., 2019), and lack of integration of technology and information asymmetry (Kant & Pal, 2017; Lusiantoro et al., 2018; Zhang et al., 2017a) are prominent contributors to significant food wastages across the globe. However, in our knowledge, there is a vast literature mostly focusing on the perishable food wastages due to improper temperature monitoring and control. Emphasizing on the global concern such as wastages of perishable food and product (Badia-Melis et al., 2018) studied key factors that challenges contemporary cold chain systems. Their study also outlined the applicability and weaknesses of various technical trends such as IoT, RFID, thermal imaging technology, wireless sensor network (WSN) that cab be suitably integrated in various phases of cold chain, to monitor and control the temperature to minimize the wastages of perishable products. Monitoring the supply chain through auctiontics (AL) is also very popular in supply chain literature and quite relevant in handling perishable supply chain (Kong et al., 2018; Kong et al., 2017). As traditional logistics approach (Liu et al., 2018; Villalobos et al., 2019) is difficult to handle the supply chain of perishables (Pérez Mesa & Gómez, 2015), in their paper, (Kong et al., 2017) proposed an AL model integrated with IOT to simulate the concept of cloud auction robot (CAR) to manage the distribution process, duties, and the practices of CAR-Agents (robots) in managing live events and to minimize the waiting time for the robots. This auction execution model was being simulated under a laboratory setting, where findings show a significant reduction in waiting time of activities though study did not conduct any experiments with real life supply chain of the perishable products. With the similar backdrop of AL as a pivotal monitoring agent for perishable cold supply chain, (Kong et al., 2018) conducted a literature review analysis on the best ways of auctions with an effective physical-internet enabled-logistics practices under the settings of a perishable supply chain trading (PSCT) and highlighted the existing deficiencies and potential research directions. Traceability of shelf life of perishable food products during the logistics process is also important towards food safety (La Scalia et al., 2016) as there is a high correlation between volatile organic compound (VOC) of food and growth rate of bacterial contamination (Mayr et al., 2003). Therefore, La Scalia et al. (2017) suggested a correlation model between VOC and bacterial growth to develop a monitoring system based on smart logistics unit (SLU) which incorporates both information and communication to ensure the food safety and its shelf life. This smart logistic unit is a remote system that monitors the quality of perishable products through identification, detect and data transfer of required information during it supply chain process.

As discussed, its importance earlier in this chapter, RFID technology plays a major role while monitoring various activities of a cold chain (Shi et al., 2010). For example, Montanari (2008) proposed a RFID enabled structural framework to monitor temperature and navigation of the product condition in the cold chain during its movement. These two monitoring systems are known as Eulerian and Lagrangian scenarios respectively that ensure minimization of management costs in the cold chain while meeting all the requirements.

Besides RFID technology, the application of WSN is also extensively reported in the literature as a monitoring tool. To monitor environmental aspects of a cold chain system such as temperature, humidity, vibration, light intensity, and carbon emission, (Wang et al., 2015) conceptualized a WSN based on Zigbee standard. This WSN architecture delivers important prior information regarding the quality of the perishable product and predictability of its self-life with highly significant real time data transmission through a proper selection of WSN topology. In an equivalent approach for improving the post-harvest quality of freshly cut branches of North American holly (winterberry), Wang et al. (2018) presented a wireless multi sensor system that captures important information during transportation period of its cold chain. Recognizing the fast-deteriorating nature of berries, the wireless multi sensor system found to be beneficial in providing prior and cautionary information regarding condition of the perishable product in the cold chain by monitoring and transferring data of important quality parameters such as temperature, relative humidity, carbon, and ethylene emission. Jedermann et al. (2017) demonstrated a prototype of an "Intelligent Container" to provide real-time remote monitoring of temperature, carbon emission, and ethylyne effects to predict hot spots that is essential for bananas during road and ocean transportation. However, this intelligent container system needs proper packaging and stowage process as a pre-requirement to maintain cooling and lacks integration of any WSN. Though applications of artificial intelligence (AI) (Singh, 2024), WSN (Feng et al., 2019) and IoT (Salah et al., 2020; Zhang et al., 2017b) are widely used technologies in the literature to monitor the activities of cold chain systems, (Tsang et al., 2018) addressed occupational safety risk such as cold related hazards via an IoT based risk management system (IoTRMS). This risk management system also addressed other common risk of cold chain system like various integration in cold chain logistics (Ali et al., 2018) and environmental conditions by incorporating a low powered WSN through real time cloud database to monitor the hazards and product quality in the cold products. Furthermore, considering the safety of frozen shellfish, (Feng et al., 2019) proposed a quality control system namely a knowledge-based hazard analysis and critical control point (HACCP) that combine WSN to monitor dynamic quality indicators such as humidity, oxygen, carbon dioxide and most importantly temperature to improve the shelf life of shellfish.

It is evident that the integration of various cutting-edge technologies to monitor or trace the products are plenty in the cold chain literature. But, (Óskarsdóttir & Oddsson, 2019) reasoned the lack of any standardized traceability techniques that is available in the existing body of knowledge. Therefore, the authors mapped different static and dynamic traceability technologies such as barcode application, RFID,

WSN etc. through a decision support framework (DSF) to avail flexible options of technologically feasible, practical, economical approaches to trace perishable products. Comparable to DSF approach, (Luo et al., 2016) proposed a fast and live intelligent tracking system with a ZigBee-based wireless sensor network that consolidates important data and information from IoT enabled different tracking technologies. This system monitors crucial activities like live temperature, humidity and physical where about of the cold chain products at a minimum cost with a large number of network capacity and simple protocols. The WSN technology is also helpful in logistic tracking management system (Jijun, 2018) to monitor transportation of time sensible perishable goods in the cargo. In (Ashok et al., 2017; Chaudhuri et al., 2018; Kumar et al., 2020), the authors reviewed various data collection methods, technology, and mechanisms on the context of cold chain infrastructure with a content analysis methodology and highlighted various challenges faced by practitioners during decision making and adoption of technologies. Temperature monitoring is very essential in most of the cold chain of refrigerated foods (Kuo & Chen, 2010; Zhang et al., 2009) as the quality and shelf life of the perishable products largely affected by many microbial activities. Therefore, finding an optimal temperature to maintain the quality is a salient challenge (Sow et al., 2018) for the refrigerated products and this was addressed by (Aung & Chang, 2014) through smart cold chain management system with integration of WSN technology. In addition to temperature, duration of transportation f perishable products plays a key role in product decaying. To address challenges of non-uniform product decay due to temperature and duration (de Keizer et al., 2017) outlined a logistic network distribution model positioned with a customer order decoupling point (CODP) that helps in deciding the hub location, extent of decay and various quality requirement. Similar work on optimal hub location in the context of agricultural product also addressed in (Wang et al., 2021a). In their paper, (Alfian et al., 2017) proposed a big data management system to assimilate data like temperature, humidity, image etc. to monitor food quality in real time. This IoT-enabled and smartphone-based sensor was experimented in Korean kimchi supply chain and is capable of detect abnormality in the sensor even though connected with number of other sensors.

In various research from (Aiello et al., 2012; Marr, 2018; Ma et al., 2020; Montanari, 2008; Koshiya et al., 2023; Pal & Kant, 2019; Sathiya et al., 2024; Sergi et al., 2021; Zhang et al., 2024) researchers have shown ways of predictive data transmission technology to track the freshness during the logistics of perishable items in real time and also reduces the overall cost. Based on the results state of hub type and width type of transportation link have strong benefits. In their work, (Sergi et al., 2021) aimed to provide support in prototyping end-to-end security monitoring the condition of the perishable goods during logistic operation by IoT and Azure Sphere platform. Using Edge computing they propose an easy way of remote and real time monitoring of goods in movement. Tang et al. (2021) designed a compact supervisory system for cold chain logistics practitioner to have a track on their shipments real time and detect any abnormalities with temperature data alone. Xiao (2021) developed a well-built traceability process constitutes of temperature monitoring and shell life prediction of tilapia in order to reduce or to eliminate the wastes, improve the transparency and traceability and to guarantee the quality of the perishable in cold chain system.

8.4 Critical Appraisals

In this study, the inclusion criteria are articles wholly related to cold chain logistics and its problems (i.e. shell life of perishables, distribution system, timely delivery, infrastructure of cold chain, temperature prediction, re-routing decisions, real-time monitoring, risks/challenges while implementing). Whereas the exclusion criteria are: Articles which mention the way of cold chain logistics design for a natural disaster or also called as humanitarian logistics (like flood, earthquake, tsunami, hurricane etc.,). Therefore, a total of 83 articles were considered for the research after screening and exclusion. These 83 articles have been grouped based on their research objectives, type of perishable product, method of study they conducted, and their approach towards the problem statement. Articles that have been grouped based on research questions into five are

1. Distribution/Transportation Problems—articles that address distribution and transportation issues in the supply chain of perishables. For example, studies focusing delivering fresh produce to customers such as supermarkets through route optimization and other cost reducing approaches.
2. Risks/Challenges of Cold chain—articles addressing the challenges and risks associated while processing in the cold chain system. For example, power outages in refrigeration system in the vaccine or other sensitive warehouses or during transportation of such items.
3. Sustainability in Cold Chain—articles addressing how to manage energy-efficient technologies for sustainability reduce cost, carbon footprints etc.
4. Monitoring the Cold Chain—articles addressing the ways to monitor the whole cold chain system through technology and digital transformation such as IOT-based and real-time monitoring systems.
5. Efficiency of Cold Chain—articles addressing the ways to improve the efficiency of the cold chain system to improve operational efficiency through various measures such as integrating AI for predictive temperature control in warehouse and refrigerated vehicles.

The articles were further classified based on criterions such as perishable product types that includes fruits, vegetables, dairy, meat, vaccines, flowers, grains, and multi-category perishables. Various methodologies namely literature reviews, simulations, and conceptual analyses were used in these studies. It is found that out of 22 different approaches as shown in the Fig. 8.8 that address various research problems in cold chain logistics area, 7 critical factors are mostly dominant as shown in the Table 8.4.

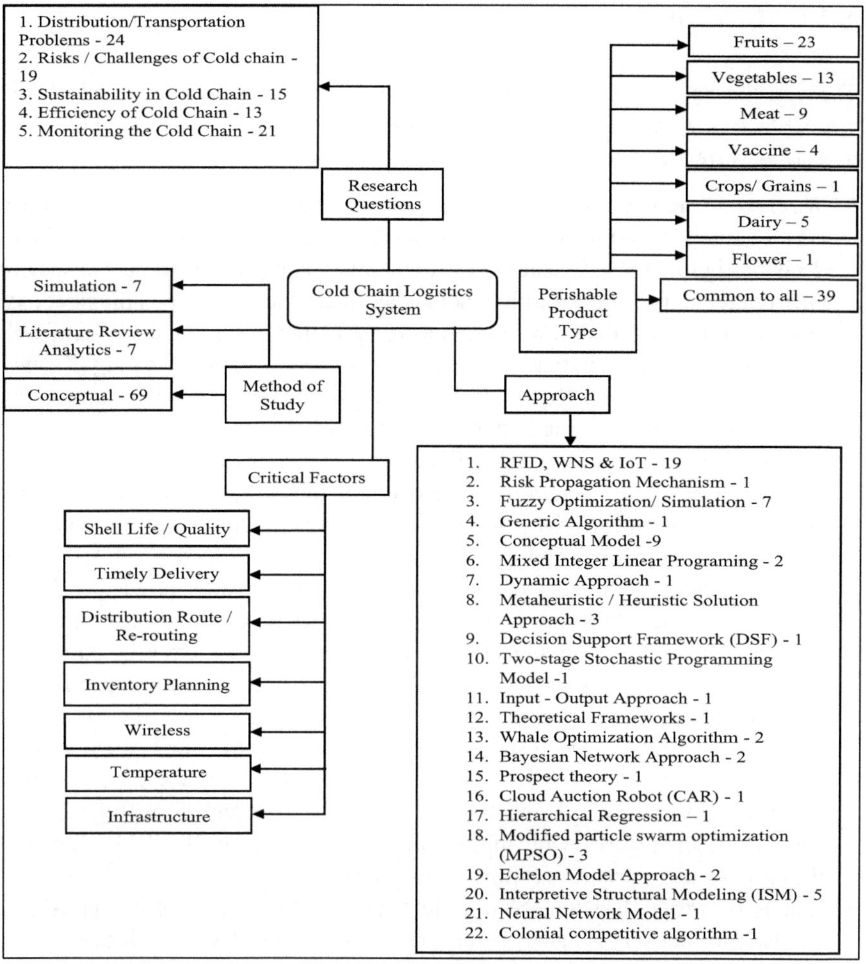

Fig. 8.8 Thematic mapping

Table 8.4 Critical Factors identified from the review

Critical Factors Identified	Number of articles that feature those critical factors
Shell Life/Quality	55
Timely Delivery	14
Distribution Route/Re-routing	30
Inventory planning	9
Wireless Technology	44
Temperature Predictions	44
Infrastructure	46

8.4.1 *Critical Factors*

Through the above comprehensive review based on interrelation, a total of nine critical factors were identified and finally narrowed down to seven paramount factors. They were discussed as given below.

- Wireless Technology—To predict the real-time location and temperature of perishable goods. Smart technologies will always be advantageous in making better decisions before the any issue arises. To maintain a perishable product at specific environmental parameters which include air quality levels (carbon dioxide, oxygen, humidity and others), wireless sensors and other technologies are extremely beneficial during transit. For example, as per Amazon Web Service blog (Koshiya et al., 2023), the use of RFID and IoT in Amazon's cold chain does a real-time location track and the temperature levels of goods.
- Temperature Predictions—The perishable products always needed to be preserved at a certain temperature. The temperature levels and requirements need to be monitored frequently. The outside temperature conditions can cause damage to the shell life/quality of perishable goods. For example, as per web articles by Pharmaceutical Commerce (Singh, 2024) and the research article by Clénet (2018) it is evident that how the predictive analytics are used by various pharmaceutical companies to maintain vaccines at required temperatures during the transportation and delivery under various situations including extreme climates.
- Shell life/Quality—For perishable goods, the quality gets deprived over a period of time, so it is required to ensure the quality throughout the supply chain process from vendor to customer. For example, the Walmart's retailing process utilize advanced sensors to check and maintain quality of meat and dairy products from suppliers to their stores while minimizing spoilage (Patel et al., 2023). Similar is the case for dairy products (Prakash et al., 2017).
- Timely delivery—It is the speed of delivery or time taken to deliver perishable goods. If the delivery is made on time, then customer satisfaction will be entrusted with the service (Sathiya et al., 2024; Wang et al., 2021b; Yan, 2017). Also, the quality of the perishable will also be in good state. For example, as per Financial Express Business report (FE Business, 2024), Food delivery services such as Zomato known for implementing time-bound logistics to ensure the freshness of foods that are to be delivered.
- Distribution Route/Re-routing—It is an important factor because all temperature sensitive goods need to be delivered without any delay. Proper routing (Biuki et al., 2020; Blackburn & Scudder, 2009; Govindan et al., 2014; Hsiao et al., 2017, 2018; Zhang et al., 2021) has to be made based on the quality of perishable products and prioritization has to be made to ensure the products or products are in a good state. In case any uncertain situation arises, there needs to be an alternate plan (re-routing) to reduce the loss. As per reports published in MIT Technology review (Woyke, 2018) and Forbes (Marr, 2018), it was exemplary to note that how UPS employs AI-driven dynamic routing for perishable deliveries, ensuring minimum delays and re-routing during unexpected traffic conditions.

- Inventory planning—Less inventory is preferable for perishable goods to reduce inventory cost (Fichtinger et al., 2015; Soysal et al., 2015; van Donselaar et al., 2006; Violi et al., 2020). The warehouse needs to be at multiple centers instead of being located at one location. Then, it will be easy to deliver the goods in less time from the nearest warehouse without any degrade in the shell life of perishable product. For example, Grocery chains like Tesco takes the advantage of decentralized warehouses to stock optimal quantities perishables, allowing faster delivery in local destinations to reducing inventory costs (Mkansi & Nsakanda, 2021; Seifert & Markoff, 2024).
- Infrastructure—In order to transport or store perishable goods, cold storage such as refrigerated trucks, cold boxes etc. are very much essential. For example, FedEx deploys refrigerated to transport seafood and flowers across long distances and hence maintains their freshness.

Figure 8.9 represents the frequency distribution of critical factors over the research questions. Shell life/Quality, Infrastructure, Temperature predictions and Wireless Technology has major proportion of all kinds of research questions as these seven critical factors are most crucial points to be looked when trying to have competitive advantage over other players. This analysis of finding above critical factors will be beneficial for new entrepreneurs who considering their business in cold chain logistic service. For maintaining a risk-free logistics service in a cold chain sector, one need to look after all these factors closely.

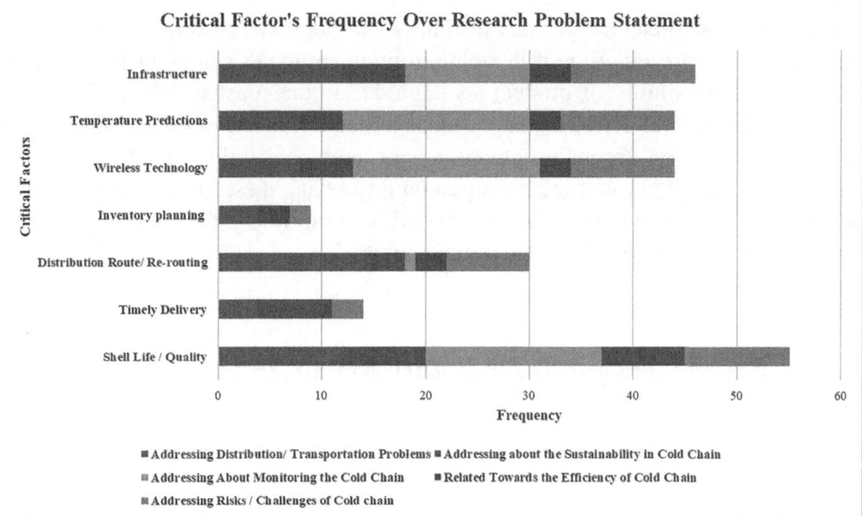

Fig. 8.9 Distribution of Critical factors over research questions

8.5 Conclusions

The digital transformation of the cold supply chain marks a significant impact on the cold chain logistic process where temperature-sensitive goods are mostly managed, transported, and delivered. The integration of technologies such as blockchain, AI, IoT and automation, the cold chain industry shows paradigm improvements in efficiency, transparency, and resiliency it its operations and service delivery. These innovative measures also instrumental in achieving goals like real-time monitoring, predictive analytics, and better decision-making. This ensures process of safe handling and delivery of perishable goods with minimised cost, waste and temperature abuse. Keeping all these in mind, this study has identified seven critical factors from an elaborative literature review. And these critical factors are essential for creating and managing an effective cold supply chain. From this study, it was evident that factor such as maintenance of the shelf life and quality of perishable goods exhibits the top priority. Many literatures argued in the support of above critical factor as the failure in preserving product quality leads to financial losses consumer trust simultaneously. It is important to the cold chain managers need to revisit possible solutions such as refrigerated vehicles, cold storage units, and advanced temperature -controlled containers.

Similarly, it is equally important to monitor peripheral conditions such as temperature levels, humidity, and gas levels frequently in order to maintain the quality of the product in the cold supply chain. Therefore, adopting wireless technologies like IoT sensors, RFID, and GPS, cold chain systems can collect information in real-time regarding perishable goods and can be responsive to any risks when required. Additionally, the delivery route-optimization will be very helpful to reduce transit times and intact the product freshness that improves customer's satisfaction and trust. Also, the adoption of flexible rerouting strategies can be instrumental in reducing the possibility of product spoilage due to unexpected delays. The other important factor is effective inventory management. This ensures cost reduction or profit maximization through optimal stocks and availability of the products. Moreover, by aligning with global sustainability goals, these critical factors when adopted can reduce food waste and lower carbon footprints in the cold supply chain by fulfilling broader social and environmental goals.

8.5.1 Recommendations for Future Research

Based on above discussions and review process, our work sees the following recommendations that are feasible to adopt or innovate for future research works.

1. *Blockchain for Enhanced Traceability*: A study can be undertaken on how blockchain can improve traceability, transparency, and data security in the cold chain logistic system to ensure product authenticity with compliance of regulatory bodies.

2. *Adapting to Climate Change*: a study can explore strategies to reduce global warming through development of energy-efficient storage systems that depends on renewable energy sources.
3. *Advanced Predictive Analytics Models*: A study can explore the advantages of various AI-driven predictive models that can anticipate future disruptions in the cold supply chain and take suitable decision to optimize logistics in real-time environment.
4. *Socioeconomic Impacts*: Another study can be undertaken to investigate the effects of digital transformation on various stake holders of cold supply chain such as small-scale farmers, local service providers particularly in developing economies where the cold chain infrastructure is rapidly evolving.
5. *Cold Chain Integration with Smart Cities*: A study can examine the integration of how smart city initiatives with cold supply chain to ensure urban food security and reduce waste.
6. *Regulatory Framework Development*: A study can be proposed to develop various policy at global and regional levels to facilitate the adoption of digital technologies keeping while privacy, security, and ethical considerations in mind.

References

Aiello, G., La Scalia, G., & Micale, R. (2012). Simulation analysis of cold chain performance based on time-temperature data. *Production Planning & Control, 23*(6), 468–476. https://doi.org/10.1080/09537287.2011.564219

Al Theeb, N., Smadi, H. J., Al-Hawari, T. H., & Aljarrah, M. H. (2020). Optimization of vehicle routing with inventory allocation problems in Cold Supply Chain Logistics. *Computers & Industrial Engineering, 142*, 106341. https://doi.org/10.1016/j.cie.2020.106341

Alfian, G., Syafrudin, M., & Rhee, J. (2017). Real-time monitoring system using smartphone-based sensors and NoSQL database for perishable supply chain. *Sustainability (Switzerland), 9*(11). https://doi.org/10.3390/su9112073

Ali, I., Nagalingam, S., & Gurd, B. (2018). A resilience model for cold chain logistics of perishable products. *International Journal of Logistics Management, 29*(3), 922–941. https://doi.org/10.1108/IJLM-06-2017-0147

Ashok, A., Brison, M., & LeTallec, Y. (2017). Improving cold chain systems: Challenges and solutions. *Vaccine, 35*(17), 2217–2223. https://doi.org/10.1016/j.vaccine.2016.08.045

Aung, M. M., & Chang, Y. S. (2014). Temperature management for the quality assurance of a perishable food supply chain. *Food Control, 40*(1), 198–207. https://doi.org/10.1016/j.foodcont.2013.11.016

Babagolzadeh, M., Shrestha, A., Abbasi, B., Zhang, Y., Woodhead, A., & Zhang, A. (2020). Sustainable cold supply chain management under demand uncertainty and carbon tax regulation. *Transportation Research Part D: Transport and Environment, 80*, 102245. https://doi.org/10.1016/j.trd.2020.102245

Badia-Melis, R., Mc Carthy, U., Ruiz-Garcia, L., Garcia-Hierro, J., & Robla Villalba, J. I. (2018). New trends in cold chain monitoring applications – A review. *Food Control, 86*, 170–182. https://doi.org/10.1016/j.foodcont.2017.11.022

Bishara, R. H. (2006). Cold chain management – An essential component of the global pharmaceutical supply chain. *American Pharmaceutical Review, 9*(1), 105–109. www.americanpharmaceuticalreview.com.

Biuki, M., Kazemi, A., & Alinezhad, A. (2020). An integrated location-routing-inventory model for sustainable design of a perishable products supply chain network. *Journal of Cleaner Production, 260*, 120842. https://doi.org/10.1016/j.jclepro.2020.120842

Blackburn, J., & Scudder, G. (2009). Supply chain strategies for perishable products: The case of fresh produce. *Production and Operations Management, 18*(2), 129–137. https://doi.org/10.1111/j.1937-5956.2009.01016.x

Bogataj, M. (2005). Stability of perishable goods in cold logistic chains. *International Journal of Production Economics, 93–94*(SPEC.ISS), 345–356. https://doi.org/10.1016/j.ijpe.2004.06.032

Bogataj, D., Bogataj, M., & Hudoklin, D. (2017). Reprint of "Mitigating risks of perishable products in the cyber-physical systems based on the extended MRP model". *International Journal of Production Economics, 194*, 113–125. https://doi.org/10.1016/j.ijpe.2017.11.004

Chaudhuri, A., Dukovska-Popovska, I., Subramanian, N., Chan, H. K., & Bai, R. (2018). Decision-making in cold chain logistics using data analytics: A literature review. *International Journal of Logistics Management, 29*(3), 839–861. https://doi.org/10.1108/IJLM-03-2017-0059

Chen, Y., Chen, M., & Hu, T. (2025). Designing a sustainable-resilient vaccine cold chain network in uncertain environments. *Computers and Chemical Engineering, 194*, 108936. https://doi.org/10.1016/j.compchemeng.2024.108936

Chermala, A., Acharya, P., & Singh, R. K. (2025). Cold chain logistics performance: A framework. *Measuring Business Excellence, 29*, 55–75. https://doi.org/10.1108/MBE-11-2023-0174

Clénet, D. (2018). Accurate prediction of vaccine stability under real storage conditions and during temperature excursions. *European Journal of Pharmaceutics and Biopharmaceutics, 125*, 76–84. https://doi.org/10.1016/j.ejpb.2018.01.005

Dai, J., Che, W., Lim, J. J., & Shou, Y. (2020). Service innovation of cold chain logistics service providers: A multiple-case study in China. *Industrial Marketing Management, 89*, 143–156. https://doi.org/10.1016/j.indmarman.2019.08.002

Dasaklis, T. K., Pappis, C. P., & Rachaniotis, N. P. (2012). Epidemics control and logistics operations: A review. *International Journal of Production Economics, 139*(2), 393–410. https://doi.org/10.1016/j.ijpe.2012.05.023

Deng, X., Yang, X., Zhang, Y., Li, Y., & Lu, Z. (2019). Risk propagation mechanisms and risk management strategies for a sustainable perishable products supply chain. *Computers and Industrial Engineering, 135*(January), 1175–1187. https://doi.org/10.1016/j.cie.2019.01.014

van Donselaar, K., van Woensel, T., Broekmeulen, R., & Fransoo, J. (2006). Inventory control of perishables in supermarkets. *International Journal of Production Economics, 104*(2), 462–472. https://doi.org/10.1016/j.ijpe.2004.10.019

van Eck, N. J., & Waltman, L. (2014). Visualizing bibliometric networks. In *Measuring scholarly impact* (pp. 285–320). Springer. https://doi.org/10.1007/978-3-319-10377-8_13

FAO. (2014). *Food losses and waste in the context of sustainable food systems A report by The High Level Panel of Experts on Food Security and Nutrition*. www.fao.org/cfs/cfs-hlpe

FAO. (2019). *State of food and agriculture 2019: Moving forward on food loss and waste reduction*. Food & Agriculture Org. https://openknowledge.fao.org/server/api/core/bitstreams/11f9288f-dc78–4171-8d02-92235b8d7dc7/content

FE Business. (2024, November 11). *Zomato launches new feature called 'Food Rescue' – Here's how it works for customers and restaurants*. Financial Express Business.

Feng, H., Chen, J., Zhou, W., Rungsardthong, V., & Zhang, X. (2019). Modeling and evaluation on WSN-enabled and knowledge-based HACCP quality control for frozen shellfish cold chain. *Food Control, 98*, 348–358. https://doi.org/10.1016/j.foodcont.2018.11.050

Fichtinger, J., Ries, J. M., Grosse, E. H., & Baker, P. (2015). Assessing the environmental impact of integrated inventory and warehouse management. *International Journal of Production Economics, 170*, 717–729. https://doi.org/10.1016/j.ijpe.2015.06.025

Govindan, K., Jafarian, A., Khodaverdi, R., & Devika, K. (2014). Two-echelon multiple-vehicle location-routing problem with time windows for optimization of sustainable supply chain network of perishable food. *International Journal of Production Economics, 152*(2009), 9–28. https://doi.org/10.1016/j.ijpe.2013.12.028

Gupta, V. K., Chaudhuri, A., & Tiwari, M. K. (2019). Modeling for deployment of digital technologies in the cold chain. *IFAC-Papers OnLine, 52*(13), 1192–1197. https://doi.org/10.1016/j.ifacol.2019.11.360

Heising, J. K., Claassen, G. D. H., & Dekker, M. (2017). Options for reducing food waste by quality-controlled logistics using intelligent packaging along the supply chain. *Food Additives and Contaminants – Part A Chemistry, Analysis, Control, Exposure and Risk Assessment, 34*(10), 1672–1680. https://doi.org/10.1080/19440049.2017.1315776

Hsiao, Y. H., Chen, M. C., & Chin, C. L. (2017). Distribution planning for perishable foods in cold chains with quality concerns: Formulation and solution procedure. *Trends in Food Science and Technology, 61*, 80–93. https://doi.org/10.1016/j.tifs.2016.11.016

Hsiao, Y. H., Chen, M. C., Lu, K. Y., & Chin, C. L. (2018). Last-mile distribution planning for fruit-and-vegetable cold chains. *International Journal of Logistics Management, 29*(3), 862–886. https://doi.org/10.1108/IJLM-01-2017-0002

Jabbour Al Maalouf, N., Elia, J., Boutros, F., Balouza, M., & Sawaya, C. (2024). Sustainable practices and their impact on the operations of fresh fruits and vegetables market in the cold chain. *Cleaner Logistics and Supply Chain, 13*, 100196. https://doi.org/10.1016/j.clscn.2024.100196

Jedermann, R., Praeger, U., & Lang, W. (2017). Challenges and opportunities in remote monitoring of perishable products. *Food Packaging and Shelf Life, 14*(May), 18–25. https://doi.org/10.1016/j.fpsl.2017.08.006

Jiang, S., Jia, S., & Guo, H. (2024). Internet of Things (IoT)-enabled framework for a sustainable Vaccine cold chain management system. *Heliyon, 10*(7), e28910. https://doi.org/10.1016/j.heliyon.2024.e28910

Jijun, X. (2018). An intelligent logistics tracking system based on wireless sensor network. *International Journal of Online and Biomedical Engineering (IJOE), 14*(01), 17–28. https://doi.org/10.3991/ijoe.v14i01.8063

Joshi, R., Banwet, D. K., & Shankar, R. (2009). Indian cold chain: Modeling the inhibitors. *British Food Journal, 111*(11), 1260–1283. https://doi.org/10.1108/00070700911001077

Kant, K., & Pal, A. (2017). Internet of perishable logistics. *IEEE Internet Computing, 21*(1), 22–31. https://doi.org/10.1109/MIC.2017.19

de Keizer, M., Akkerman, R., Grunow, M., Bloemhof, J. M., Haijema, R., & van der Vorst, J. G. A. J. (2017). Logistics network design for perishable products with heterogeneous quality decay. *European Journal of Operational Research, 262*(2), 535–549. https://doi.org/10.1016/j.ejor.2017.03.049

Kong, X. T. R., Zhong, R. Y., Xu, G., & Huang, G. Q. (2017). Robot-enabled execution system for perishables auction logistics. *Industrial Management and Data Systems, 117*(9), 1954–1971. https://doi.org/10.1108/IMDS-03-2016-0114

Kong, X., Huang, G. Q., Luo, H., & Yen, B. P. C. (2018). Physical-internet-enabled auction logistics in perishable supply chain trading: State-of-the-art and research opportunities. *Industrial Management and Data Systems, 118*(8), 1671–1694. https://doi.org/10.1108/IMDS-10-2017-0486

Koshiya, N., Dryden, D., Cravinho, J. & Swagle, J. (2023). *Enhancing retail store inventory management through RFID technology & AWS*. AWS for industries. https://aws.amazon.com/blogs/industries/enhancing-retail-store-inventory-management-through-rfid-technology-aws/

Kumar, A., Mangla, S. K., Kumar, P., & Karamperidis, S. (2020). Challenges in perishable food supply chains for sustainability management: A developing economy perspective. *Business Strategy and the Environment, 29*(5), 1809–1831. https://doi.org/10.1002/bse.2470

Kuo, J. C., & Chen, M. C. (2010). Developing an advanced Multi-Temperature Joint Distribution System for the food cold chain. *Food Control, 21*(4), 559–566. https://doi.org/10.1016/j.foodcont.2009.08.007

La Scalia, G., Aiello, G., Miceli, A., Nasca, A., Alfonzo, A., & Settanni, L. (2016). Effect of vibration on the quality of strawberry fruits caused by simulated transport. *Journal of Food Process Engineering, 39*(2), 140–156. https://doi.org/10.1111/jfpe.12207

La Scalia, G., Nasca, A., Corona, O., Settanni, L., & Micale, R. (2017). An innovative shelf life model based on smart logistic unit for an efficient management of the perishable food supply chain. *Journal of Food Process Engineering, 40*(1), 1–13. https://doi.org/10.1111/jfpe.12311

Li, Y., Lim, M. K., & Tseng, M. L. (2019). A green vehicle routing model based on modified particle swarm optimization for cold chain logistics. *Industrial Management and Data Systems, 119*(3), 473–494. https://doi.org/10.1108/IMDS-07-2018-0314

Lin, Q., Zhao, Q., & Lev, B. (2020). Cold chain transportation decision in the vaccine supply chain. *European Journal of Operational Research, 283*(1), 182–195. https://doi.org/10.1016/j.ejor.2019.11.005

Liu, H., Pretorius, L., & Jiang, D. (2018). Optimization of cold chain logistics distribution network terminal. *EURASIP Journal on Wireless Communications and Networking, 2018*(1). https://doi.org/10.1186/s13638-018-1168-4

Luo, H., Zhu, M., Ye, S., Hou, H., Chen, Y., & Bulysheva, L. (2016). An intelligent tracking system based on internet of things for the cold chain. *Internet Research, 26*(2), 435–445. https://doi.org/10.1108/IntR-11-2014-0294

Lusiantoro, L., Yates, N., Mena, C., & Varga, L. (2018). A refined framework of information sharing in perishable product supply chains. *International Journal of Physical Distribution and Logistics Management, 48*(3), 254–283. https://doi.org/10.1108/IJPDLM-08-2017-0250

Ma, X., Wang, J., Bai, Q., & Wang, S. (2020). Optimization of a three-echelon cold chain considering freshness-keeping efforts under cap-and-trade regulation in Industry 4.0. International Journal of Production Economics, 220(July), 0–1. doi:https://doi.org/10.1016/j.ijpe.2019.07.030.

Marr, B. (2018). *The brilliant ways UPS uses Artificial Intelligence, machine learning and big data*. Forbes, Innovation, Enterprise Tech.

Mayr, D., Margesin, R., Klingsbichel, E., Hartungen, E., Jenewein, D., Schinner, F., & Märk, T. D. (2003). Rapid detection of meat spoilage by measuring volatile organic compounds by using proton transfer reaction mass spectrometry. *Applied and Environmental Microbiology, 69*(8), 4697–4705. https://doi.org/10.1128/AEM.69.8.4697-4705.2003

Mejjaouli, S., & Babiceanu, R. F. (2018). Cold supply chain logistics: System optimization for real-time rerouting transportation solutions. *Computers in Industry, 95*, 68–80. https://doi.org/10.1016/j.compind.2017.12.006

Mercier, S., Villeneuve, S., Mondor, M., & Uysal, I. (2017). Time–temperature management along the food cold chain: A review of recent developments. *Comprehensive Reviews in Food Science and Food Safety, 16*(4), 647–667. https://doi.org/10.1111/1541-4337.12269

Mkansi, M., & Nsakanda, A. L. (2021). Leveraging the physical network of stores in e-grocery order fulfilment for sustainable competitive advantage. *Research in Transportation Economics, 87*, 100786. https://doi.org/10.1016/j.retrec.2019.100786

Montanari, R. (2008). Cold chain tracking: A managerial perspective. *Trends in Food Science and Technology, 19*(8), 425–431. https://doi.org/10.1016/j.tifs.2008.03.009

Mustafa, M. F. M. S., Navaranjan, N., & Demirovic, A. (2024). Food cold chain logistics and management: A review of current development and emerging trends. In *Journal of Agriculture and Food Research* (Vol. 18). Elsevier B.V. https://doi.org/10.1016/j.jafr.2024.101343

Óskarsdóttir, K., & Oddsson, G. V. (2019). Towards a decision support framework for technologies used in cold supply chain traceability. *Journal of Food Engineering, 240*, 153–159. https://doi.org/10.1016/j.jfoodeng.2018.07.013

Pal, A., & Kant, K. (2019). Internet of perishable logistics: Building smart fresh food supply chain networks. *IEEE Access, 7*, 17675–17695. https://doi.org/10.1109/ACCESS.2019.2894126

Patel, A. S., Brahmbhatt, M. N., Bariya, A. R., Nayak, J. B., & Singh, V. K. (2023). Blockchain technology in food safety and traceability concern to livestock products. In *Heliyon* (Vol. 9, Issue 6). Elsevier. https://doi.org/10.1016/j.heliyon.2023.e16526

Pérez Mesa, J. C., & Gómez, E. G. (2015). Collaborative firms managing perishable products in a complex supply network: An empirical analysis of performance. *Supply Chain Management, 20*(2), 128–138. https://doi.org/10.1108/SCM-06-2014-0185

du Plessis, F., Goedhals-Gerber, L., & van Eeden, J. (2024). The impacts of climate change on marine cargo insurance of cold chains: A systematic literature review and bibliometric analysis.

Transportation Research Interdisciplinary Perspectives, 23, 101018. https://doi.org/10.1016/j.trip.2024.101018

Prakash, S., Soni, G., Rathore, A. P. S., & Singh, S. (2017). Risk analysis and mitigation for perishable food supply chain: A case of dairy industry. *Benchmarking: An International Journal, 24*(1), 2–23. https://doi.org/10.1108/BIJ-07-2015-0070

Principato, L., Secondi, L., & Pratesi, C. A. (2015). Reducing food waste: An investigation on the behaviour of Italian youths. *British Food Journal, 117*(2), 731–748. https://doi.org/10.1108/BFJ-10-2013-0314

Principato, L., Ruini, L., Guidi, M., & Secondi, L. (2019). Adopting the circular economy approach on food loss and waste: The case of Italian pasta production. *Resources, Conservation and Recycling, 144*, 82–89. https://doi.org/10.1016/j.resconrec.2019.01.025

Principato, L., Mattia, G., Di Leo, A., & Pratesi, C. A. (2021). The household wasteful behaviour framework: A systematic review of consumer food waste. *Industrial Marketing Management, 93*, 641–649. https://doi.org/10.1016/j.indmarman.2020.07.010

Rahbari, A., Nasiri, M. M., Werner, F., Musavi, M., & Jolai, F. (2019). The vehicle routing and scheduling problem with cross-docking for perishable products under uncertainty: Two robust bi-objective models. *Applied Mathematical Modelling, 70*, 605–625. https://doi.org/10.1016/j.apm.2019.01.047

Raut, R. D., Gardas, B. B., Narwane, V. S., & Narkhede, B. E. (2019). Improvement in the food losses in fruits and vegetable supply chain – A perspective of cold third-party logistics approach. *Operations Research Perspectives, 6*(January), 100117. https://doi.org/10.1016/j.orp.2019.100117

Salah, K., Alfalasi, A., Alfalasi, M., Alharmoudi, M., Alzaabi, M., Alzyeodi, A., & Ahmad, R. W. (2020). IoT-enabled shipping container with environmental monitoring and location tracking. *2020 IEEE 17th Annual Consumer Communications & Networking Conference (CCNC)*, 1–6. https://doi.org/10.1109/CCNC46108.2020.9045495

Sathiya, V., Nagalakshmi, K., Raju, K., & Lavanya, R. (2024). Tracking perishable foods in the supply chain using chain of things technology. *Scientific Reports, 14*(1). https://doi.org/10.1038/s41598-024-72617-3

Schanes, K., Dobernig, K., & Gözet, B. (2018). Food waste matters – A systematic review of household food waste practices and their policy implications. *Journal of Cleaner Production, 182*, 978–991. https://doi.org/10.1016/j.jclepro.2018.02.030

Sebatjane, M. (2024). A sustainable inventory model for a two-echelon cold chain with green technology investments and stock-dependent demand under carbon emissions tax regulation. *Cleaner Logistics and Supply Chain, 13*, 100173. https://doi.org/10.1016/j.clscn.2024.100173

Seifert, R. W., & Markoff, R. (2024). *Tesco and Ocado: Contrasting online grocery supply chain models*. IMD Research & Knowledge Case Study.

Sergi, I., Montanaro, T., Benvenuto, F. L., & Patrono, L. (2021). A smart and secure logistics system based on IoT and cloud technologies. *Sensors, 21*(6), 2231. https://doi.org/10.3390/s21062231

Sharma, H. B., Vanapalli, K. R., Cheela, V. S., Ranjan, V. P., Jaglan, A. K., Dubey, B., Goel, S., & Bhattacharya, J. (2020). Challenges, opportunities, and innovations for effective solid waste management during and post COVID-19 pandemic. *Resources, Conservation and Recycling, 162*, 105052. https://doi.org/10.1016/j.resconrec.2020.105052

Shi, J., Zhang, J., & Qu, X. (2010). Optimizing distribution strategy for perishable foods using RFiD and sensor technologies. *Journal of Business & Industrial Marketing, 25*(8), 596–606. https://doi.org/10.1108/08858621011088338

Silvia La Face. (2024, December 13). *What to expect in 2025–5 upcoming cold chain trends*. Maersk Growth Insights. https://www.maersk.com/insights/growth/2024/12/13/what-to-expect-in-2025-upcoming-cold-chain-trends

Singh, J. (2024). *Pharmaceutical cold chain logistics in the age of artificial intelligence*. Pharmaceutical Commerce. https://www.pharmaceuticalcommerce.com/view/pharmaceutical-cold-chain-logistics-artificial-intelligence

Singh, R., & Shabani, A. (2016). The identification of key success factors in sustainable cold chain management: Insights from the Indian food industry. *Journal of Operations and Supply Chain Management, 9*(2), 1–16. https://doi.org/10.12660/joscmv9n2p1-16

Sow, C., Sanou, C., Medah, C., Schlumberger, M., Mireux, F., Ouédraogo, I., Ouédraogo, S. M., & Betsem, E. (2018). Challenges of cold chain quality for routine EPI in south-west Burkina-Faso: An assessment using automated temperature recording devices. *Vaccine, 36*(26), 3747–3755. https://doi.org/10.1016/j.vaccine.2018.05.062

Soysal, M., Bloemhof-Ruwaard, J. M., Haijema, R., & Van Der Vorst, J. G. A. J. (2015). Modeling an Inventory Routing Problem for perishable products with environmental considerations and demand uncertainty. *International Journal of Production Economics, 164*, 118–133. https://doi.org/10.1016/j.ijpe.2015.03.008

Tang, J., Zou, Y., Xie, R., Tu, B., & Liu, G. (2021). Compact supervisory system for cold chain logistics. *Food Control, 126*, 108025. https://doi.org/10.1016/j.foodcont.2021.108025

Tang, Y. M., Chau, K. Y., Kuo, W. T., & Liu, X. X. (2024). IoT-Based Information System on Cold-Chain Logistics Service Quality (ICCLSQ) Management in Logistics 4.0. *Information Systems Frontiers, 26*(2), 689–708. https://doi.org/10.1007/s10796-023-10393-7

Taylor, D. H., & Fearne, A. (2006). Towards a framework for improvement in the management of demand in agri-food supply chains. *Supply Chain Management: An International Journal, 11*(5), 379–384. https://doi.org/10.1108/13598540610682381

Tsang, Y. P., Choy, K. L., Wu, C. H., Ho, G. T. S., Lam, C. H. Y., & Koo, P. S. (2018). An Internet of Things (IoT)-based risk monitoring system for managing cold supply chain risks. *Industrial Management and Data Systems, 118*(7), 1432–1462. https://doi.org/10.1108/IMDS-09-2017-0384

Villalobos, J. R., Soto-Silva, W. E., González-Araya, M. C., & González-Ramirez, R. G. (2019). Research directions in technology development to support real-time decisions of fresh produce logistics: A review and research agenda. *Computers and Electronics in Agriculture, 167*, 105092. https://doi.org/10.1016/j.compag.2019.105092

Violi, A., Laganá, D., & Paradiso, R. (2020). The inventory routing problem under uncertainty with perishable products: An application in the agri-food supply chain. *Soft Computing, 24*(18), 13725–13740. https://doi.org/10.1007/s00500-019-04497-z

Vrat, P., Gupta, R., Bhatnagar, A., Pathak, D. K., & Fulzele, V. (2018). Literature review analytics (LRA) on sustainable cold-chain for perishable food products: Research trends and future directions. *Opsearch, 55*(3–4), 601–627. https://doi.org/10.1007/s12597-018-0338-9

Wang, J., Wang, H., He, J., Li, L., Shen, M., Tan, X., Min, H., & Zheng, L. (2015). Wireless sensor network for real-time perishable food supply chain management. *Computers and Electronics in Agriculture, 110*, 196–207. https://doi.org/10.1016/j.compag.2014.11.009

Wang, X., Li, L., Moga, L. M., Zhang, X., & Zhang, Y. (2018). Development and evaluation on a wireless multi-sensors system for fresh-cut branches of the North American holly cold chain. *Computers and Electronics in Agriculture, 148*(March), 132–141. https://doi.org/10.1016/j.compag.2018.03.011

Wang, M., Cheng, Q., Huang, J., & Cheng, G. (2021a). Research on optimal hub location of agricultural product transportation network based on hierarchical hub-and-spoke network model. *Physica A: Statistical Mechanics and its Applications, 566*, 125412. https://doi.org/10.1016/j.physa.2020.125412

Wang, M., Wang, Y., Liu, W., Ma, Y., Xiang, L., Yang, Y., & Li, X. (2021b). How to achieve a win–win scenario between cost and customer satisfaction for cold chain logistics? *Physica A: Statistical Mechanics and its Applications, 566*, 125637. https://doi.org/10.1016/j.physa.2020.125637

Woyke, E. (2018). *How UPS uses AI to deliver holiday gifts in the worst storms*. MIT Technology Review.

Xiao, X. (2021). Improved traceability process for frozen tilapia waste elimination in cold chain. *Cleaner Engineering and Technology, 4*, 100148. https://doi.org/10.1016/j.clet.2021.100148

Yan, R. (2017). Optimization approach for increasing revenue of perishable product supply chain with the Internet of Things. *Industrial Management & Data Systems, 117*(4), 729–741. https://doi.org/10.1108/IMDS-07-2016-0297

Yavari, M., & Zaker, H. (2020). Designing a resilient-green closed loop supply chain network for perishable products by considering disruption in both supply chain and power networks. *Computers & Chemical Engineering, 134*, 106680. https://doi.org/10.1016/j.compchemeng.2019.106680

Zhang, J., Lu, L., Song, M. W., Moga, L. M., & Shuan, Z. X. (2009). Development of temperature-managed traceability system for frozen and chilled food during storage and transportation. *Journal of Food, Agriculture & Environment, 7*(3/4), 28–31.

Zhang, H., Qiu, B., & Zhang, K. (2017a). A new risk assessment model for agricultural products cold chain logistics. *Industrial Management & Data Systems, 117*(9), 1800–1816. https://doi.org/10.1108/IMDS-03-2016-0098

Zhang, Y., Zhao, L., & Qian, C. (2017b). Modeling of an IoT-enabled supply chain for perishable food with two-echelon supply hubs. *Industrial Management and Data Systems, 117*(9), 1890–1905. https://doi.org/10.1108/IMDS-10-2016-0456

Zhang, S., Chen, N., She, N., & Li, K. (2021). Location optimization of a competitive distribution center for urban cold chain logistics in terms of low-carbon emissions. *Computers & Industrial Engineering, 154*, 107120. https://doi.org/10.1016/j.cie.2021.107120

Zhang, L., Zhang, M., Mujumdar, A. S., Wu, C., & Wang, D. (2024). Advanced model predictive control strategies for nondestructive monitoring quality of fruit and vegetables during supply chain processes. *Computers and Electronics in Agriculture, 225*, 109262. https://doi.org/10.1016/j.compag.2024.109262

Chapter 9
An Analysis of Critical Success Factors for Digital Transformation in Agri-food Supply Chains

Manish B. R. Veeramani and Marcelo Martins de Sá

Abstract This chapter identifies and evaluates critical success factors (CSFs) for digital transformation in agri-food supply chains, focusing on the Brazilian and Indian craft beer industry. Expert insights were collected using the Fuzzy DEMATEL method to assess the causal relationships among 14 CSFs related to sustainability, resilience, and technological adoption. Key findings indicate that digital logistics infrastructure, e-governance, trust and collaboration, and social sustainability are essential for achieving resilient supply chain performance. In addition, the chapter contrasts the Brazilian and Indian craft beer industries, outlining similarities and differences in raw material sourcing, production, and supply chain structures. These comparative insights provide contextual depth for emerging market strategies. The findings offer practical recommendations for improving traceability, reducing food waste, and enabling effective digital adoption, particularly for small and medium-sized enterprises (SMEs). Stakeholders can enhance supply chain adaptability and sustainability by prioritizing strategic enablers such as smart farming and AI-driven analytics.

Keywords Digital technology · Critical success factors · Fuzzy DEMATEL · Emerging economies · Craft brewery industry

---

M. B. R. Veeramani · M. M. de Sá (✉)
Newcastle Business School, Northumbria University, Newcastle upon Tyne, UK
e-mail: marcelo.sa@northumbria.ac.uk

Y. Khojasteh, S.M. M. Jahan Abadi (eds.), *Digital Transformation of Supply Chains*, International Series in Operations Research & Management Science 372, https://doi.org/10.1007/978-3-032-01206-7_9

9.1 Introduction

Agri-food supply chains (AFSCs) are critical to feeding billions of people and underpinning national economies (FAO, 2023a). Historically, AFSCs have been a fundamental pillar of human civilization, integrating activities such as agricultural production, manufacturing, distribution, and the interconnected flow of information, financial transactions, and contractual agreements (Kumar & Singh, 2022; Kumar & Agrawal, 2023). As global food demand rises, these networks play a central role in ensuring food security, addressing malnutrition, and improving equitable food distribution (FAO, 2023a; Godfray et al., 2018).

However, traditional supply networks are facing increasing pressures from climate change, leading to declining agricultural yields, expansion into ecologically vulnerable areas, degradation of staple crop nutrition, and reductions in biodiversity and livestock productivity (IPCC, 2022). Rising temperatures and extreme weather events have further disrupted agricultural output, while shifts in precipitation patterns have intensified soil degradation and water scarcity (FAO, 2021). Additionally, elevated atmospheric CO_2 levels have been linked to nutrient depletion in staple crops, such as wheat and rice, reducing their protein, zinc, and iron concentrations (Myers et al., 2014). The continued expansion of agricultural production into preserved natural areas has accelerated biodiversity loss and ecosystem degradation, placing further strain on global food systems (Springmann et al., 2018).

These increase the uncertainties that affect the coordination of quality and resilience throughout the downstream supply network to avoid food waste (de Moraes et al., 2020; Costa et al., 2022) or the upstream supply network to mitigate risks and losses (de Sá et al., 2020). For example, the perishable nature of agri-food products results in significant losses throughout the supply network, highlighting the urgent need for improved food waste management and supply chain resilience (Costa et al., 2021). Strengthening these supply networks is also essential for achieving the United Nations Sustainable Development Goals (SDGs), particularly Goal 2 (Zero Hunger) and Goal 12 (Responsible Consumption and Production), which emphasize sustainable agricultural practices, food accessibility, and waste reduction (United Nations, 2023). Advances in digital technologies and policy-driven interventions offer promising pathways to enhance sustainability, mitigate climate risks, and ensure long-term food security (FAO, 2022).

The broader operational landscape further complicates these challenges. Global supply networks face escalating disruptions, ranging from geopolitical conflicts and rising energy costs to environmental crises and pandemics. Their global, resource-intensive nature makes them highly vulnerable to systemic shocks, such as the COVID-19 pandemic, severely affecting food security, economic stability, and vulnerable populations. In the post-pandemic era, concerns over food safety, security, and quality have grown, prompting stakeholders to adopt more transparent, traceable, and circular economy-focused approaches (Kersten et al., 2024).

The COVID-19 pandemic revealed major vulnerabilities in supply networks, as facility shutdowns and supply chain disruptions led to material shortages, labor constraints, and operational bottlenecks (Choi, 2021; Ivanov, 2022). These disruptions underscored the need for resilient supply systems that can withstand, adapt to, and recover from crises (OECD, 2020b; Altay et al., 2018). However, resilience strategies must also be sustainable, as disruptions intensify resource consumption and waste (Sarkis et al., 2020; Ivanov & Dolgui, 2020). In this context, digital technologies are essential for enhancing AFSC resilience, efficiency, and security, particularly in mitigating food safety risks that became more pronounced during the pandemic.

To effectively align sustainability with resilience, AFSCs (historically designed for economic efficiency) must be reassessed. Strengthening production capabilities, optimizing sourcing strategies, and improving distribution networks are essential for creating agile, cost-effective, and crisis-resistant supply systems (Joshi & Sharma, 2022). Integrating digital technologies enables real-time connectivity, advanced analytics, and improved decision-making, leading to greater economic returns, reduced environmental impact, and better working conditions.

For SMEs, emerging digital technologies offer significant opportunities to enhance competitiveness by streamlining operations, integrating supply chains, and scaling production efficiently (Queiroz et al., 2020). However, the growing complexity of global trade, heightened pressure for supply chain transparency, and stricter sustainability regulations are reshaping the agri-food landscape. To remain competitive, SMEs must adopt digitally enabled business models that enhance traceability, compliance, and decision-making in response to environmental changes and regulatory pressures (UNCTAD, 2020). Technologies like Internet of Things (IoT), Artificial Intelligence (AI)-driven analytics, and blockchain improve supply chain visibility and enable SMEs to optimize logistics, reduce costs, and enhance risk management (Queiroz & Wamba, 2024). Successful implementation of these technologies requires strategic planning and tailored digital transformation strategies to ensure operational efficiency and long-term resilience in agri-food supply networks.

This chapter explores the Brazilian craft beer supply chain's CSFs for achieving digital transformation. This rapidly evolving system exemplifies how global supply networks embrace strategic, operational, and sustainable innovations to adapt to shifting market demands, consumer preferences, and regulatory frameworks. Beer is not only the most widely consumed alcoholic beverage worldwide but also holds deep cultural significance in Brazil. According to the Brazilian Beer Association (CervBrasil), beer is as integral to Brazilian culture as Carnival, reflecting its economic and social prominence. In 2023, the Brazilian beer industry continued to expand, with 1847 registered breweries, marking a 6.8% growth compared to the previous year. As one of the world's largest beer consumers, Brazil's brewing sector plays a significant role in the national economy, contributing to GDP growth and job creation, particularly in the rapidly expanding craft beer segment (AGENCIAGOV, 2024). As the third-largest beer consumer globally, Brazil increasingly integrates digital technologies to enhance efficiency, transparency, and competitiveness in its beer supply network. A comprehensive understanding of Brazil's "Cerveja" culture (beer in Brazilian Portuguese) necessitates an examination of the distinct segments within its supply network, from raw material sourcing to production, distribution, and retailing (see Fig. 9.1).

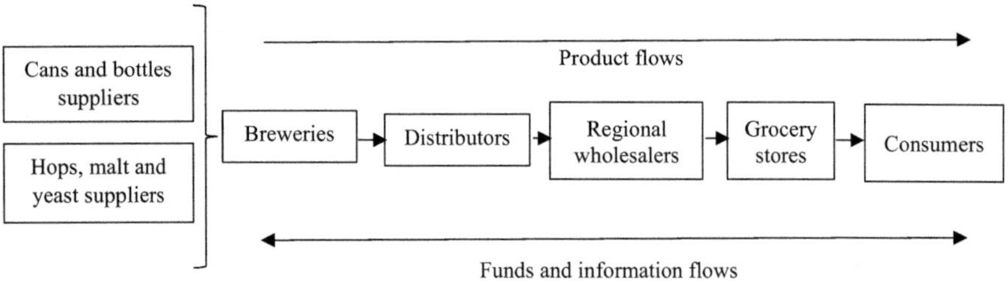

Fig. 9.1 Simplified beer AFSC

For the Brazilian beer AFSC's success, post-consumption factors and pre-production activities must be well-determined to ensure beer product quality in Brazil (Violino et al., 2020). Morado (2009) states that quality-conscious production and distribution are vital to Brazil's beer enthusiasts. The study "To Love Beer Above All Things: An Analysis of Brazilian Craft Beer Subculture of Consumption" (Koch & Sauerbronn, 2019) discusses the rising popularity of craft beers, highlighting that this subculture, driven by the motto "drink less, drink better," is poised to capture a significant future market share. This trend is further supported by data from the Brazilian Ministry of Agriculture, Livestock, and Supply (MAPA), which reports a sharp rise in craft beer microbreweries, increasing from 33 in 1999 to 1549 in 2022, reflecting the sector's rapid expansion (MAPA, 2023).

The growth of microbreweries could drive digital transformation among these entities to achieve faster, more flexible, and cost-effective AFSC activities, such as sourcing unique raw materials. For large-scale Brazilian beer producers, adopting Industry 4.0 technologies and digital supply networks can enhance operational efficiency, foster better supplier collaboration, and drive revenue growth. Innovations such as the Internet of Beer (IoB) (e.g., including cost-effective sensors and cloud-based ERP systems) are increasingly used at every stage of the beer supply chain, from sourcing and brewing to distribution and retailing (Violino et al., 2020).

Adopting digital technologies shifts AFSC management from traditional supply chain planning to demand-driven planning based on analytical demand forecasting, enhancing operational effectiveness. One of the primary objectives of a digital supply network is to enable accurate demand forecasting, leading to higher operational efficiency and enhanced resilience. By implementing demand-driven operations planning, digital technologies offer real-time, end-to-end supply chain transparency and visibility. This transparency empowers decision-makers with greater confidence in managing real-time disruptions related to regulations, market shifts, and logistical constraints. Furthermore, adopting digital environments can collectively reduce transaction costs, enhancing overall supply chain agility.

However, several factors hamper digital transformation in the AFSC:

1. Resistance to process changes required for new digital technologies.
2. Lack of adequate data capturing, analysis, and management frameworks to align with ESG commitments.

3. The impact of digital transformation on long-term business growth and employee morale, particularly without proper change management strategies.

Understanding these factors is essential for addressing the barriers to digital transformation. Effective communication, environmental and social governance (ESG) alignment, and real-time management of AFSC processes are critical for improving environmental sustainability and reducing food wastage and carbon emissions. The recent COVID-19 outbreak has highlighted flaws in existing AFSC management practices and underscored the need for research in AFSC resilience, sustainability, and digital transformation for building resilient and sustainable AFSCs.

This chapter begins with a comprehensive literature review covering critical aspects of AFSCs, the Brazilian craft beer industry, and its digital transformation strategies. It explores similarities and differences between Indian AFSCs previously informed by the literature (Joshi & Sharma, 2022), assesses the impact of COVID-19 on supply network recovery and resilience, and analyzes the drivers and barriers influencing digital technology adoption. The research framework and methodology section details using the Fuzzy Delphi method to identify CSFs. The chapter concludes by presenting the results and offering insights for academics and practitioners.

9.2 Literature Review

The Web of Science and Scopus databases were used for the literature review. With a timeframe of 2015–2022, the search terms "digital technologies" or "digital technologies adoption" and "agri-food supply chain" were used. The search yielded 343 results. After removing duplicates, initial exclusion criteria such as record type (conference papers, book chapters) and ease of access were used. This study includes peer-reviewed academic papers from leading operations and supply chain management journals selected based on their theoretical depth, empirical rigor, and contribution to the field. Emphasis was placed on research published in recognized, high-impact journals, such as the Journal of Operations Management and the Journal of Supply Chain Management, as well as recent studies offering novel insights into digital transformation and supply network resilience. NVivo was used to analyse the articles.

After reviewing the abstract and full text, the articles were excluded for relevance to the current study. Articles were included based on inclusion criteria such as use in the original study and relevance to research objectives, and the Brazilian and Indian beer AFSC context to better illustrate the context of emerging markets. Additional papers were included during the revision stage of this chapter in 2025. Finally, using NVivo coding, articles were grouped based on prevalent themes, which identified past studies and research gaps in digital transformation adoption. The literature review framework followed the steps below (Fig. 9.2).

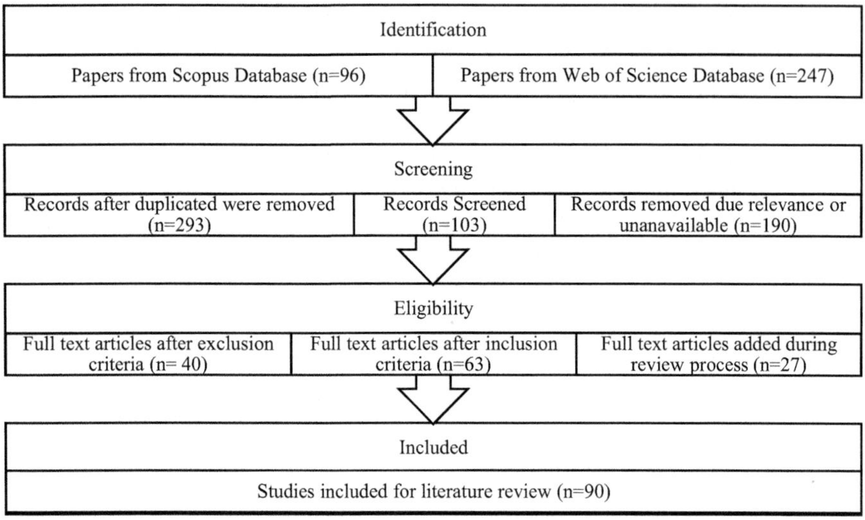

Fig. 9.2 Literature review framework

9.2.1 Agri-food Supply Chain and Its Characteristics

The agri-food supply networks differ from a company's other manufacturing or service supply chain structure, focusing on core activities such as supplying raw materials needed to produce finished products. Each actor in the AFSC adds differentiated value to the final product delivered to consumers. For example, farmers in a modern AFSC produce raw materials, which are processed into finished products and distributed to regional distributors. These distributors store or, in some cases, use cross-docking facilities, especially for perishable goods, to bypass storage and expedite the supply chain. The finished products are then delivered to retailers for consumer purchase. Key stakeholders in the AFSC include farmers, suppliers, distributors, consumers, NGOs, regulatory bodies, and other relevant parties (Yadav et al., 2022).

The demand for agricultural supply is steadily rising, so a resilient AFSC is expected to provide consumers with various options involving both fresh and processed food (Zhu et al., 2018). However, it has been argued that different stakeholders in the AFSC face several challenges, making the process more complicated. Farmers, for example, face a variety of challenges, including a lack of resources and difficulty being profitable. Furthermore, uncertain climate change affects farmers in a variety of ways. For example, adverse weather conditions resulted in inconsistent crop yields. Aside from these, intermediaries must deal with product perishability, supply and demand balancing, and warehouse management. Also, AFSC is plagued by a lack of real-time or near real-time data and ineffective resource management.

Ultimately, these problems increase the waste of products that could have fed many people. Thus, because of all these issues/challenges faced in the AFSC, it is crucial to transform the traditional AFSC into a resilient and sustainable AFSC (Yadav et al., 2022; Gardas et al., 2019).

9.2.1.1 The Brazilian Agri-food Supply Chain and the Digital Transformation

Agribusiness remains a cornerstone of the Brazilian economy, contributing 24.2 percent of the country's GDP in 2022 (CNA and CEPEA, 2025; Statista, 2024). Beyond primary agricultural production, the sector encompasses agro-industry, agricultural services, and input production, forming a highly integrated and dynamic supply network. Over the last three decades, technological advancements, mechanization, and sustainable farming practices have significantly boosted agricultural productivity. Between 1975 and 2016, Brazil's multifactor agricultural productivity more than tripled, with an annual growth rate of approximately 3.1 percent (OECD-FAO, 2023). This growth trajectory is expected to continue, with the Brazilian Ministry of Agriculture projecting that grain production will rise from 250.9 million tons in 2020 to 370 million tons by 2032, representing a 47.4 percent increase (MAPA, 2023).

Given this rapid expansion, enhancing supply chain efficiency and resilience is essential for ensuring long-term profitability and sustainability in an era of increasing climate, geopolitical, and economic disruptions. Adopting digital technologies, including blockchain, IoT, and AI-driven analytics, has played a transformative role in streamlining logistics, optimizing inventory management, and reducing food waste and operational costs (FAO, 2023b).

Digital start-ups are disrupting established markets in Brazil, driven by the government's "Going Digital in Brazil" policy (OECD, 2020a), which promotes an integrated approach to digital transformation. This strategy emphasizes key enablers of digital growth and the shift towards a data-driven economy, with potential in the AFSC context. However, Brazil's complex business environment and burdensome tax regulations create significant challenges, limiting profitability and innovation. To address these issues, the government introduced the "E-Digital" policy (CGEE-MCTI, 2022) to harmonize and coordinate governmental initiatives on digital issues to advance sustainable development. This strategy, updated for the 2022–2026 cycle, outlines a comprehensive approach to leveraging digital technologies for economic and social development.

9.2.2 Similarities and Differences

Regarding the current competitive growth in the craft beer industry, the supply chain processes in Brazil and India include similarities and differences in raw material sourcing, production, distribution processes, and other relevant activities. As a result, the following are a few key similarities and differences discussed in the current supply chain process of the beer AFSC in these two countries.

9.2.2.1 Similarities Between Brazilian and Indian Craft Beer Industry Contexts

According to a market analysis by the Brazilian Ministry of Agriculture, Livestock, and Supply (MAPA), Brazil remains highly dependent on imports to supply key ingredients for beer production despite its strong agricultural capacity and extensive farmland (MAPA, 2024). The country primarily imports wheat, hops, and malt, as domestic production of these ingredients remains limited or underdeveloped. Hop production in Brazil is still in its early stages, requiring the industry to rely heavily on imports from countries such as Argentina, the United States, and Germany (MAPA, 2022). Similarly, insufficient domestic wheat and barley production continues to challenge meeting demand in the local beer sector, leading to price fluctuations and supply chain vulnerabilities (EMBRAPA, 2015, 2023).

The steady growth of beer consumption in Brazil directly correlates with the increasing demand for imported raw materials and advanced brewing equipment. As Brazil's beer industry expands, investments in modern brewing technology, processing machinery, and ingredient supply chains have become critical to sustaining production efficiency and meeting consumer preferences (Euromonitor International, 2024, 2025a, 2025b; Euromonitor International and MAPA, 2025).

Dias and Falconi (2018) highlight that in Brazilian beer supply networks, packaging, and distribution processes are frequently outsourced to third-party collaborators, a strategy widely adopted by domestic and foreign beer producers. This outsourcing model allows international brewing companies to partner with local firms, leveraging regional expertise and supply chain efficiencies. As a result, foreign beer producers operating in Brazil are more likely to collaborate with local brewers to manufacture beer while relying on independent packaging companies to handle bottling and distribution. Overseas brewing companies establish long-term partnerships with regional players, enabling them to reduce sourcing, production, packaging, and distribution costs in both Brazil and export markets (Dias & Falconi, 2018).

A comparable collaboration strategy has been observed in India's beer industry (Vashishth & Tripathi, 2016). Foreign beer producers entering the Indian market commonly have contractual agreements with local brewers. This strategy helps lower operational costs, including raw material sourcing, skilled labor acquisition, and production and distribution logistics within India and other foreign markets. Furthermore, despite India's significant domestic production of key raw materials such as wheat and barley, the country's beer industry continues to import essential ingredients like hops and yeast to meet the rising consumption demands and ensure the quality and consistency of beer production.

In this vein, the craft beer industries in Brazil and India have experienced significant growth, driven by shifting consumer preferences, increased investment in

Table 9.1 Key trends in craft beer supply networks in Brazil and India

Aspect	Brazil	India
Key imported raw materials	Hops, malt, yeast, wheat and barley	Hops, malt, yeast, wheat and barley
Main raw material supplier countries	*Hops:* Argentina, Germany, United States *Malt & Barley:* Canada, Belgium, Germany *Wheat:* Argentina, Canada *Yeast:* Europe and North America	*Hops:* Germany, United States, New Zealand *Malt & Barley:* Belgium, Germany, UK, Australia, France *Wheat:* Australia, France *Yeast:* Europe and North America
Dependency on imports High-quality imports dependence	Domestic hop production is still in the early stages, leading to high import reliance Although some *malt* production exists locally, *high-quality* malt for craft beer is often imported	Domestic hop cultivation is limited to regions like Himachal Pradesh but not sufficient for large-scale craft brewing Despite India producing a significant amount of *barley*, *high-quality* malting barley is still imported
Craft-beer industry growth and supply implications	Expanding market: the third-largest beer consumer globally, with a growing craft beer segment driven by consumer preference for premium and artisanal products	Experiencing double-digit growth, driven by urbanization, rising disposable incomes, and an evolving drinking culture
Production and supply network management	Foreign breweries *often collaborate* with local brewers to reduce costs in sourcing, production, and packaging	Foreign breweries *enter contractual agreements* with Indian brewers to lower operational costs
Logistics, packaging and warehousing	Frequently outsourced to third-party collaborators in supply networks	Frequently outsourced to third-party collaborators in supply networks

Sources: adapted from (EMBRAPA, 2015, 2023; Euromonitor International, 2024, 2025a, 2025b; Euromonitor International and MAPA, 2025; Dias & Falconi, 2018; MAPA, 2022, 2024; Statista, 2024; Vashishth & Tripathi, 2016)

brewing technology, and evolving supply chain strategies. However, both countries remain highly dependent on imported raw materials, particularly hops, malt, yeast, wheat, and barley, essential for high-quality beer production. Table 9.1 provides a comparative overview of key trends in Brazil and India's craft beer supply networks, highlighting their import dependencies, supply chain strategies, and industry growth patterns.

9.2.2.2 Differences Between the Brazilian and Indian Craft Beer Industry Contexts

There are a few significant differences in the supply chain processes of the Indian and Brazilian beer industries. In this context, a significant difference has been recognized in raw material sourcing. According to Knoth and Borrit (2014), brewing companies in India significantly achieve more opportunities due to increased consumption and the availability of locally sourced raw materials. India is one of the world's leading producers of wheat and barley. Nonetheless, the country increased the production of other ingredients such as malt, hops, and yeast. As a result, when it comes to differentiating the supply chain of the beer industry in Brazil from the Indian beer industry, sourcing raw materials can be considered a significant factor (Knoth & Borrit, 2014). Brazil has limited domestic capacity for producing key ingredients essential for beer production, mainly hops, yeast, and wheat. As a result, the country relies heavily on imports to meet the demands of its growing beer industry. Hops are primarily sourced from Argentina, Germany, and the United States, while high-quality wheat and yeast are imported from Canada and Europe (MAPA, 2023; EMBRAPA, 2023). This import dependency increases production costs and supply chain vulnerabilities, making the industry susceptible to global market fluctuations and logistical disruptions.

Foreign breweries in Brazil and India adopt collaborative strategies with local brewers to reduce operational costs and improve supply chain efficiency. However, the nature of these partnerships differs significantly due to variations in market structure, regulatory frameworks, and supply chain constraints. In Brazil, foreign breweries often outsource sourcing, production, and packaging to local brewing partners, leveraging existing supply chain infrastructure to minimize costs and streamline operations. This model allows multinational beer companies to reduce reliance on imported raw materials, benefiting from established local expertise in production and distribution. Additionally, partnerships with local brewers help navigate complex logistics in Brazil's large and geographically diverse market.

Conversely, foreign breweries primarily engage in contractual agreements with local brewers in India, which serve as a cost-effective strategy to handle state-specific alcohol regulations and taxation policies. Unlike in Brazil, where collaboration often extends across multiple stages of the supply network, foreign beer producers in India primarily license production to domestic companies, reducing the need for direct investments in brewing facilities and ingredient sourcing. This approach helps multinational brewers overcome market entry barriers while ensuring greater flexibility in adapting to India's highly fragmented beer market. Despite these differences, both markets benefit from foreign-local collaborations, allowing international brands to expand their presence, optimize distribution networks, and reduce overhead costs while adapting to local supply chain realities.

9.2.3 *Impact of COVID-19 on Agri-food Supply Chains*

Before COVID-19, organizations associated with AFSCs accepted lean and JIT activities to reduce stock and expenses. However, the pandemic exposed vulnerabilities in these approaches, leading to inflation, price volatility, disruptions in traceability, and weakened supply chain resilience.

However, the pandemic exposed vulnerabilities in these approaches, leading to inflation, price volatility, disruptions in traceability, and weakened supply chain resilience. The supply side of AFSCs faced significant challenges in ensuring the accessibility and availability of agricultural products, particularly as lockdowns, labor shortages, and logistical constraints disrupted food production and distribution (Joshi & Sharma, 2022).

Beyond supply disruptions, consumer behavior changed dramatically. The pandemic reduced workforce availability, increased physical distancing requirements, and shifted toward contactless product deliveries (Ciruela-Lorenzo et al., 2020). Additionally, home consumption preferences intensified, prompting AFSCs to restructure operations to meet these evolving market demands. As production and consumption patterns fluctuated, enhanced coordination and collaboration among AFSC stakeholders became crucial for developing resilient and adaptive distribution channels to sustain food security and accessibility.

Additionally, the pandemic underscored the urgent need for digital transformation in AFSCs to address challenges related to traceability, transparency, and efficiency. However, digital inequalities in emerging markets present a barrier to adoption, particularly for smallholder farmers who lack access to real-time market data and digital infrastructure. Digital transformation has the potential to revolutionize AFSCs by creating novel information-sharing platforms, optimizing resource allocation, and enabling more inclusive market participation. Without strategic development efforts, digitalization risks widening the gap between technologically advanced stakeholders and those with limited access to digital tools (Medina & Thomé, 2021).

According to Bolfe et al. (2020), social sustainability is one of the most influential aspects of AFSCs, and the pandemic amplified socio-economic disparities within supply networks. Labor shortages, physical distancing mandates, and travel restrictions weakened workforce security, exacerbating AFSC vulnerabilities and increasing the disconnect among supply chain partners (Bolfe et al., 2020; Romani et al., 2020).

Furthermore, growing price elasticity and heightened consumer anxiety led to shifts in purchasing behavior, necessitating continuous monitoring and adaptation of AFSC processes to improve responsiveness and supply chain efficiency.

Transparency is another crucial aspect of AFSC. The pandemic heightened food safety concerns, prompting organizations to implement stricter food processing controls to prevent contamination and virus transmission (Menon & Jain, 2024). Blockchain technology, AI-driven analytics, and digital ledgers have become valuable tools for enhancing transparency, reducing transaction costs, and ensuring

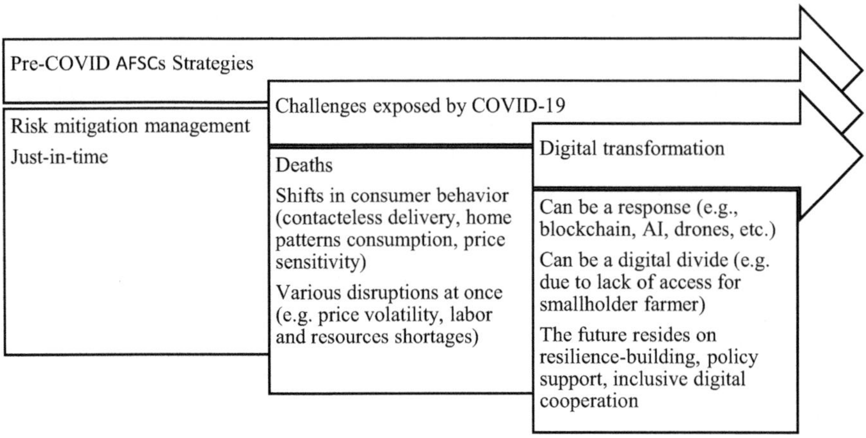

Fig. 9.3 The impact of COVID-19 and digital transformation on agri-food supply chains

reliable traceability across AFSCs. These innovations support regulatory compliance and build consumer trust in food safety measures.

Another point to consider is that COVID-19 disproportionately impacted smallholder farmers, particularly in developing countries, where agricultural sectors rely on traditional supply networks with limited technological integration (Rama & Wilkinson, 2019). Restricted movement and physical distancing mandates disrupted market access, weakening economic activity in rural areas and threatening livelihoods. These disruptions have raised concerns about AFSC resilience, prompting discussions on the long-term sustainability of conventional agri-food networks.

Integrating advanced digital technologies has been proposed as a transformative solution for AFSCs. Technologies such as wireless sensor networks, AI, geospatial analytics, remote sensing, and drones can enhance supply chain visibility, predictive analytics, and real-time monitoring of agricultural production (Rama & Wilkinson, 2019). However, successful digital transformation requires targeted investments, policy support, and infrastructure development to ensure that all supply chain stakeholders, especially small-scale farmers, benefit from technological advancements.

Figure 9.3 illustrates the transformational impact of COVID-19 on AFSC, highlighting pre-pandemic strategies, key challenges, and digital responses. It outlines how the crisis exposed vulnerabilities in traditional models, driving the adoption of digital technologies to improve resilience, traceability, and efficiency. Additionally, the diagram addresses barriers to digitalization, particularly in emerging markets, and explores future strategies for building a more adaptive and inclusive AFSC framework.

9.2.4 Drivers and Barriers to Adoption of Digital Technology

The digital transformation of AFSC involves the integration of multiple existing and advanced technologies to optimize and modernize traditional supply chain processes. Digitalization has reshaped business operations by enhancing supply network efficiency, visibility, and automation. Technologies such as the IoT, blockchain, AI, and big data analytics have enabled improved traceability, real-time monitoring, and predictive analytics, fostering more sustainable and resilient supply chains (Cricelli et al., 2024). However, despite digital transformation's advantages, its adoption in AFSCs is often hindered by financial, technical, and organizational barriers that require strategic interventions.

One of the key drivers of digital transformation in AFSC is the growing demand for supply chain transparency and traceability. Consumers increasingly seek information on food safety, sustainability, and ethical sourcing, pushing firms to adopt technologies that enhance visibility and compliance with regulatory standards (Belhadi et al., 2024). Blockchain and IoT applications, for instance, facilitate real-time tracking of food products, reducing fraud and contamination risks (Ali et al., 2023).

Other critical drivers are operational efficiency and risk mitigation. Digital tools help businesses streamline inventory management, optimize logistics, and automate administrative tasks, reducing human errors and improving supply chain resilience (Dadsena et al., 2024). AI-powered analytics allow firms to predict demand fluctuations, mitigate disruptions, and enhance decision-making processes. The push towards sustainable agriculture and resource optimization also plays a vital role in digital adoption, as technologies like precision farming and smart irrigation reduce food waste and improve productivity (Panetto et al., 2020).

Government initiatives and industry regulations further accelerate digitalization efforts. Policies supporting Industry 4.0 adoption (i.e., E-digital policies), subsidies for agri-tech innovations, and compliance requirements for food traceability are pushing agri-food businesses toward digital transformation (Dixit et al., 2024). Additionally, disruptions such as the COVID-19 pandemic and geopolitical tensions have highlighted the need for resilient supply chains, prompting companies to invest in digital solutions to manage uncertainties (Gupta et al., 2022).

Despite these drivers, several barriers hinder the implementation of digital technologies in AFSC. One of the most significant obstacles is the high financial investment required for digital transformation. Small and medium-sized enterprises (SMEs), which dominate the agri-food sector, often lack the capital to invest in expensive infrastructure, software, and employee training (Dadsena et al., 2024). The uncertain return on investment (ROI) further discourages businesses from adopting digital technologies (Dixit et al., 2024).

Many firms struggle with a shortage of skilled personnel capable of managing and integrating advanced digital tools. The complexity of technologies such as blockchain, AI, and IoT requires specialized knowledge, which is often lacking among traditional supply chain professionals (Agrawal et al., 2019). Without proper training programs, businesses face difficulties in adapting to digital workflows (Holmström et al., 2019).

Organizational inertia and resistance to change pose another major barrier to digital transformation. Many stakeholders, including farmers, suppliers, and logistics partners, are hesitant to adopt digital solutions due to concerns about disruptions to existing workflows and job displacement (Belhadi et al., 2024). A lack of strategic leadership and commitment further delays digital initiatives (Dixit et al., 2024).

Poor digital infrastructure, particularly in rural and developing regions, limits the effectiveness of digital solutions. Many agricultural areas suffer from unstable internet connectivity and inadequate IT infrastructure, making real-time data exchange and automation challenging (Gupta et al., 2022). Without reliable network systems, the benefits of digitalization cannot be fully realized.

As digital supply chains rely on extensive data sharing, cybersecurity risks such as hacking, data breaches, and unauthorized access pose critical threats. Businesses must invest in robust cybersecurity frameworks to protect sensitive supply chain information, but many lack the resources to do so (Ali et al., 2023). Addressing regulatory compliance and data privacy laws further adds complexity to digital adoption (Cricelli et al., 2024).

In sum, digital transformation in AFSCs involves the integration of Industry 4.0 technologies, including blockchain, AI, IoT, and big data analytics, to enhance efficiency, transparency, and resilience (Ali & Govindan, 2023). These advancements improve traceability, real-time monitoring, and predictive analytics, making supply chains more responsive to disruptions and sustainability challenges (Brintrup et al., 2024).

However, despite the numerous benefits, organizations face significant barriers to implementation, including high costs, lack of digital expertise, and cybersecurity risks (Cichosz et al., 2020). While technological advancements drive efficiency, transparency, and risk management improvements, barriers such as financial constraints, workforce skill gaps, and cybersecurity threats must be proactively managed.

Companies that develop strategic digital roadmaps, invest in workforce training, and implement robust cybersecurity frameworks will be better positioned to maximize the benefits of digitalization in the agri-food supply chain (Ali & Govindan, 2023). By addressing these challenges through targeted investments, education, and policy alignment, AFSCs can achieve greater resilience, sustainability, and competitive advantage. Table 9.2 provides a summary of the key drivers and barriers influencing digital transformation, highlighting the importance of some potential solutions for strategic planning and investment to navigate these challenges effectively.

Table 9.2 Drivers and barriers to digital transformation

Drivers	Barriers	Potential Solutions
Increased supply network traceability	Workforce digital skills gap in AI, blockchain, and cybersecurity	Change management strategies: to promote a culture of digital transformation through incentives and support. Investment in training programs for digital literacy and workforce development.
Increased supply network transparency	Employees' and leaders' resistance	
Easier identification and monitoring of supply disruption risks	Fragmented Digital Infrastructure: Lack of standardization across supply chain partners inhibits interoperability	Standardization and interoperability frameworks: Adoption of global standards for digital platforms can enhance integration while maintaining flexibility
Elimination of repetitive tasks through Industry 4.0 implementation	Lack of time, budget, or resources. Poor digital transformation strategy	

9.2.5 Synthesis of CSF Criteria

This section brings on the CSFs identified by Joshi and Sharma (2022) in the Indian agri-food supply chain digital adoption study. Rather than modifying these CSFs, the approach involved validating them within the context of the Brazilian craft beer industry to ensure their relevance and applicability. The original study employed the Fuzzy Delphi method, allowing an expert panel to confirm the significance of each CSF in driving sustainable and efficient AFSC operations. Similarly, in this study, specialist validation was conducted with professionals from the Brazilian craft beer industry, ensuring the selected CSFs aligned with local supply network dynamics and market conditions.

The validation process involved a panel of specialists with an average of over 15 years of experience in the market, working across different segments of the Brazilian craft beer industry. These professionals were primarily entrepreneurs and functional managers in operations, procurement, logistics, and production management. Their extensive industry expertise allowed a practical assessment of how digital transformation influences sustainability and efficiency in craft beer supply networks. Once validated, the CSFs were analyzed using the DEMATEL (Decision-Making Trial and Evaluation Laboratory) approach to assess their interdependencies and impact on AFSC sustainability and performance.

The validated CSFs provide a structured framework for evaluating sustainability and resilience in AFSCs. These factors address key social, economic, environmental, technological, and governance-related challenges, ensuring a comprehensive approach to digital transformation. The sustainability of AFSCs, including those in the craft beer industry, depends on how effectively these factors are integrated into strategic decision-making, operational efficiency, and technological advancements. Table 9.3 presents a structured synthesis of the validated CSFs, their implications for AFSC sustainability, and academic literature supporting their relevance.

Table 9.3 Critical success factors identified in the literature

Critical success factors		Implications on the sustainability of agri-food supply networks	References
CSF01	Social Sustainability Performance	It focuses on fair resource distribution, food security, and improving community well-being. Enhancing labor conditions, fair wages, and stakeholder engagement strengthens supply chain stability and resilience.	Toussaint et al. (2022), Silvestri et al. (2024)
CSF02	Economic Sustainability Performance	It ensures economic viability by enhancing productivity, reducing costs, and improving access to affordable, safe food. It optimizes resources, minimizes waste, and drives innovation for long-term sustainability and competitiveness.	Joshi and Sharma (2022), Hassoun et al. (2023)
CSF03	Environmental Sustainability Performance	It aims to reduce environmental impacts by cutting greenhouse gas emissions, conserving resources, using renewable energy, and adopting a circular economy and sustainable farming practices to ensure ecological balance and combat climate change.	Sezer et al. (2024)
CSF04	Transparency	It enhances traceability, and information sharing promotes trust and accountability across stakeholders. Transparency reduces errors, ensures compliance, and supports ethical and sustainable practices throughout the supply chain.	Gligor et al. (2022), Yadav et al. (2022)
CSF05	Traceability	It ensures accountability and efficiency by tracking products, ensuring safety and quality compliance. Digital tools like blockchain and IoT enable real-time monitoring, reducing risks and supporting sustainable practices.	Vern et al. (2024), Feng et al. (2020)
CSF06	Technical Training and Skill Upgradation for Stakeholders	It supports adopting sustainable practices and technologies, fosters innovation, improves efficiency, and promotes eco-friendly methods.	Kovács et al. (2024)
CSF07	Digital Logistics and Technology Infrastructure	It enables better inventory management, energy optimization, and real-time monitoring while addressing barriers like high implementation costs and skill gaps.	Fernandez-Vidal et al. (2022), Ali and Phan (2022)
CSF08	E-governance	It supports policy implementation and compliance through digital platforms that enhance transparency, stakeholder engagement, and accountability. E-governance improves decision-making, reduces inefficiencies, and ensures fair resource distribution, promoting sustainability and resilience in the supply chain.	Jagtap et al. (2024)

(continued)

Table 9.3 (continued)

Critical success factors		Implications on the sustainability of agri-food supply networks	References
CSF09	Sustainable Food Security decisions	It focuses on ensuring consistent food availability, accessibility, and quality while minimizing environmental impacts. Decision-making processes supported by data analytics and forecasting tools enhance food distribution efficiency, reduce waste, and address supply-demand imbalances.	Giongo et al. (2025), Awokuse et al. (2024)
CSF10	Trust and Collaboration	It strengthens partnerships to reduce risks, improve decision-making, and encourage adopting sustainable and ethical practices. Collaborative efforts drive innovation, promote resource efficiency, and support long-term resilience.	Barbosa and Cansino (2024), Gajdić et al. (2023), Grohmann et al. (2023)
CSF11	Operational Effectiveness and Scalability	It streamlines processes, optimizes resources, and reduces waste to improve supply chain efficiency. Its scalability allows adaptation to demand changes and expansion while maintaining sustainability.	Bechtsis et al. (2022), Mishra et al. (2025)
CSF12	Standardization of Sustainable Practices	It ensures consistency, efficiency, and accountability in supply chains. It encourages best practices, ensures regulatory compliance, reduces variability, and improves resource management, supporting long-term sustainability and stakeholder collaboration.	Ariadi et al. (2024), Silvestri et al. (2024)
CSF13	Security	It protects data, prevents disruptions, and ensures food safety. Security measures reduce risks like fraud and cyberattacks, enhance resilience, build trust, and support sustainable operations with reliable food access.	El Jaouhari et al. (2024), Bechtsis et al. (2022)
CSF14	Smart Farming	It supports precision agriculture, real-time monitoring, and data-driven decision-making, enhancing productivity and sustainability. These innovations help farmers adapt to climate change, reduce waste, and meet growing food demand sustainably.	Kazancoglu et al. (2024), Yadav et al. (2022), Musa and Basir (2021), Musa et al. (2021)

Social sustainability is fundamental in ensuring fair resource distribution, equitable wages, and improved labor conditions. The panel of specialists emphasized that in the Brazilian craft beer industry, strengthening stakeholder relationships and ensuring fair working conditions directly contributes to long-term supply network resilience (Toussaint et al., 2022; Silvestri et al., 2024). Similarly, economic sustainability is critical for enhancing productivity, optimizing resource utilization, and reducing costs, ensuring that small and medium-sized breweries remain competitive (Joshi & Sharma, 2022; Hassoun et al., 2023).

Environmental sustainability remains a top priority in craft beer production, with strategies to reduce waste, greenhouse gas emissions, and energy consumption. Sustainable brewing techniques, circular economy models, and renewable energy adoption have been identified as critical for long-term ecological balance (Sezer et al., 2024). The validation panel reinforced the importance of wastewater treatment, energy-efficient brewing processes, and responsible ingredient sourcing in enhancing environmental performance.

Transparency emerged as a key driver of trust and accountability, particularly in ensuring quality control, traceability, and ethical business practices. Blockchain and IoT solutions have been instrumental in enhancing product tracking, minimizing fraud risks, and ensuring regulatory compliance (Yadav et al., 2022; Gligor et al., 2022). Complementing this, traceability systems enable real-time beer production and distribution monitoring, reducing supply network risks and ensuring product authenticity (Vern et al., 2024; Feng et al., 2020).

Digital logistics and technology infrastructure were also identified as transformational factors for optimizing inventory management, production efficiency, and supply network coordination. The panel of experts confirmed that the integration of IoT, AI, and blockchain has significantly enhanced decision-making capabilities, improved demand forecasting, and minimized inefficiencies despite facing barriers such as high implementation costs and skill shortages (Ali & Phan, 2022; Fernandez-Vidal et al., 2022).

The study's expert panel emphasized the importance of technical training and skill upgradation to ensure effective digital adoption and innovation in the beer AFSC. Research has shown that brewers and supply network professionals who receive structured training in digital tools and sustainable practices are more successful in adapting to emerging technologies and improving operational efficiency (Kovács et al., 2024; Joshi & Sharma, 2022). Similarly, e-governance is critical in shaping industry policy compliance, taxation regulations, and digital transparency. Digital governance platforms enhance stakeholder engagement, regulatory enforcement, and sustainable resource distribution (Jagtap et al., 2024).

Sustainable food security decisions remain an integral part of supply network resilience. Data analytics and AI-driven forecasting tools help breweries optimize ingredient sourcing, reduce waste, and align production with market demand (Giongo et al., 2025; Awokuse et al., 2024). The ability to make data-driven decisions supports cost-effective production while minimizing environmental impacts.

Trust and collaboration among supply network stakeholders have been identified as key elements in mitigating risks, fostering innovation, and improving market positioning. The craft beer industry thrives on strong partnerships between producers, suppliers, and distributors, which ensures greater resilience to market fluctuations and supply network disruptions (Gajdić et al., 2023; Barbosa & Cansino, 2024; Grohmann et al., 2023).

Operational effectiveness and scalability enable beer supply networks to expand sustainably while maintaining efficiency. The craft beer industry faces unique challenges in scaling operations while maintaining product differentiation, requiring innovative approaches to resource optimization and waste reduction (Bechtsis et al.,

2022; Mishra et al., 2025). Likewise, the standardization of sustainable practices plays a significant role in regulatory compliance, process consistency, and resource efficiency, ensuring alignment with industry-wide sustainability goals (Ariadi et al., 2024; Silvestri et al., 2024).

Security concerns were also raised during expert validation, particularly regarding cybersecurity, data protection, and supply network integrity. The increasing use of digital transactions and online supply network platforms has prioritized security for mitigating fraud risks, cyberattacks, and data breaches (El Jaouhari et al., 2024; Bechtsis et al., 2022).

Finally, innovative farming technologies have emerged as a key enabler of sustainable brewing ingredients sourcing, with AI-driven precision agriculture helping brewers optimize the quality and sustainability of raw materials (Kazancoglu et al., 2024; Yadav et al., 2022; Musa & Basir, 2021; Musa et al., 2021).

The synthesis of validated CSFs highlights their relevance in agri-food supply networks, particularly in the Brazilian beer industry. Experienced craft beer entrepreneurs and operational specialists reinforce the practical significance of these factors in supply network management.

Table 9.3 summarizes each CSF, its sustainability implications, and the supporting literature. This validated framework forms a basis for research and practical application, offering insights into the barriers and enablers of digital transformation in AFSCs. Table 9.3 illustrates the critical areas necessary for sustainable agri-food supply chains, serving as a foundation for future research.

9.3 Research Methodology and Analysis

This section outlines the research design and methodology for identifying CSFsfor digital transformation in the Brazilian craft beer AFSC. It details Fuzzy Decision-Making Trial and Evaluation Laboratory (DEMATEL) method, the data collection process, and the subsequent analysis to derive meaningful insights.

The methodology was designed to ensure a systematic and comprehensive evaluation of CSFs, integrating literature findings, expert opinions, and quantitative analysis to generate actionable results.

9.3.1 Research Framework and Method Selection

The fuzzy DEMATEL method was chosen for this study because it effectively establishes cause-and-effect relationships in complex systems (Chang et al., 2011; Yadav et al., 2022). DEMATEL transforms interdependent relationships among multiple variables into causal-effect matrices, allowing for a structured approach to problem-solving. The study followed a step-by-step approach, including CSF selection, expert validation, fuzzy analysis, and DEMATEL-based prioritization.

Two essential methods considered were Fuzzy DEMATEL and Analytic Network and Process (ANP). The Fuzzy DEMATEL method was selected over ANP due to its effectiveness in evaluating the structure and relationships among alternative components. While both methods facilitate decision-making by prioritizing indicators and incorporating qualitative and quantitative factors, Fuzzy DEMATEL was selected due to its ability to analyze complex interdependencies more effectively. ANP, despite its strengths, presents significant challenges in managing a large number of variables, making it less suitable for this study. The main limitation of Fuzzy DEMATEL is its reliance on expert judgment, which introduces subjectivity (Joshi & Sharma, 2022). To mitigate this, convenience sampling was used to recruit an expert panel, ensuring a diverse representation of industry professionals with practical insights (Joshi & Sharma, 2022).

The selected Fuzzy DEMATEL approach provides a robust cause-effect analysis, identifying which CSFs drive digital transformation and which are more reactive to changes. The research framework is presented in Fig. 9.4.

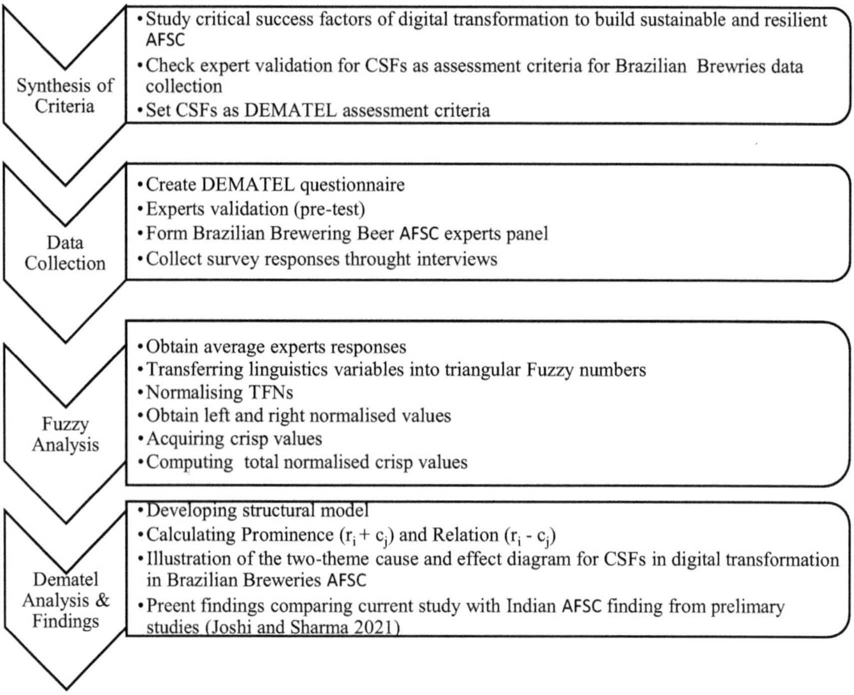

Fig. 9.4 Research methodology design

9.3.2 Data Collection and Survey Design

To ensure the validity and industry relevance of the selected CSFs for digital transformation in the Brazilian craft beer AFSC, this study employed an expert panel with an average of over 10 years of industry experience. The panel consisted of 10 professionals from various sectors of the beer industry, including craft brewery entrepreneurs, supply chain managers, procurement and logistics specialists, packaging and sustainability managers, and hop and wheat farmers.

These experts had extensive backgrounds in supply chain operations, digital transformation, procurement strategies, and environmental sustainability. Many held graduate and postgraduate degrees in Supply Chain Management, Operations Management, Logistics, and Food and Beverage Business and had professional certifications in Lean Six Sigma, Supply Chain Digitalization, and Food Safety Compliance.

The selection of these experts was based on their practical experience in managing real-world challenges related to beer production, distribution, procurement, and technology integration. The diversity of their roles provided a holistic perspective on the barriers and enablers of digital transformation within the Brazilian craft beer AFSC. Their direct involvement in decision-making ensured that the CSFs were validated against actual operational conditions, regulatory constraints, and technological advancements.

The experts were drawn primarily from Brazil, with some possessing international experience in global beer supply chains. Their combined expertise comprehensively validated the study's findings, ensuring that the proposed CSFs were theoretically sound and practically applicable to industry challenges. This expert validation process was conducted through a structured DEMATEL questionnaire, allowing for a systematic assessment of CSF interdependencies and digital transformation impact in the Brazilian craft beer AFSC.

Data were collected from both primary and secondary sources. Secondary data included academic literature, industry reports, and case studies, ensuring the study built upon established research. Primary data were collected via a structured DEMATEL questionnaire survey in 2021 to capture expert insights. The questionnaire consisted of four sections:

1. Instructions—Provide guidelines for respondents by sharing participant information sheets and asking for a consent form to record the interviews.
2. Definitions—Clear descriptions of CSFs and their relevance to digital transformation were sent before the interview process and clarified before starting recording.
3. Linguistic Scale and Measurement—The Fuzzy Linguistic Scale (Table 9.4) for evaluating relationships between CSFs.
4. Expert Background Information—Capturing respondent expertise and industry experience.

Convenience sampling was employed to select experts from various craft beer supply network segments to ensure adequate representation. Their insights were crucial

Table 9.4 Specialists' panel demographics data

Expert	Supply Node	Experience (Years)	Job Role
Expert 1	Craft Beer Micro Brewery	9	Supply And Logistics Manager
Expert 2	Craft Beer Micro Brewery	18	Entrepreneur
Expert 3	Craft Beer Micro Brewery	17	Entrepreneur
Expert 4	Craft Beer Micro Brewery	6	Procurement Manager
Expert 5	Craft Beer Micro Brewery	8	Procurement Manager
Expert 6	Packaging Supplier (SME)	5	Entrepreneur
Expert 7	Logistics provider	12	CEO
Expert 8	Wheat Farmer	35	Farming
Expert 9	Hop Farmer	9	Farming
Expert 10	Hop Farmer	9	Farming, Procurement, Import and Export

Table 9.5 DEMATEL survey measurement scale

Measurement Scale	Terms to Fill in Matrix	Explanation	Triangular Fuzzy Numbers
Very high influence	VH	It represents a very high correlation between the two CSFs.	(0.75, 1.0, 1.0)
High influence	H	It represents a high correlation between the two CSFs and each other.	(0.5, 0.75, 1.0)
Low influence	L	It represents a low correlation between the two CSFs and each other.	(0.25, 0.5, 0.75)
Very low influence	VL	It represents a very low correlation between the two CSFs.	(0, 0.25, 0.5)
No influence	No	It represents no correlation between two CSFs with each other.	(0, 0.25, 0.5)

in establishing the connections between CSFs, offering practical validation of theoretical frameworks.

Data for this study were gathered from both primary and secondary sources. Secondary data encompassed information from relevant journals, literature, and online platforms, with reputable journals being prioritized in this research. Primary data were collected via a DEMATEL questionnaire survey, which was crafted to gather fundamental information about the industry. It consisted of four sections: instructions for completion, definitions, the main questionnaire, and basic respondent information. The definitions section featured the fuzzy linguistic scale (Table 9.5), and a sample of 10 participants was surveyed using the DEMATEL questionnaire (Table 9.4), following CSF definitions and pertinent examples (Table 9.3).

Table 9.6 Steps for processing expert responses in fuzzy DEMATEL analysis

Steps	Description	Equations
Convert Expert Responses into Fuzzy Triangular Numbers	Expert opinions are mapped to fuzzy triangular numbers using a predefined linguistic scale.	
Normalize Fuzzy Triangular Numbers	Convert fuzzy numbers into a standardized range using the following transformations:	$xl_{ij}^{k}=\frac{l_{ij}^{k}-\min l_{ij}^{k}}{\Delta_{\min}^{\max}}$ (9.1) $xm_{ij}^{k}=\frac{m_{ij}^{k}-\min l_{ij}^{k}}{\Delta_{\min}^{\max}}$ (9.2) $xr_{ij}^{k}=\frac{r_{ij}^{k}-\min l_{ij}^{k}}{\Delta_{\min}^{\max}}$ (9.3) Where: $\Delta_{\min}^{\max}=\max r_{ij}^{k}-\min l_{ij}^{k}$ (9.4)
Calculate Left and Right Normalized Values	Obtain refined normalized values using:	$x1s_{ij}^{k}=\frac{xm_{ij}^{k}}{1+xm_{ij}^{k}-xl_{ij}^{k}}$ (9.5) $xrs_{ij}^{k}=\frac{xr_{ij}^{k}}{1+xr_{ij}^{k}-xm_{ij}^{k}}$ (9.6)
Derive Crisp Values	Convert fuzzy values into crisp values using:	$x_{ij}^{k}=\frac{x1s_{ij}^{k}\left(1-x1s_{ij}^{k}\right)+xrs_{ij}^{k}xrs_{ij}^{k}}{1-x1s_{ij}^{k}+xrs_{ij}^{k}}$ (9.7)
Compute Total Normalized Crisp Values	Compute absolute crisp values by transforming back:	$Z_{ij}^{k}=\min l_{ij}^{k}+x_{ij}^{k}\Delta_{\min}^{\max}$ (9.8)
Integrate Crisp Values	Aggregate all crisp values for final computations:	$Z_{ij}=\frac{1}{p}\left(Z_{ij}^{1}+Z_{ij}^{2}+Z_{ij}^{p}\right)$ (9.9)

9.3.3 *Fuzzy Analysis*

The fuzzy DEMATEL method is well-suited for defining structural relationships when decision-makers clearly understand the research problem. This approach combines the fuzzy set concept with the DEMATEL methodology to manage the complexities of segmenting multiple factors (Chang et al., 2011). Using a linguistic scale, decision-makers can express their insights in simple, everyday terms. This method has been widely applied to establish cause-and-effect relationships in models, helping to reduce ambiguity and uncertainty in decision-making.

Defuzzification is the process of transforming fuzzy values into precise, crisp values. The fuzzy analysis in this study was conducted using Eqs. (9.1)–(9.9), as outlined in the base paper by Joshi and Sharma (2022), along with other significant Fuzzy DEMATEL studies (Erkan et al., 2021; Altuntas & Yilmaz, 2016). The equations have been broken down into clear steps to simplify understanding. Additionally, the research framework section demonstrates the application of these equations using expert responses collected in this study.

At this stage, the expert responses are fully defuzzified, and the next step is to construct the normalized direct-relation matrix N, which consists of values between 0 and 1. Once this matrix is prepared, the Decision-Making Trial and Evaluation Laboratory (DEMATEL) analysis can proceed.

9.3.3.1 DEMATEL Analysis and Illustration of Findings Using Cause-Effect Diagram

The DEMATEL analysis constructs a structural model to identify and visualize the relationships between factors using a directed graph (cause-and-effect diagram). The equations are streamlined into steps below for more straightforward comprehension. Additionally, the following section applies these equations to the expert responses from this study, demonstrating their implementation in developing the total relation matrix.

Step A: Development of the Total Relation Matrix

The process begins with constructing the total relation matrix, which captures the direct and indirect influences among factors. This matrix serves as the foundation for subsequent calculations in the DEMATEL method.

$$T = N\left(I - N\right)^{-1} \tag{9.10}$$

Step B: Calculation of Prominence and Relation

Let I be the identity matrix. Calculate r (sum of rows) and c_j (sum of columns) from the total relation model. The sum of r and c_j (i.e., $r_i + c_j$) is called prominence, represents the overall importance of a factor, and is plotted on the horizontal (x-axis) of the cause-effect diagram. The difference between r_i and c_j (i.e., $r_i - c_j$) is called relation, indicates whether a factor is more of a cause or an effect, and is plotted on the vertical (y-axis). This information is illustrated in the Fig. 9.5 below.

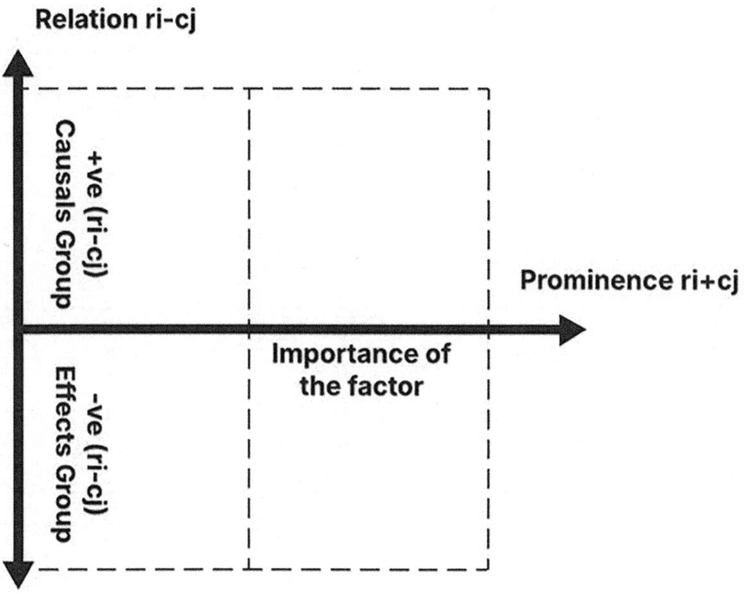

Fig. 9.5 Causal and effect group classification in DEMATEL

Step C: Illustration of the Cause-Effect Diagram
A cause-effect scatter plot uses the prominence and relation values as x and y coordinates. This visualization categorizes factors into causal and effect groups, aiding decision-making and strategic prioritization. The CSFs positioned above the x-axis belong to the causal group, while those below the x-axis form the effect group. The x-axis coordinates indicate the importance of each CSF. Using Fuzzy DEMATEL, this structured approach effectively identifies cause-effect relationships among the CSFs, offering valuable insights to enhance the resilience and sustainability of the Brazilian craft beer AFSC through digital transformation. The study's Fuzzy DEMATEL analysis systematically categorizes and ranks the 14 CSFs for digital transformation in the Brazilian craft beer AFSC using a cause-effect scatter plot diagram. This classification visually distinguishes the CSFs into causal and effect groups, following steps A to C.

Table 9.7 categorizes the 14 CSFs for digital transformation in the Brazilian craft beer AFSC into causal and effect groups based on their relation ($r_i - c_j$) values. Among these factors, nine CSFs (Economic Sustainability Performance, Environmental Performance, Technical Training and Skill Upgradation, E-Governance, Digital Logistics and Technology Infrastructure, Sustainable Food Security Decisions, Standardization of Sustainable Practices, Security, and Smart Farming) fall into the causal group. This classification indicates that these factors play a leading role in influencing other variables within the system, making them key drivers of digital transformation.

Table 9.7 Illustration of cause-and-effect group

Success Factor	CSF	Prominence (ri + cj)	Relation (ri − cj)	Identity
Social sustainability performance	CSF1	6.326	−1.326	Effect
Economic sustainability performance	CSF2	4.663	0.455	Causal
Environmental performance	CSF3	4.709	0.509	Causal
Transparency	CSF4	5.938	−1.106	Effect
Traceability	CSF5	5.804	−0.620	Effect
Technical training and skill up-gradation for stakeholders	CSF6	4.234	0.344	Causal
E-governance	CSF7	4.531	0.935	Causal
Digital logistics and technology infrastructure	CSF8	4.747	0.951	Causal
Sustainable food security decisions	CSF9	5.421	0.925	Causal
Trust and collaboration	CSF10	6.911	−0.233	Effect
Operational effectiveness and scalability	CSF11	5.827	−0.627	Effect
Standardization of sustainable practices	CSF12	5.541	0.840	Causal
Security	CSF13	6.106	0.141	Causal
Smart farming	CSF14	3.457	0.501	Causal

On the other hand, five CSFs (Social Sustainability Performance, Transparency, Traceability, Trust and Collaboration, and Operational Effectiveness and Scalability) are classified as effect factors, meaning they are influenced by other CSFs rather than driving the system themselves. These factors typically reflect the outcomes of improvements in the causal factors and serve as indicators of successful digital technology integration. The results suggest that strategic focus should be placed on strengthening the causal factors, as they will directly impact improving the effect factors, thereby enhancing the overall resilience and sustainability of the Brazilian craft beer AFSC.

In cause-effect diagrams, the causal group refers to a group of factors that, when increasing or decreasing, increases or decreases in another group of factors, which is referred to as the effects group. Table 9.3 displays the cause-and-effect groups obtained from this study's fuzzy-DEMATEL analysis and compares the base paper's cause-and-effect group.

A negative correlation, or inverse correlation, indicates that an increase/decrease in one CSF inversely affects another CSF. This negative correlation and interrelationship among the 14 CSFs are consistent with the base paper's cause-effect diagram. Following this classification, Fig. 9.6 visually represents these relationships through a cause-effect scatter plot. The x-axis (prominence) reflects the overall importance of each CSF within the system, while the y-axis (relation) determines whether a factor is more influential (causal) or influenced (effect). CSFs above the x-axis belong to the causal group, meaning they drive changes in the system. In contrast, those below the x-axis form the effect group, indicating that other factors influence them. The trend line illustrates a medium, negative correlation with no significant outliers. Each CSF is represented by a number to enhance clarity, making

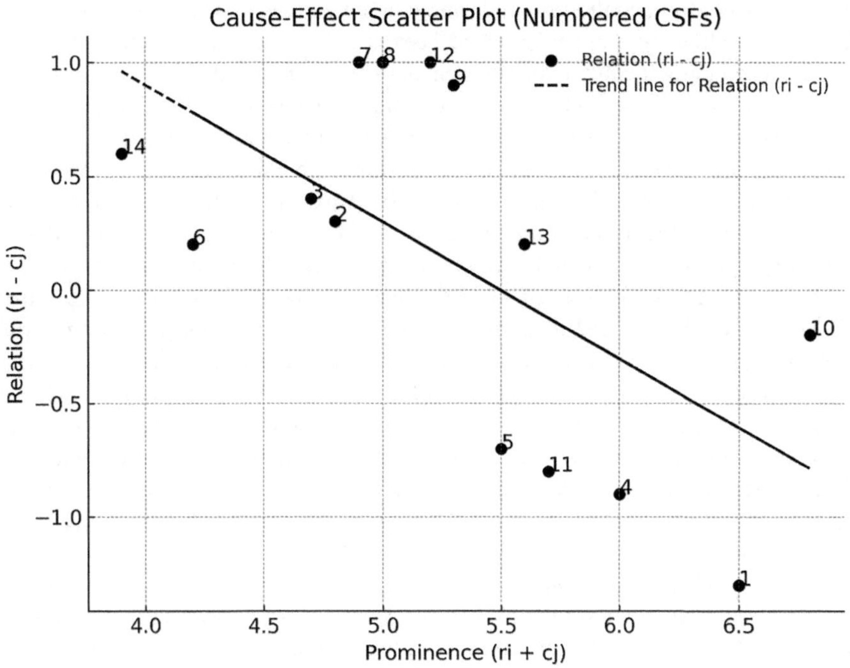

Fig. 9.6 Cause-effect two-themed scatter plot with negative correlation

it easier to identify their respective positions and roles in the digital transformation process.

The study was conducted when the global supply chain was experiencing supply and demand disruptions due to the COVID-19 pandemic, the Russia-Ukraine war, and fluctuations in commodity prices and workforce availability. AFSC has faced challenges in various areas, including transportation constraints, limited food supply, and stock heaping. Furthermore, strict lockdown guidelines imposed following the pandemic have increased the costs of managing the supply and accessibility of agricultural raw materials, products, labor, and workforce. These difficulties existed during the observation period of Joshi and Sharma (2022).

From Fig. 9.6 and Table 9.6, trust and collaboration have the highest prominence (ri – cj) value, while digital logistics and technology infrastructure have the highest relational (ri – cj) value. As a result, AFSC can generate value by implementing digital transformation and developing new frameworks, improving its social sustainability performance.

This interrelationship is evident from the analysis that CSF 8, i.e.digital logistics and technology infrastructure, strongly influences CSF10, i.e., trust and collaboration. A possible explanation for this influence is that CSF 8 can help increase cooperation among various supply chain players and thus help generate CSH 10, i.e., trust and collaboration. Collaboration strategies are essential for brewing companies to have inexpensive sourcing, production, packaging, and distribution. Craft beer producers form contracts with farming organizations to obtain raw materials at a low cost, allowing them to produce beer at a low cost (Vashishth & Tripathi, 2016). A modern, resilient distribution network requires digital transformation strategies such as cloud ERP systems for omnichannel product distribution.

Digital transformation can benefit competitively when organizations actively learn, support, and engage in supply chain resilience-building techniques that adjust to market requirements (Govindan et al., 2020). Adopting digital transformation can overcome the problems that limit market access for beer producers in novel ways by extending support services and developing novel methods of enhancing AFSC.

According to an analysis of the Brazilian craft beer consumption subculture (Koch & Sauerbronn, 2019), craft beers produced by microbreweries and craft breweries have limited access to unique raw materials and consumers. For these organizations, digital transformation could promote business speed and customer experience through effective beer raw material sourcing, production, and delivery methods supporting internal establishments for originality. At an affordable one-time investment or through monthly payment plans, digital transformation can help develop robust and effective beer AFSC for microbreweries and craft breweries in today's cost-focused technology environment.

The second highest prominence (ri + cj) value is social sustainability and performance. The second highest relational (ri − cj) value is e-governance. From the analysis, it is clear that both factors strongly influence building a resilient and suitable AFSC. Based on this, governance is the set of rules, policies, and systems to direct and control individual and collective actions of the organization and its supply chain partners to improve its ESG responsibility while maximizing efficiency and profit. When enabled by information and communication digital transformation, this governance is called supply chain E-governance. E-governance direct the decision-makers in ESG-compliant supply chain planning.

For example, during uncertain conditions such as a pandemic, SC planners must source raw materials without affecting the people's right to nutritious food, which could positively contribute to social sustainability and performance. From a resilience perspective, this ESG-compliant decision-making requires SC visibility and significant commitment from top-level management to ensure appropriate decisions throughout challenging times.

The third-highest prominence (ri + cj) value of CSF is security (CSF13), and the third-highest relational (ri − cj) value of CSF is sustainable food security decisions (CSF9). Security in the supply chain means monitoring and controlling mechanisms safeguarding the organization against supply chain security concerns such as data risks, fraud prevention, and third-party risks. In contrast, sustainable food security decisions ensure the consistent availability of agricultural raw materials for production. Both factors are strongly related and strongly influence building a resilient and suitable AFSC.

Digital Transformation can provide data protection solutions to create secure and resilient beer AFSCs while reducing third-party risks (Yadav et al., 2022). In this way, digital technology can help make sustainable food safety decisions. Adopt multi-tier food security decisions, such as investments in end-to-end supply chain visibility and real-time risk monitoring, have proven to reduce the Forrester effect/bullwhip, i.e., demand distortion caused by limited information sharing among supply chain partners, especially downstream supply chain partners.

Digital transformation can empower new intelligent agriculture practices that secure raw beer materials such as hops and water. COVID-19 has probed the promptness of the international supply chain and demonstrated a low level of contingency planning. Due to the low level of digital transformation in Brazilian beer, AFSCs' transparency, secure communication, and information exchanges among supply chain partners are challenging and required to successfully build a sustainable and resilient beer AFSC.

9.4 Final Considerations

The study examined the cause-effect relationship of the CSFs for digital transformation in the Brazilian craft beer agri-food supply chain, focusing on enhancing resilience and sustainability. The findings underscore the strategic importance of digital transformation, providing actionable insights for industry practitioners, policymakers, and researchers. This section highlights the importance of the identified CSFs, their practical applications, and recommendations for effective digital transformation implementation.

9.4.1 Importance and Practical Utility of Critical Success Factors

The validated CSFs offer a structured framework to guide decision-makers, supply chain managers, and brewers in navigating digital transformation. The study demonstrated that nine CSFs (e.g., digital logistics, economic sustainability, smart farming) function as causal drivers, meaning they influence other factors and play a critical role in shaping the resilience and efficiency of the AFSC. Meanwhile, five CSFs (e.g., social sustainability, transparency, trust, and collaboration) represent effect factors, meaning they reflect the overall success of digital integration.

Adopting digital transformation can improve traceability, inventory management, and operational efficiency for industry practitioners. E-governance mechanisms can streamline regulatory compliance, reducing barriers for small breweries to access markets. Additionally, technical training and skill development are essential to overcoming digital adoption barriers, equipping professionals with the expertise to leverage AI, blockchain, and IoT in AFSC management.

9.4.2 Enhancing Supply Network Resilience Through Digital Transformation

The COVID-19 pandemic and global supply chain disruptions highlighted the need for flexible, data-driven decision-making frameworks. The study confirmed that digital transformation fosters agility and adaptability, allowing breweries to mitigate risks and enhance supply network resilience.

Implementing the IoB offers microbreweries a competitive advantage by improving real-time demand tracking, optimizing supply routes, and enhancing consumer engagement. However, the study also emphasizes that cost-effective solutions must be developed to ensure widespread digital adoption, particularly for small-scale beer producers.

9.4.3 Strategic Implications for Policymakers and Industry Leaders

The study aligns with Brazil's "Going Digital in Brazil" strategy, reinforcing the need for policy-driven incentives to promote technology adoption in AFSCs. Policymakers should consider (1) Regulatory frameworks that support digital transformation while ensuring fair market access, (2) Investment in digital infrastructure, particularly in underserved regions where microbreweries operate, and (3) Public-private partnerships to facilitate training programs and funding opportunities for breweries transitioning to digitized supply network operations.

The study provides industry leaders with a set of CSFs for integrating digital transformation while balancing sustainability, resilience, and efficiency. Standardizing digital practices, such as blockchain-enabled traceability and AI-powered supply chain analytics, can drive efficiency and reduce operational risks.

9.4.4 Recommendations and Future Research Directions

This study offers a comprehensive framework for the CSFs of digital transformation in the Brazilian craft beer industry, reinforcing its role in enhancing resilience, efficiency, and sustainability. By concentrating on strategic CSFs, industry stakeholders can proactively adapt to evolving market challenges while promoting long-term growth and innovation. Future research should investigate comparative AFSC studies in other developing regions to validate the applicability of digital strategies across diverse economic contexts. Integrating smart technologies, policy interventions, and industry collaboration will be essential in ensuring a robust and future-ready beer industry. Expand research on digital transformation in developing economies to enable cross-country comparisons and tailor digital strategies to emerging markets.

The study offers the following recommendations as action points to maximize the benefits of digital transformation: (i) prioritize CSFs that serve as enablers of digital transformation, such as digital logistics, smart farming, and economic sustainability (ii) adopt AI-driven analytics for demand forecasting and risk assessment, allowing breweries to optimize production and resource allocation and (iii) create financial incentives for small brewers to adopt cost-effective digital solutions, bridging the gap between technology accessibility and industry adoption.

References

AGENCIAGOV. (2024). *Com 1.847 cervejarias registradas no Brasil, setor cresce 6,8% em 2023*. May 8, 2024. https://agenciagov.ebc.com.br/noticias/202405/com-1-847-cervejarias-registradas-no-brasil-setor-cresce-6-8-em-2023

Agrawal, P., Narain, R., & Ullah, I. (2019). Analysis of barriers in implementation of digital transformation of supply chain using interpretive structural modelling approach. *Journal of Modelling in Management, 15*(1), 297–317. https://doi.org/10.1108/JM2-03-2019-0066

Ali, I., & Govindan, K. (2023). Extenuating operational risks through digital transformation of agri-food supply chains. *Production Planning & Control, 34*(12), 1165–1177. https://doi.org/10.1080/09537287.2021.1988177

Ali, I., & Phan, H. M. (2022). Industry 4.0 technologies and sustainable warehousing: A systematic literature review and future research agenda. *The International Journal of Logistics Management, 33*(2), 644–662. https://doi.org/10.1108/IJLM-05-2021-0277

Ali, I., Sadiddin, A., & Cattaneo, A. (2023). Risk and resilience in agri-food supply chain SMEs in the pandemic era: A cross-country study. *International Journal of Logistics Research and Applications, 26*(11), 1602–1620. https://doi.org/10.1080/13675567.2022.2102159

Altay, N., Gunasekaran, A., Dubey, R., & Childe, S. J. (2018). Agility and resilience as antecedents of supply chain performance under moderating effects of organizational culture within the humanitarian setting: A dynamic capability view. *Production Planning and Control, 29*(14), 1158–1174. https://doi.org/10.1080/09537287.2018.1542174

Altuntas, S., & Yilmaz, M. K. (2016). Fuzzy DEMATEL method to evaluate the dimensions of marketing resources: An application in SMEs. *Journal of Business Economics and Management, 17*(3), 347–364. https://doi.org/10.3846/16111699.2015.1068220

Ariadi, G., Werastuti, D. N. S., & Inggarwati, K. (2024). The effect of sustainable supply chain toward sustainability performance mediated by joint efforts and sharing activities: Evidence from Bali Farmer Groups. *Measuring Business Excellence, 28*(3/4), 366–380. https://doi.org/10.1108/MBE-12-2023-0195

Awokuse, T., Lim, S., Santeramo, F., & Steinbach, S. (2024). Robust policy frameworks for strengthening the resilience and sustainability of agri-food global value chains. *Food Policy, 127*, 102714. https://doi.org/10.1016/j.foodpol.2024.102714

Barbosa, M. W., & Cansino, J. M. (2024). The impacts of environmental collaboration on the environmental performance of agri-food supply chains: A Mediation-moderation analysis of external pressures. *International Journal of Logistics Research and Applications, 27*(12), 2712–2736. https://doi.org/10.1080/13675567.2024.2310024

Bechtsis, D., Tsolakis, N., Iakovou, E., & Vlachos, D. (2022). Data-driven secure, resilient and sustainable supply chains: Gaps, opportunities, and a new generalised data sharing and data monetisation framework. *International Journal of Production Research, 60*(14), 4397–4417. https://doi.org/10.1080/00207543.2021.1957506

Belhadi, A., Kamble, S., Subramanian, N., Singh, R. K., & Venkatesh, M. (2024). Digital capabilities to manage agri-food supply chain uncertainties and build supply chain resilience during compounding geopolitical disruptions. *International Journal of Operations & Production Management, 44*(11), 1914–1950. https://doi.org/10.1108/IJOPM-11-2022-0737

Bolfe, É. L., Jorge, L. A. d. C., Sanches, I. D., Júnior, A. L., da Costa, C. C., Victoria, D. d. C., Inamasu, R. Y., Grego, C. R., Ferreira, V. R., & Ramirez, A. R. (2020). Precision and digital agriculture: Adoption of technologies and perception of Brazilian farmers. *Agriculture (Switzerland), 10*(12), 1–16. https://doi.org/10.3390/agriculture10120653

Brintrup, A., Kosasih, E., Schaffer, P., Zheng, G., Demirel, G., & MacCarthy, B. L. (2024). Digital supply chain surveillance using artificial intelligence: Definitions, opportunities and risks. *International Journal of Production Research, 62*(13), 4674–4695. https://doi.org/10.1080/00207543.2023.2270719

CGEE-MCTI. (2022). *Brazilian digital transformation strategy (e-digital) 2022–2026 cycle*. Brasília. https://www.gov.br/mcti/pt-br/acompanhe-o-mcti/transformacaodigital/arquivosestrategiadigital/digitalstrategy_2022-2026.pdf

Chang, B., Chang, C., & Wu, C. (2011). Fuzzy DEMATEL method for developing supplier selection criteria. *Expert Systems with Applications, 38*(3), 1850–1858. https://doi.org/10.1016/j.eswa.2010.07.114

Choi, T.-M. (2021). Risk analysis in logistics systems: A research agenda during and after the COVID-19 pandemic. *Transportation Research Part E: Logistics and Transportation Review, 145*(January), 102190. https://doi.org/10.1016/j.tre.2020.102190

Cichosz, M., Wallenburg, C. M., & Knemeyer, A. M. (2020). Digital transformation at logistics service providers: Barriers, success factors and leading practices. *The International Journal of Logistics Management, 31*(2), 209–238. https://doi.org/10.1108/IJLM-08-2019-0229

Ciruela-Lorenzo, A. M., Del-Aguila-Obra, A. R., Padilla-Meléndez, A., & Plaza-Angulo, J. J. (2020). Digitalization of agri-cooperatives in the smart agriculture context. Proposal of a digital diagnosis tool. *Sustainability, 12*(4), 1325. https://doi.org/10.3390/su12041325

CNA, and CEPEA. (2025). *PIB do Agronegócio*. https://cepea.esalq.usp.br/upload/kceditor/files/CT-PIB-AGRO_23.JAN.25.pdf

Costa, F. H. d. O., de Moraes, C. C., da Silva, A. L., Delai, I., Chaudhuri, A., & Pereira, C. R. (2022). Does resilience reduce food waste? Analysis of Brazilian Supplier-Retailer Dyad. *Journal of Cleaner Production, 338*(March), 130488. https://doi.org/10.1016/j.jclepro.2022.130488

Costa, F. H. d. O., Moraes, C. C., da Silva, A. L., Pereira, C. R., Delai, I., & Jabbour, A. B. L. d. S. (2021). Does resilience impact food waste? Moving the debate on. *RAE - Revista de Administração de Empresas (Journal of Business Management), 5*(61), 1–21. https://doi.org/10.1590/S0034-759020210506

Cricelli, L., Mauriello, R., & Strazzullo, S. (2024). Technological innovation in agri-food supply chains. *British Food Journal, 126*(5), 1852–1869. https://doi.org/10.1108/BFJ-06-2022-0490

Dadsena, K. K., Pant, P., Paul, S. K., & Pratap, S. (2024). Overcoming strategies for supply chain digitization barriers: Implications for sustainable development goals. *Business Strategy and the Environment, 33*(5), 3887–3910. https://doi.org/10.1002/bse.3681

Dias, M. d. O., & Falconi, D. (2018). The evolution of craft beer industry in Brazil. *Journal of Economics and Business, 1*(4), 618–626. https://doi.org/10.31014/aior.1992.01.04.55

Dixit, V. K., Malviya, R. K., Kumar, V., & Shankar, R. (2024). An analysis of the strategies for overcoming digital supply chain implementation barriers. *Decision Analytics Journal, 10*, 100389. https://doi.org/10.1016/j.dajour.2023.100389

El Jaouhari, A., Arif, J., Jawab, F., Samadhiya, A., & Kumar, A. (2024). Unfolding the role of metaverse in agri-food supply chain security: Current scenario and future perspectives. *International Journal of Food Science & Technology, 59*(5), 3451–3460. https://doi.org/10.1111/ijfs.16972

EMBRAPA. (2015). *Cultivo de Cevada – Dados Sistema de Produção*. https://www.infoteca.cnptia.embrapa.br/infoteca/bitstream/doc/1155367/1/Cultivo-de-Cevada.pdf

EMBRAPA. (2023). *Indicações Técnicas para a Produção de Cevada Cervejeira nas Safras 2023 e 2024*. Brasília. https://www.embrapa.br/fale-conosco.

Erkan, E. F., Uygun, Ö., Kiraz, A., & Canpolat, O. (2021). An integrated fuzzy DEMATEL and fuzzy cognitive maps approach for the assessing of the industry 4.0 model. *Journal of Engineering Research, 11*(2), 236–251. https://doi.org/10.36909/jer.12303

Euromonitor International. (2024). *Beer in Brazil*. https://www.portal.euromonitor.com/Analysis/Tab

Euromonitor International. (2025a). *Beer in Brazil – Context*. https://www.portal.euromonitor.com/Analysis/Tab

Euromonitor International. (2025b). *Beer in Brazil – Datagraphics*. https://www.portal.euromonitor.com/Analysis/Tab

Euromonitor International, and MAPA. (2025). *Beer in Brazil – Analysis*. https://www.portal.euromonitor.com/Analysis/Tab

FAO. (2021). *The State of Food and Agriculture 2021 – Making Agrifood Systems More Resilient to Shocks and Stress*. Edited by Food and Agriculture Organization of the United Nations. *The State of Food and Agriculture 2021*. Rome: FAO. https://doi.org/10.4060/cb4476en

FAO. (2022). *The future of food and agriculture – Drivers and triggers for transformation*. FAO. https://doi.org/10.4060/cc0959en

FAO. (2023a). *Technical Webinar Series on Avocado and Pineapple Value Chains Webinar #13: How Can the European Green Deal Drive Sustainability in the Tropical Fruit Sector? Summary Report*. Rome. https://applications.icao.int/icec/Home/Index.

FAO. (2023b). *The State of Food and Agriculture 2023. The State of Food and Agriculture 2023*. FAO. https://doi.org/10.4060/cc7724en

Feng, H., Wang, X., Duan, Y., Zhang, J., & Zhang, X. (2020). Applying blockchain technology to improve agri-food traceability: A review of development methods, benefits and challenges. *Journal of Cleaner Production, 260*(July), 121031. https://doi.org/10.1016/j.jclepro.2020.121031

Fernandez-Vidal, J., Gonzalez, R., & Gasco, J. (2022). Managing digital transformation: The view from the top. *Journal of Business Research, 152*, 29–41. https://doi.org/10.1016/j.jbusres.2022.07.020

Gajdić, D., Kotzab, H., & Petljak, K. (2023). Collaboration, trust and performance in agri-food supply chains: A bibliometric analysis. *British Food Journal, 125*(2), 752–778. https://doi.org/10.1108/BFJ-07-2021-0723

Gardas, B. B., Raut, R. D., & Narkhede, B. (2019). Identifying critical success factors to facilitate reusable plastic packaging towards sustainable supply chain management. *Journal of Environmental Management, 236*, 81–92. https://doi.org/10.1016/j.jenvman.2019.01.113

Giongo, V., Acosta, A. d. S., Dossa, Á. A., Santi, A., do Amaral, A. J., Caierão, E., Denardin, J. E., et al. (2025). How can the environmental impacts of wheat cultivation and wheat flour production be reduced? A life cycle assessment of Brazilian Wheat. *Journal of Cleaner Production, 489*, 144650. https://doi.org/10.1016/j.jclepro.2024.144650

Gligor, D. M., Davis-Sramek, B., Tan, A., Vitale, A., Russo, I., Golgeci, I., & Wan, X. (2022). Utilizing blockchain technology for supply chain transparency: A resource orchestration perspective. *Journal of Business Logistics, 43*, 140–159. https://doi.org/10.1111/jbl.12287

Godfray, H. C. J., Aveyard, P., Garnett, T., Hall, J. W., Key, T. J., Lorimer, J., Pierrehumbert, R. T., Scarborough, P., Springmann, M., & Jebb, S. A. (2018). Meat consumption, health, and the environment. *Science, 361*(6399). https://doi.org/10.1126/science.aam5324

Govindan, K., Mina, H., Esmaeili, A., & Gholami-Zanjani, S. M. (2020). An integrated hybrid approach for circular supplier selection and closed loop supply chain network design under uncertainty. *Journal of Cleaner Production, 242*, 118317. https://doi.org/10.1016/j.jclepro.2019.118317

Grohmann, P., Dietze, V., Krämer, C., & Feindt, P. H. (2023). Governance and trust in sustainability-based agri-food value chains. A comparative analysis of five cases in Germany. *Frontiers in Sustainable Food Systems, 7*, 1–14. https://doi.org/10.3389/fsufs.2023.1130895

Gupta, H., Yadav, A. K., Kusi-Sarpong, S., Khan, S. A., & Sharma, S. C. (2022). Strategies to overcome barriers to innovative digitalisation technologies for supply chain logistics resilience during pandemic. *Technology in Society, 69*, 101970. https://doi.org/10.1016/j.techsoc.2022.101970

Hassoun, A., Marvin, H. J. P., Bouzembrak, Y., Barba, F. J., Castagnini, J. M., Pallarés, N., Rabail, R., et al. (2023). Digital transformation in the agri-food industry: Recent applications and the role of the COVID-19 pandemic. *Frontiers in Sustainable Food Systems, 7*(1217813), 1–16. https://doi.org/10.3389/fsufs.2023.1217813

Holmström, J., Holweg, M., Lawson, B., Pil, F. K., & Wagner, S. M. (2019). The digitalization of operations and supply chain management: Theoretical and methodological implications. *Journal of Operations Management, 65*(8), 728–734. https://doi.org/10.1002/joom.1073

IPCC. (2022). *Climate Change 2022: Impacts, Adaptation, and Vulnerability. Contribution of Working Group III to the Sixth Assessment Report of the Intergovernmental Panel on Climate Change*. Cambridge, UK and New York, NY, USA. https://doi.org/10.1017/9781009325844

Ivanov, D. (2022). Viable supply chain model: Integrating agility, resilience and sustainability perspectives—Lessons from and thinking beyond the COVID-19 pandemic. *Annals of Operations Research, 319*(1), 1411–1431. https://doi.org/10.1007/s10479-020-03640-6

Ivanov, D., & Dolgui, A. (2020). Viability of intertwined supply networks: Extending the supply chain resilience angles towards survivability. A position paper motivated by COVID-19 outbreak. *International Journal of Production Research, 58*(10), 2904–2915. https://doi.org/10.1080/00207543.2020.1750727

Jagtap, S., Raut, R., & Dani, S. (2024). Advancing the digital frontier in agri-food supply chains. *International Journal of Food Science & Technology, 59*(5), 3433–3435.

Joshi, S., & Sharma, M. (2022). Digital Technologies (DT) adoption in agri-food supply chains amidst COVID-19: An approach towards food security concerns in developing countries. *Journal of Global Operations and Strategic Sourcing, 15*(2), 262–282. https://doi.org/10.1108/JGOSS-02-2021-0014

Kazancoglu, Y., Lafci, C., Kumar, A., Luthra, S., Garza-Reyes, J. A., & Berberoglu, Y. (2024). The role of agri-food 4.0 in climate-smart farming for controlling climate change-related risks: A business perspective analysis. *Business Strategy and the Environment, 33*(4), 2788–2802. https://doi.org/10.1002/bse.3629

Kersten, C. C., Kerber, J. M. C., Silva, J. d. S., Bouzon, M., & Campos, L. M. D. S. (2024). Traceability in the agri-food supply chain: A new perspective under the circular economy approach. *Production, 34*(e20240009), 1–19. https://doi.org/10.1590/0103-6513.20240009

Knoth, M., & Borrit, J. B. (2014). The Indian Beer Market: The next 'China' or a hyped market with little future growth? *Scandinavian Brewers' Review, 72*(2), 1–5. https://www.niras.com/media/nqgm40rk/scandinavian-brewers-review_india.pdf

Koch, E. S., & Sauerbronn, J. F. R. (2019). 'To love beer above all things': An analysis of Brazilian craft beer subculture of consumption. *Journal of Food Products Marketing, 25*(1), 1–25. https://doi.org/10.1080/10454446.2018.1431577

Kovács, T., Botos, S., & Felföldi, J. (2024). What does digital transformation do in agriculture? A systematic literature review of agri-food sector's digitalisation. *Journal of Agricultural Informatics, 14*(2), 26–42. https://doi.org/10.17700/jai.2023.14.2.707

Kumar, A., & Agrawal, S. (2023). Challenges and opportunities for agri-fresh food supply chain management in India. *Computers and Electronics in Agriculture, 212*(108161), 1–22. https://doi.org/10.1016/j.compag.2023.108161

Kumar, P., & Singh, R. K. (2022). Strategic framework for developing resilience in agri-food supply chains during COVID 19 pandemic. *International Journal of Logistics Research and Applications, 25*(11), 1401–1424. https://doi.org/10.1080/13675567.2021.1908524

MAPA. (2022). *Lúpulo No Brasil: Perspectivas e Realidades*. https://www.gov.br/agricultura/pt-br/arquivos/livro_lupulo-no-brasil-perspectivas-e-realidade_baixa_semmarcacao.pdf

MAPA. (2023). *Projeções do Agronegócio – Brasil – 2022/23 a 2032/33*. Brasília. https://www.gov.br/agricultura/pt-br/assuntos/noticias/producao-de-graos-brasileira-devera-chegar-a-390-milhoes-de-toneladas-nos-proximos-dez-anos/ProjeesdoAgronegcio20232033.pdf

MAPA. (2024). *Anuário da Cerveja 2024 – Ano referência 2023*. Brasília. https://sindicerv.com.br/2025/wp-content/uploads/2024/05/Anuario-da-cerveja-2024-referencia-2023-MAPA-versao-web.pdf

Medina, G., & Thomé, K. (2021). Transparency in global agribusiness: Transforming Brazil's soybean supply chain based on companies' accountability. *Logistics, 58*(5), 1–15. https://doi.org/10.3390/logistics

Menon, S., & Jain, K. (2024). Blockchain technology for transparency in agri-food supply chain: Use cases, limitations, and future directions. *IEEE Transactions on Engineering Management, 71*, 106–120. https://doi.org/10.1109/TEM.2021.3110903

Mishra, R., Singh, R. K., Kumar, S., Mangla, S. K., & Kumar, V. (2025). Critical success factors of blockchain technology adoption for sustainable and resilient operations in the banking industry during an uncertain business environment. *Electronic Commerce Research, 25*(1), 595–629. https://doi.org/10.1007/s10660-023-09707-3

Morado, R. (2009). Larousse da Cerveja - A História e as curiosidades de uma das bebidas mais populares do mundo. 1st ed. São Paulo: Alaúde Editorial Ltd. Available at: www.alaude.com.br

de Moraes, C. C., Costa, F. H. d. O., Pereira, C. R., da Silva, A. L., & Delai, I. (2020). Retail food waste: Mapping causes and reduction practices. *Journal of Cleaner Production, 256*(120124), 1–16. https://doi.org/10.1016/j.jclepro.2020.120124

Musa, S. F. P. D., & Basir, K. H. (2021). Smart farming: Towards a sustainable agri-food system. *British Food Journal, 123*(9), 3085–3099. https://doi.org/10.1108/BFJ-03-2021-0325

Musa, S. F. P. D., Basir, K. H., & Luah, E. (2021). The role of smart farming in sustainable development. *International Journal of Asian Business and Information Management, 13*(2), 1–12. https://doi.org/10.4018/IJABIM.20220701.oa5

Myers, S. S., Zanobetti, A., Kloog, I., Huybers, P., Leakey, A. D. B., Bloom, A. J., Carlisle, E., et al. (2014). Increasing CO2 threatens human nutrition. *Nature, 510*(7503), 139–142. https://doi.org/10.1038/nature13179

OECD. (2020a). *OECD reviews of digital transformation: Going digital in Brazil*. OECD Reviews of Digital Transformation. Paris: OECD. https://doi.org/10.1787/e9bf7f8a-en

OECD. (2020b). Tacking Coronavirus (COVID-19): Contributing to a Global Effort. *Comparing Crises: Great Lockdown versus Great Recession*. FAO. https://doi.org/10.4060/ca8833en

OECD-FAO. (2023). *OECD-FAO Agricultural Outlook 2023–2032*. OECD-FAO Agricultural Outlook. OECD. https://doi.org/10.1787/08801ab7-en

Panetto, H., Lezoche, M., Hormazabal, J. E. H., Diaz, M. d. M. E. A., & Kacprzyk, J. (2020). Special issue on agri-food 4.0 and digitalization in agriculture supply chains – New directions, challenges and applications. *Computers in Industry, 116*(103188), 1–3. https://doi.org/10.1016/j.compind.2020.103188

Queiroz, M. M., De Bourmont, M., & Telles, R. (2020). Blockchain adoption in operations and supply chain management: Empirical evidence from an emerging economy. *International Journal of Production Research, 59*(20), 1–17. https://doi.org/10.1080/00207543.2020.1803511

Queiroz, M. M., & Wamba, S. F. (2024). A structured literature review on the interplay between emerging technologies and COVID-19 – Insights and directions to operations fields. *Annals of Operations Research, 335*(3), 937–963. https://doi.org/10.1007/s10479-021-04107-y

Rama, R., & Wilkinson, J. (2019). Innovation and disruptive technologies in the Brazilian agrofood sector. *Systèmes Alimentaires/Food Systems, 4*, 51–70. https://doi.org/10.15122/isbn.978-2-406-09829-4.p.0051

Romani, L. A. S., Bariani, J. M., Drucker, D. P., Vaz, G. J., Mondo, V. H. V., Moura, M. F., Bolfe, E. L., Oliveira, S. R. d. M., de Sousa, P. H. P., & Junior, A. L. (2020). Role of research and development institutions and AgTechs in the digital transformation of agriculture in Brazil. *Revista Ciência Agronômica, 51*(5), 1–8. https://doi.org/10.5935/1806-6690.20200082

de Sá, M. M., Miguel, P. L. d. S., de Brito, R. P., & Pereira, S. C. F. (2020). Supply chain resilience: The whole is not the sum of the parts. *International Journal of Operations and Production Management, 40*(1), 92–115. https://doi.org/10.1108/IJOPM-09-2017-0510

Sarkis, J., Cohen, M. J., Dewick, P., & Schröder, P. (2020). A brave new world: Lessons from the COVID-19 pandemic for transitioning to sustainable supply and production. *Resources, Conservation and Recycling, 159*(104894), 1–4. https://doi.org/10.1016/j.resconrec.2020.104894

Sezer, M. D., Kazancoglu, Y., Mangla, S. K., & Lafçı, Ç. (2024). Smart, sustainable, and resilient food supply chains in disruptive events context. *Business Strategy and the Environment, 33*(7), 6156–6171. https://doi.org/10.1002/bse.3801

Silvestri, C., Silvestri, L., Piccarozzi, M., & Ruggieri, A. (2024). Toward a framework for selecting indicators of measuring sustainability and circular economy in the agri-food sector: A systematic literature review. *The International Journal of Life Cycle Assessment, 29*(8), 1446–1484. https://doi.org/10.1007/s11367-022-02032-1

Springmann, M., Clark, M., Mason-D'Croz, D., Wiebe, K., Bodirsky, B. L., Lassaletta, L., de Vries, W., et al. (2018). Options for keeping the food system within environmental limits. *Nature, 562*(7728), 519–525. https://doi.org/10.1038/s41586-018-0594-0

Statista. (2024). *Contribution of the Agribusiness to the Gross Domestic Product (GDP) in Brazil from 2010 to 2023, by Segment*. December 2024. https://www.statista.com/statistics/1104407/agribusiness-share-gdp-brazil-segment/

Toussaint, M., Cabanelas, P., & Muñoz-Dueñas, P. (2022). Social sustainability in the food value chain: What is and how to adopt an integrative approach? *Quality and Quantity, 56*(4), 2477–2500. https://doi.org/10.1007/s11135-021-01236-1

UNCTAD. (2020). *Maximizing sustainable agri-food supply chain opportunities to redress COVID-19 in developing countries*. https://unctad.org/system/files/official-document/ditctabinf2020d9_en.pdf

United Nations. (2023). *The-Sustainable-Development-Goals-Report-2023: Special Edition. Towards a Rescue Plan for People and Planet*. https://unstats.un.org/sdgs/report/2023/

Vashishth, A., & Tripathi, N. (2016). Study on market analysis of Indian beer industry. *International Journal of Business and Management Invention ISSN, 5*(7), 28–32. https://doi.org/10.13140/RG.2.2.23406.20801

Vern, P., Panghal, A., Mor, R. S., Kumar, V., & Jagtap, S. (2024). Blockchain-based traceability framework for agri-food supply chain: A proof-of-concept. *Operations Management Research, 1–20*. https://doi.org/10.1007/s12063-024-00529-6

Violino, S., Figorilli, S., Costa, C., & Pallottino, F. (2020). Internet of beer: A review on smart technologies from mash to pint. *Food, 9*(950), 1–20. https://doi.org/10.3390/foods9070950

Yadav, S., Luthra, S., & Garg, D. (2022). Internet of Things (IoT) based coordination system in agri-food supply chain: Development of an efficient framework using DEMATEL-ISM. *Operations Management Research, 15*(1–2), 1–27. https://doi.org/10.1007/s12063-020-00164-x

Zhu, Z., Chu, F., Dolgui, A., Chu, C., Zhou, W., & Piramuthu, S. (2018). Recent advances and opportunities in sustainable food supply chain: A model-oriented review. *International Journal of Production Research, 56*(17), 5700–5722. https://doi.org/10.1080/00207543.2018.1425014

Chapter 10
Reducing Transaction Costs Through Digitalization: Challenges in the Order to Cash Process of Logistics Service Providers

Christiaan de Goeij and Luca Gelsomino

Abstract Logistics Service Providers (LSPs) face specific challenges in their Order-to-Cash (O2C) processes due to complex pricing structures and relatively low profit margins. In this chapter, using a multiple case study approach, we apply Transaction Cost Economics (TCE) to examine the factors leading to inefficiencies in LSPs' O2C processes. Our research is based on four in-depth case studies of LSPs in the Netherlands, combining 19 semi-structured interviews with employees from different departments and secondary data on orders, invoices, and payments. The findings reveal that the main O2C inefficiencies stem from a lack of an integrated vision of the process as a whole, insufficient internal coordination and collaboration between departments, and challenges related to digitalization and automation, including limitations in employees' IT skills. These insights highlight the need for LSPs to optimize their O2C processes to improve financial performance and operational efficiency.

Keywords Order-to-Cash Process · O2C · Logistics service providers · Transaction cost economics

C. de Goeij (✉)
Faculty of Business, Media and Law, Windesheim University of Applied Sciences, Zwolle, The Netherlands
e-mail: caj.de.goeij@windesheim.nl

L. Gelsomino
Faculty of Economics and Business, University of Groningen, Groningen, The Netherlands

Y. Khojasteh, S.M. M. Jahan Abadi (eds.), *Digital Transformation of Supply Chains*, International Series in Operations Research & Management Science 372, https://doi.org/10.1007/978-3-032-01206-7_10

10.1 Introduction

The Order-to-Cash (O2C) process can be explained briefly as the process starting from the receipt of an order, to the acceptance of an order, to the delivery of a product or service, to the sending of the invoice, to the approval of the invoice by the buyer and to the final payment. If suppliers shorten their O2C process, they receive cash faster, resulting in a better working capital position (Hofmann & Kotzab, 2010). Furthermore, suppliers can achieve advantages in terms of transaction costs and workforce productivity by limiting the time spent on order and invoice management (Perego & Salgaro, 2010), and can improve customer relations by limiting the amount of disputed invoices (Jimenez-Martinez & Polo-Redondo, 2004).

Since the Covid-19 crisis, payment terms have been extended on a large scale. According to Intrum (2024) the average payment term in the EU increased from 40 to 62 days from 2019 to 2024.

Logistics Service Providers (LSPs) are particularly vulnerable to extended payment terms due to the structural complexities of their operations. According to De Goeij (2021), LSPs in the Netherlands send out invoices 3.7 days later and receive payments 10.6 days later after invoice approval compared to the average across industries. Although they process and fulfill customer orders relatively quickly, invoicing is often delayed. This is partly due to the complexity of logistics pricing structures, which involve variable costs such as waiting times, currency fluctuations, and additional fees. Contractual payment terms frequently extend beyond a month, with actual payments often exceeding agreed deadlines. Additionally, the sector's strong customer focus leads to a preference for personal follow-ups over automated reminders for late payments, further increasing administrative workload. These factors underscore the urgent need for LSPs to optimize their O2C processes, not only by accelerating invoicing but also by improving internal coordination and digital capabilities.

Even though both the academic literature and industry reports focus primarily on payment terms, the O2C process comprises many other steps. The buyer is responsible for approving the invoice and making the final payment, while the supplier is responsible for most of the other steps in the process. For suppliers it's therefore important to optimize these steps to avoid liquidity problems and high transaction costs. Since LSPs often face low margins (Elliot et al., 2020), a long and inefficient O2C process has a significant impact on profitability.

10.2 Literature Review

10.2.1 The Order-to-Cash Process

In Fig. 10.1 a typical O2C process is mapped, based on earlier literature (Hofmann & Kotzab, 2010; Perego & Salgaro, 2010; Silvestro & Lustrato, 2014). A typical O2C process consists of the following phases:

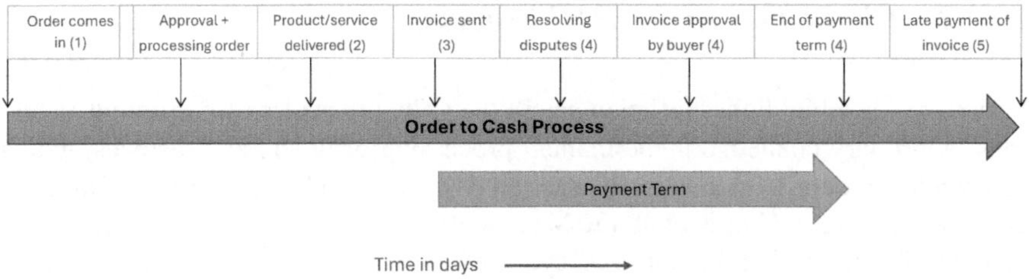

Fig. 10.1 O2C process steps

1. *Order acceptance and approval.* The supplier receives and accepts the order from the buyer, after which it is processed internally.
2. *Delivery of goods/services.* The supplier delivers the agreed-upon goods or services to the buyer.
3. *Invoicing.* Suppliers usually issue invoices after delivery, although in some cases invoices are sent beforehand.
4. *Payments according to agreed terms.* The payment takes place after approval of the invoice by the buyer. In case there are mistakes in the invoice, this can lead to disputes resulting in increased invoice approval time and late payment.
5. *Payments (after contractually agreed payment term).* Buyers do not always adhere to agreed payment terms, for example due to liquidity issues or inefficiencies in their payment processes, requiring follow-up communication from suppliers.

Even though most O2C processes consist of these five phases, many variations exist across industries. For instance, in large construction projects, multiple payments may be scheduled over time, with an initial payment often made before project commencement.

In many companies, O2C is not managed as an integrated process (Korotina et al., 2015). Different departments are often involved in various phases. Sales or customer service typically handle negotiations with customers about orders, while finance or administration follows up on invoices (De Goeij et al., 2016). Because the O2C process is not managed as an integrated process, it is uncommon for companies to measure and monitor the length and effects of the O2C process on liquidity or costs. The literature is mostly focused on specific parts of the O2C process, especially payment terms (e.g. De Goeij et al., 2021; Van der Vliet et al., 2015), or on the digitalization of the invoicing process (e.g. Fairchild, 2004; Fairchild, 2005; Koch, 2019), and therefore, offers limited insights into how companies can manage O2C as an integrated process.

10.2.2 Challenges in the O2C Process

Various studies highlight inefficiencies in the O2C process across different industries. A primary challenge is the manual processing of orders and invoices, which contributes to delays, increased transaction costs, and higher error rates (Fairchild, 2004; Koch, 2019). Cronie et al. (2024) note that suppliers often delay sending out invoices, further prolonging the O2C cycle. Moreover, the absence of customer segmentation strategies and the existence of multiple invoice types per company (e.g. invoices per order, weekly invoices or monthly invoices) add to process complexity, leading to longer lead times and higher administrative costs (De Goeij et al., 2016).

Another major issue is late payments, often linked to disputes over invoice accuracy (Silvestro & Lustrato, 2014). Without a standardized dispute resolution process, these delays become more frequent and costly (De Goeij et al., 2016). Furthermore, internal goal conflicts, particularly between sales and finance departments, exacerbate inefficiencies in O2C management (Dekkers et al., 2020). Finance teams tend to focus on liquidity and cash flow, while sales teams, prioritizing customer relationships, may be more inclined to agree to extended payment terms. Such misalignments negatively impact overall profitability and working capital efficiency.

The logistics sector faces additional complexities due to cross-border transactions, variable fuel costs, waiting times, and currency fluctuations, which make pricing structures more intricate and invoicing more prone to errors (De Goeij et al., 2016). Given that logistics operates on low profit margins, inefficiencies in the O2C process can significantly impact financial performance (Elliot et al., 2020).

10.2.2.1 Digitalization Challenges in the O2C Process

One of the most discussed solutions to O2C inefficiencies is digitalization, which can help reduce transaction costs and improve process visibility (Caniato et al., 2016; Fairchild, 2005). Studies indicate that e-invoicing, automated order processing, and digital payment reminders can significantly reduce O2C lead times, error rates, and payment delays (Dello Iacono et al., 2015; Koch, 2019; Nienhuis et al., 2013). However, adoption levels vary widely, particularly among small and medium-sized enterprises (SMEs), where manual invoicing processes remain common (Cronie et al., 2024).

Digitalization also affects real-time process monitoring, enabling companies to track invoice submission times, outstanding payments, and approval bottlenecks. This might enhance managerial control over the O2C process and can allow firms to implement data-driven strategies for process improvement (Hu, 2015). In particular, automation of invoice reminders has the potential to improve payment behavior among buyers, yet many companies still rely on manual follow-ups due to customer relationship concerns (Cronie et al., 2024).

Despite these benefits, LSPs face unique challenges in digitalizing their O2C processes. The heterogeneous IT landscape across supply chain partners makes seamless system integration difficult, particularly for companies using outdated ERP or accounting systems (Cronie et al., 2023). Logistics transactions frequently involve complex pricing structures, such as fuel surcharges, waiting times, and multi-leg shipments, which add layers of complexity to automation efforts. LSP staff often need to manually adjust invoices due to last-minute order changes, fluctuating fees, or customer-specific billing agreements (Cronie et al., 2023; De Goeij et al., 2016).

10.2.3 Transaction Cost Economics

Transaction Cost Economics (TCE) is a theoretical framework that explains how firms seek to minimize the costs associated with economic exchanges beyond direct production costs (Williamson, 1975). Transaction costs arise from activities such as searching for business partners, negotiating agreements, monitoring performance and enforcing contracts. TCE emphasizes the costs of governing transactions and the structures firms use to mitigate inefficiencies (Williamson, 1975, 1985), In buyer-supplier relationships, transaction costs are particularly relevant in complex processes such as O2C, where multiple parties interact, and inefficiencies can lead to financial and operational losses.

TCE theory explains that organizations choose among three governance modes to minimize transaction costs: market, hierarchy and hybrid (Williamson, 1979). The market governance mode relies on open competition and price mechanisms, making it suitable for standardized goods and services. Hierarchy governance modes involve managing transactions within a firm through direct control, reducing reliance on external partners but increasing coordination costs. Hybrid governance modes combine elements of both, using contracts, partnerships, or alliances to balance stability and flexibility. The governance structure influences how firms manage O2C processes since it determines the degree of control, coordination, and contractual reliance in managing orders, invoicing, and payments.

The main factors influencing transaction costs and the choice of governance mode are *uncertainty, asset specificity,* and *frequency* (Williamson, 1975). *Uncertainty* refers to "unanticipated changes in circumstances surrounding a transaction" (Grover & Malhotra, 2003, p. 460). TCE literature distinguishes between *environmental* and *behavioral uncertainty.* Environmental uncertainty relates to potential economic, political, or regulatory changes, increasing the difficulty of drafting sufficient agreements in advance. It is often operationalized by examining *business context uncertainty,* in terms of *macro-economic changes* and *institutional changes,* as well as the *market complexity* (Rindfleisch & Heide, 1997).

Behavioral uncertainty is associated with bounded rationality and opportunism. *Bounded rationality* refers to actors' *restrictions in using available information* due to cognitive or time-related limitations in processing available information. Since O2C processes are cross-functional, *intra-firm collaboration* between sales, finance, and operations departments is crucial in managing the process. Weak intra-firm and interdepartmental coordination can further amplify bounded rationality (Dekkers et al., 2020).

Opportunism, defined by Williamson (1985, p. 47) as "self-interest seeking with guile," often occurs when one party in a transaction has leverage due to an *imbalance in bargaining power* (Ireland & Webb, 2007) or when actors provide *incomplete or misleading information* (Hobbs, 1996). Digitalization might limit opportunism by increasing transaction transparency (De Goeij et al., 2021).

Asset specificity refers to "the transferability of assets that support a given transaction" (Grover & Malhotra, 2003, p. 459). *Investments* in highly specific assets tend to have limited value outside of a particular exchange relationship. In the context of LSPs, investments in O2C digitalization may require buyer-specific adaptations, such as custom integrations with clients' ERP systems or compliance with client-specific invoicing requirements, increasing asset specificity. Specific customer wishes in orders or invoices, such as the use of specific IT, might lead to a need for specific *skills and resources*.

The *frequency* of transactions is the third concept influencing transaction costs (Williamson, 1975). Williamson (1985) explains that larger transactions of a recurring kind give an incentive to apply hierarchical governance, since the costs of this governance structure will be easier to recover in this case. Digitalization can play a role in reducing transaction costs for high-frequency transactions which result in a high *number of invoices* by automating routine tasks, such as invoicing and payment reminders (Cronie et al., 2024; Table 10.1).

10.3 Methodology

This research is exploratory and examines the degree to which LSPs face transaction costs in their O2C processes as a consequence of uncertainty, asset specificity and frequency. Due to the exploratory nature of the research and the complexity of the subject of O2C processes, with multiple departments and companies involved, case study methodology is a good fit (Yin, 2009).

We chose a theoretical sampling approach by selecting cases that either extend or replicate theory (Eisenhardt, 1989). The case study sample is summarized in Table 10.2. The cases are homogenous in terms of geography and industry, since the companies are all based in the Netherlands and they are all LSPs, allowing for cross-case comparability. At the same time there is heterogeneity in the sample, in terms of yearly turnover, the main activities of LSPs and the industries of their clients, which enables the investigation of factors that can explain differences between cases.

Data triangulation is achieved in our research by conducting 19 semi-structured interviews with employees from different company departments. In all cases at least

Table 10.1 TCE concepts and their impact on the O2C process

TCE concept	Definition	Examples of impact on O2C process
Behavioral uncertainty	Uncertainty arising from bounded rationality (limited ability to process information) and opportunism (self-interest seeking with guile) (Williamson, 1985)	• Bounded rationality can lead to errors in invoicing, increasing approval time and the likelihood of disputes • Opportunistic behavior from buyers, such as delaying payments beyond agreed terms, extends O2C lead times and increases transaction monitoring efforts
Environmental uncertainty	Uncertainty caused by external factors such as regulatory changes, economic conditions, and market complexity (Rindfleisch & Heide, 1997)	• Geopolitical tensions, macroeconomic fluctuations and regulatory changes, might increase transaction costs by making payment behaviors and pricing structures more unpredictable • This can complicate order and invoicing processes, requiring frequent adjustments and extending O2C lead times
Asset specificity	Degree to which investments are specialized and have little value outside of a specific exchange relationship (Grover & Malhotra, 2003)	• Custom IT integrations and client-specific invoicing requirements raise transaction costs by limiting standardization and automation potential • Investments in buyer-specific adaptations may increase financial risk if the relationship ends or changes significantly
Frequency	The number of times a transaction occurs (Williamson, 1985)	• High invoicing volumes incentivize automation to reduce transaction costs but require strong process controls to prevent errors • Without digitalization, repetitive invoicing might increase administrative burdens

both sales/customer service and financial/administrative employees are interviewed, enabling analysis of both financial/transactional aspects and sales-linked customer relationship aspects (Dekkers et al., 2020). In this study, secondary data plays a crucial role, especially the extensive data on orders, invoices and payments, which was extracted from internal systems. All LSPs had interns as research assistants on the topic of O2C process optimization inside the company for 5 months, during a time frame between 2016 and 2021. Combining secondary data with interview information increased construct validity (Yin, 2009), and enabled the assessment of both the length of different steps in the O2C process and the main inefficiencies per step.

After collection, the data were first reviewed by the main contact person at each of the four companies, who provided additional information or corrections where necessary. Subsequently, the researchers conducted a thorough review to ensure consistency and accuracy.

Table 10.2 Case study sample

	LSP 1	LSP 2	LSP 3	LSP 4
Size (yearly turnover in €)	25-50 million	50-100 million	100-150 million	25-50 million
Main activities	(Sea and Road) Forwarding, Warehousing, 3PL	(Road) Transport, Warehousing	(Sea and Road) Transport, Forwarding, Warehousing	(Road) Transport, Warehousing, Forwarding
Most important industries of clients	Fashion	Interior design, construction	Food & Beverages, retail	Larger LSPs + luxury goods, food & beverages, chemical products
Interviews	CEO, administrative employees (2x), customer service employee	Sales manager, finance manager, head of customer service, head of credit control	CFO (3x), financial/ administrative employee, sales managers (2x), CEO, logistics engineer	Sales manager, customer service employees (2x), account manager
Secondary data	In all cases: Research assistant inside company for 5 months tracking extensive order, invoicing and payment data (MS excel-extractions from internal IT-systems), sales volumes and payment terms for all customers. Financial ratios (e.g. DPO, DSO, current ratio) and annual reports			

For analysis, we applied a thematic coding approach using a deductive framework based on established TCE factors—uncertainty, asset specificity and frequency (Williamson, 1975). This framework enabled the systematic categorization of key factors contributing to O2C process inefficiencies. Guided by the literature, these factors were further refined into sub-factors and specific items, all of which were introduced in the literature review in italics (Sect. 10.2.3). The coding process was conducted in multiple stages. Initially, each researcher independently coded the data according to the predefined framework. This was followed by a round of inter-coder agreement, in which discrepancies were discussed until consensus was reached. To ensure robustness, the final coding results were validated through iterative review and discussions.

Appendix 1 presents the complete coding structure, detailing all factors, sub-factors, and items, along with examples of coded results and corresponding interview quotes. These coded insights were crucial for the within- and cross-case analysis, providing a structured way to compare findings across the four LSPs. In the discussion, we contextualize these findings by highlighting the peculiarities of our specific sample, explaining the extent to which they may also apply to other LSPs or industries beyond logistics.

10.4 Results

The results are discussed case by case after which the cross-case analysis is done in the discussion and conclusion section.

10.4.1 Case of LSP 1

As an expeditor LSP 1 arranges sea transportation from many different parts of the world, especially from Asia, to Europe. Sometimes they also take care of follow-up transportation, from European ports to the final customer destination. This means LSP 1 frequently has to arrange multiple partial deliveries for one order, leading to changes in the regular order input. The CEO mentions this follow-up is not always registered in the same way, which results in a lack of clarity about when the service is actually delivered and when the invoice can be sent. Internal guidelines for the necessary internal collaboration required for this follow-up transportation are missing.

The CEO of LSP 1 mentions employee awareness about the importance of the order-to-cash process is high: "*everybody knows within our organisation that financial flows are very important. We try to really convince everybody of that*". The company has a high level of digitalization and most of its customers make use of its software for tracking the status of goods that are being shipped and stored. The CEO mentions most information about orders and invoices is linked to this software.

Despite this high level of information availability and awareness of employees, the majority of invoices are paid later than agreed upon. While the contractually agreed payment term with most customers is 30 days, especially the larger companies frequently pay later than agreed upon. In some cases disputes are a reason for late payment. LSP 1 currently doesn't register and monitor disputes, and a standardized approach for dispute resolution is missing.

The company works with credit limits since 2021 to make customers pay faster and limit the amount of outstanding receivables. The impact of outstanding receivables became much bigger, since the prices of the containers increased since 2020. A customer service employee elaborates on this: "*Nowadays customers easily exceed the credit limit, since the costs of sea containers are sky high*".

10.4.2 Case of LSP 2

Despite a growing demand for their services from customers in recent years, LSP 2 describes they are in a low-margin business where large customers dictate the contract terms. Especially larger customers pay later than agreed upon, resulting in actual payment terms being more than 10 days higher than what is contractually agreed upon. LSP 2 explains one of their core competencies is in relational resources, since they are flexible with customer wishes in orders. However, this flexibility also brings complications in invoicing and payments, as the CFO explains: "*a large part of the problem starts when agreements are made with customers. A lot of flexibility is given here, which means a growing amount of variables has to be taken into account for invoicing*". Once there are more exceptions to 'common' orders, there is a lack of standardisation possibilities. LSP 2 often makes adjustments to orders, even after they have already been approved, in response to specific customer wishes. Additionally, customers often don't use LSP 2's EDI system for order input, instead sending e-mails with order information that must be manually entered into the system. This contributes to a substantial amount of disputes, since the manual input is prone to mistakes.

LSP2 has the ambition to reduce disputes, it has a dispute registration system to track the causes. The sales manager explains this registration system hasn't been successful yet, since employees can choose out of 17 categories of disputes, many of which overlap. The company has recently adopted a new IT-system which gives many options for tracking the status of orders and invoices, however the CFO explains that one of the key inefficiencies in the process is employees not knowing how to use many of the system's functionalities.

10.4.3 Case of LSP 3

LSP 3 faces some of the same challenges in their O2C process as LSP 1 and LSP 2. Like LSP 2, being flexible in dealing with customer wishes is seen as important, leading to many exceptions in orders and invoices, which increases the complexity and number of input mistakes in the O2C process. Similar to LSP 1, LSP 3 frequently fails to internally register or communicate the delivery date to invoicing employees on time or at all. This results in internal confusion about the moment when they can send an invoice.

In addition, the CFO of LSP 3 mentions administrative employees often have to manually create or correct certain invoices when either the IT-system is not capable of dealing with customer specific variables for the invoices or the employees are not capable of effectively using the IT system. Customer service and administrative employees mention the workload is heavy, leading to even more mistakes made and additional corrections needed before the invoice is sent out. For these reasons, the time between delivery and sending the invoice is exceptionally long for LSP 3 (12,7 days on average), and the number of disputes is high. A logistics engineer working for LSP 3 explains they don't have a standardized approach for dispute resolution: "*more clarity about roles and responsibilities in case of disputes would be better*".

Table 10.3 Averages of Invoicing and payment time in case study sample

Time in days (averages)	LSP 1	LSP 2	LSP 3	LSP 4
Sending invoice after delivery	2.3	4.6	12.7	4.0
Contractually agreed payment term	34.3	33	30	38.6
Actual payment term	37.9	43.1	36.8	41.1
Difference contractual and actual payment term	3.6	10.1	6.8	2.5

10.4.4 Case of LSP 4

The CFO of LSP 4 explains many mistakes and disputes in the O2C process are a result of customers not using the EDI system of LSP 4 for order receipt, but simply sending e-mails which have to be manually put in the system. Especially for these types of customers problems arise with linking order information to invoice information. At the same time, the CFO of LSP 4 explains there is not enough knowledge in the company to use the system in the right way: *"There are certain pain points in specific situations that require a settlement, which seems difficult to accommodate within the IT-system. Personally, I do not have the knowledge to address these issues"*. In addition, miscommunication between the operations and sales department in many occasions led to wrong prices in orders, resulting in complications in invoicing and sometimes to disputes with customers. As in the case of LSP 3, administrative and customer service employees of LSP 4 experience work pressure as a result of complications in order and invoicing processes, and explain this is a factor contributing to even more mistakes in invoicing. Disputes are registered in a standardized way to track the causes of mistakes, but not resolved in a standardized way. In comparison to the other three LSPs, LSP 4 shows a relatively small difference between the average contractually agreed payment term and the actual payment term. The CFO explains the organization puts significant effort into communicating with buyers in case payments are not received on time (Table 10.3).

10.5 Discussion

Bounded rationality is a key contributor to the transaction costs associated with O2C processes in LSPs. The lack of integration and collaboration between departments exacerbates inefficiencies. Sales and operations departments often prioritize customer satisfaction by allowing last-minute changes to orders, while finance prioritizes timely invoicing and payment collection. This misalignment creates friction, delays, and increases the need for manual adjustments to orders and invoices.

Digitalization has the potential to mitigate these issues by streamlining communication and ensuring all departments have real-time access to consistent information. Modular IT solutions, for instance, can provide tailored functionalities for tracking delivery status, updating order changes, and automating invoice creation. However, the case studies reveal that the effectiveness of digital tools depends on employees' ability to use them. In LSPs 2, 3, and 4, a lack of IT skills among employees limits the potential of existing systems. Training programs and user-friendly interfaces are critical to addressing this challenge. While O2C literature highlights the benefits of digitalization in payments, it often overlooks the need for skilled human resources to leverage IT effectively.

Opportunistic behavior by buyers, manifesting as late payments or late approval of invoices, adds to transaction costs. Larger buyers, who wield significant bargaining power, often extend payment terms beyond what is contractually agreed upon, impacting the liquidity of LSPs. The hesitance of LSPs to use digital tools such as automated payment reminders reflects a fear of damaging customer relationships.

To address this, LSPs can implement practical, data-driven solutions like using monitoring tools to track payment patterns and identify consistently late-paying customers. Automated reminders can be introduced for routine follow-ups, while maintaining personalized communication for key accounts to preserve relationships.

Rapidly changing market conditions, such as fluctuating container prices and increasing demand for logistics services, exacerbate *environmental uncertainty* in O2C processes. These dynamics lead to more complex pricing structures and a higher frequency of manual invoice adjustments, particularly evident in LSPs 1 and 3. Additionally, the unpredictability of order volumes and delivery timelines complicates planning and coordination efforts. LSP 1 demonstrates a key solution might be the use of collaborative software. Their in-house-developed tool enables customers to share and access real-time information on demand expectations, order statuses, and invoices, significantly enhancing transparency and mitigating environmental uncertainty.

The high degree of customization required by some buyers, such as unique invoicing formats, contributes to *asset specificity* in O2C processes. Especially for LSPs 2 and 3, accommodating these specific demands often necessitates manual interventions, leading to longer lead times and increased errors. There are many O2C solution providers offering modular IT solutions that enable customer-specific adaptations without overhauling the entire system. For instance, customizable invoicing templates or integration capabilities with client ERP systems can automate complex processes, reducing manual workloads. Training employees to leverage such modular functionalities can enhance their effectiveness.

Weekly or monthly invoicing reduces the *frequency* of invoicing, which can help limit transaction costs by decreasing administrative workloads. However, since all four LSPs combine different invoicing types, such as per-order, weekly, and monthly invoicing, opportunities for standardization are reduced. Monthly invoices consolidate all services within a month, making them larger and more detailed, which increases complexity and the potential for mistakes and disputes compared to per-order invoices. From a liquidity perspective, per-order invoicing is advantageous, as weekly or monthly invoices delay payments for services delivered early in the invoicing period. To balance these trade-offs, it is crucial for LSPs to regularly monitor and evaluate the impact of different invoicing methods and frequencies (Table 10.4).

Table 10.4 Summary of main inefficiencies and potential solutions using digitalization

TCE Factor	Main inefficiencies identified	Potential solutions using digitalization
Bounded rationality (Behavioral uncertainty)	• Lack of integration and collaboration between departments (sales, finance, operations) • Frequent manual adjustments to orders and invoices due to last-minute changes, resulting in increased mistakes • Employees struggle to use existing IT systems effectively	• Provide employee training on IT tools to make better use of functionalities and reduce manual work • Use (AI-powered) validation tools to reduce invoicing errors before submission • Introduce role-based dashboards to enhance clarity on responsibilities and workflows
Opportunism (behavioral uncertainty)	• Larger customers frequently pay later than agreed upon • Buyers provide incorrect invoice data or delay approvals • Hesitation to use automated payment reminders due to concerns over customer relationships	• Use automated payment tracking tools to monitor late-paying customers • Implement tiered payment reminder systems (e.g., automated reminders for routine follow-ups, personal communication for key accounts)
Environmental uncertainty	• Fluctuating container prices and demand complicate pricing structures and invoicing • Unpredictability in order volumes and delivery timelines affects planning	• Use collaborative software to allow real-time data sharing between LSPs and customers • Deploy predictive analytics to anticipate demand fluctuations and adjust pricing dynamically
Asset specificity	• Custom invoicing formats and buyer-specific delivery requirements increase manual interventions • Specific IT functionalities requiring advanced skills that employees lack	• Use modular IT solutions that enable customer-specific adaptations • Standardize the more frequent customer-specific requirements through configurable system settings • Provide employee training on IT tools to make better use of functionalities
Frequency of invoicing	• Combining different invoicing types (per-order, weekly, monthly) reduces standardization opportunities • Monthly invoices consolidate administrative tasks, but increase complexity and potential for errors in invoices, while per-order invoicing improves cash flow but might increase administrative workload	• Implement tools to monitor and evaluate the impact of different invoicing methods and frequencies on cash flow • Use automated reconciliation tools to match invoices with orders, reducing disputes

Transaction costs resulting from bounded rationality, opportunism, environmental uncertainty, asset specificity and the frequency of invoicing are significant in all cases. There is a high volume of manual order and invoicing input for employees, which is prone to mistakes. Correcting internal mistakes made in orders or invoices requires cross-functional coordination. When mistakes lead to disputes, this necessitates additional communication with buyers and extra invoice processing work. Since the majority of invoices are paid late for all LSPs, considerable follow-up activities on late payments are required. The rapidly changing environment, especially in terms of prices and demand, requires substantial extra negotiation with customers.

In theory, higher transaction costs favour hierarchies as a *governance mode*. However, the organization of orders and payments in the O2C process of LSPs shows a different picture. Vertical integration and cooperative behaviour in the supplier-buyer relation are relatively low, which can be partly explained by opportunistic behaviour of buyers. However, internally, it seems like rules, procedures and vertical commands are largely lacking, leading to many inefficiencies. An explanation could be that the majority of the transaction costs mentioned are a consequence of frictions, what Langlois (2006, p. 1389) describes as mundane transaction costs and "seemingly dull categories of costs". These types of costs do not appear to be a top priority for managers, and therefore often remain largely unaddressed.

10.6 Conclusions

Our results support managerial and theoretical contributions in three categories. First, there is a lack of studies researching the O2C process as a whole, while our results show the value of doing so. LSPs appear to be relatively quick with order approval and order input. However, our case study results show that there are often mistakes made in the order phase, which contribute to transaction costs in later stages.

Second, our research provides insights into the TCE factors leading to higher transaction costs in the context of O2C processes of LSPs. Opportunism of buyers contributes to longer payment times. Asset specificity manifests itself in the form of buyers having very specific wishes, which lead to complications for the invoicing process. Environmental uncertainty is evident in the form of price and demand uncertainty and pricing complexity. Frequency of invoicing is generally high in logistics. However, we show that especially bounded rationality, particularly a lack of knowledge and internal collaboration between departments involved in the O2C process, has a substantial impact on transaction costs. The LSPs explain that they need greater clarity on roles and responsibilities, especially in relation to dispute management. In addition, the contrast between sales and finance goals, described by Dekkers et al. (2020, p. 7) as 'keeping the customer happy' vs. 'getting paid', is very relevant in our context. The friction created leads to 'mundane' transaction costs, which are often overlooked by managers (Langlois, 2006). LSPs can benefit from carefully monitoring the impact of long O2C processes on working capital and transaction costs, and by discussing the impact internally to create the right balance between sales and financial goals.

Our third contribution relates to digitalisation and IT usage in O2C processes, a topic that receives significant attention in the literature. Our results confirm the importance of well-functioning IT systems for the optimization of the process. LSPs could benefit significantly from solutions to reduce manual workloads and streamline complex processes. Digital solutions can enhance supply chain transparency and reduce environmental uncertainty by providing real-time visibility into order statuses, demand forecasts, and payment schedules. However, the case studies reveal that bounded rationality, stemming from employees' limited skills or knowledge, often undermines the effectiveness of these digital tools. When employees lack sufficient training or user-friendly interfaces are absent, the potential of such tools remains underutilized, leading to inefficiencies, errors, and higher transaction costs. By emphasizing both the technological and human aspects of digitalization, our findings contribute to a more comprehensive understanding of how digital tools can address inefficiencies in O2C processes.

Future research could benefit from exploring the management of the O2C process as an integrated system, rather than focusing solely on isolated components such as payment terms or invoicing digitalization. By examining how companies can holistically align cross-departmental collaboration, IT integration, and performance monitoring across all O2C stages, valuable insights could be gained to reduce inefficiencies and transaction costs. Detailed perspectives from different industries, supported by empirical data, are needed to achieve this. Our research relies on an extensive amount of primary and secondary data from LSPs in the Netherlands. However, future research might also benefit from considering data from other countries, especially those with higher average payment terms.

Acknowledgements This research was conducted in the context of two NWO SIA RAAK projects. These are "Order-to-Cash Process Optimization for SMEs", under grant number "RAAK.MKB12.028", and "Supply Chain Finance for SMEs", under grant number "RAAK.MKB19.015".

Appendix 1 Coding structure with example results

Factor	Sub-factor	Item	Examples
Behavioural uncertainty	Bounded rationality	Restriction in using available information	• Lack of employee capabilities in using O2C related IT-functionalities. Example quote of customer service employee 1 of LSP 4: *"There are certain pain points in specific situations that require a settlement, which seems difficult to accommodate within the IT-system. Personally, I do not have the knowledge to address these issues."* • Not having sufficient IT resources to efficiently manage O2C process steps. Example quote of sales manager of LSP 2: *"The automatic pricing in the IT system is static, resulting in various exceptions for each customer. Checking and billing these exceptions takes a lot of time and is very error-prone."*
		Intra-firm collaboration	• Lack of clarity of roles and responsibilities. Example quote of logistics engineer of LSP 3: *"more clarity about roles and responsibilities in case of disputes would be better"* • Sales and finance not aligned. Example quote of customer service employee of LSP 1: *"Sales makes the agreements with the customer, who also determines the rates. However, sometimes when a shipment arrives, you need to "fill your contract," as it is called, which includes all the terms and conditions. But sometimes, this doesn't happen. In that case, you have to report it to sales, but by then, the shipment has already been in transit for 3 to 4 weeks."* • Delivery moment not registered, resulting in confusion about when invoice can be sent. Example quote of customer service employee 2 of LSP 4: *"For most customers, the unloading date is not recorded in the system."*
	Opportunism	Buyer uses bargaining power	• Large buyers dictate terms and often don't live up to agreed upon payment terms, or take little effort in approving invoices faster. Example quote of CFO of LSP 3: *"For one of our top-3 customers the invoice approval time is very unpredictable. Sometimes it takes 20 days, sometimes 25 days and sometimes even 40 days."* • Buyers requiring invoicing on a monthly or weekly basis instead of per order introduces greater variability in the administrative process for each customer, leading to longer average payment times. *Example:* ERP data analysis reveals that all included LSPs manage a mix of customers with different invoicing preferences—Some requiring invoices per order, others on a weekly basis, and others on a monthly cycle

		Incomplete/ misleading information	• Buyer not providing correct invoice information or not providing information on when invoices are approved or paid. Example quote of sales manager of LSP 2: *"The information received from the warehouse is also crucial. It happens that an order does not match what is delivered. Either because the customer did not provide the correct information, or because it got lost somewhere. This then requires repair, which takes time."* • Buyer gives limited room for communication possibilities about late payments. Example quote of finance manager of LSP 2: *"A major client of ours has a payment term of 60 days. It sometimes happens that it is unclear who within the company is responsible for the invoice. As a result, the invoice is returned, the issue needs to be investigated again, and the payment term starts over. This costs a tremendous amount of time and money."* • Goal incongruence: Buyer wants to pay late, supplier wants early payment. Example quote of finance manager of LSP 3: *"Buyers postpone payment terms, arbitrarily extending them from 30 to 90 days (…) and the buyers promise that they will approve invoices faster, but it doesn't happen—It's nonsense. In the end, it costs me money."*
Environmental uncertainty	Business context uncertainty	Macro-economic changes	• Inflation and rapidly changing container prices. Example quote of customer service employee LSP 1: *"Nowadays customers easily exceed the credit limit, since the costs of sea containers are sky high"* • Changes in demand. Example quote of customer service employee of LSP 1: *"My customer is currently so unpredictable that things can still change at the very last minute. The delivery date might change, or even what we're supposed to deliver. Sometimes we get a message from the warehouse saying, "we don't have this in stock anymore, but we do have that." then we have to rearrange everything."*
		Institutional changes	• Take-over of companies changes administrative organization. Example quote of financial/administrative employee of LSP 3: *"Over the past four years, we've taken over multiple companies, and each one had its own way of invoicing. That made our O2C process a real challenge!"* • Changes of personnel in the organization. Example quote of customer service employee of LSP 4: *"A few months ago, (name employee) was let go. (name employee) was extremely fast and had a lot of experience. The replacement is simply slower, which increases the pressure on him and places a larger burden on others."*
	Market complexity		• Complex pricing structure in industry (e.g. multiple partial deliveries for one order is common - giving complications for invoicing). Example quote of customer service employee LSP 4: *"Shipments need to be consolidated and billed together. The system currently can't handle this, so it has to be done manually."*

(continued)

Factor	Sub-factor	Item	Examples
Asset specificity	Investment		• Being flexible with customer wishes, leading to many exceptions to 'common' orders and a lack of standardisation possibilities. Example quote of financial manager of LSP 2: *"A large part of the problem starts when agreements are made with customers. A lot of flexibility is given here, which means a growing amount of variables has to be taken into account for invoicing."*
	Skills and resources		• Specific IT system functionalities requiring specific skills. Example quote of sales manager of LSP 4: *"The system is so complicated that a certain level of expertise is needed for such pain points. This expertise is required to support me in addressing the various issues."*
Frequency	# of invoices		• Transactions are recurring in every case and are relatively frequent. Example: For the top-25 customers LSP 1 sent out 4439 invoices in 2020

References

Caniato, F., Gelsomino, L. M., Perego, A., & Ronchi, S. (2016). Does finance solve the supply chain financing problem? *Supply Chain Management: An International Journal, 21*(5), 534–549.

Cronie, G., De Goeij, C., Holman, R., & Hoek, N. (2023). *Surviving and thriving in times of uncertainty: order-to-cash process optimization for SMEs*. Windesheim White Paper Series, Vol. 1. Retrieved January 6, 2025, from: WZ OND SCF - Lead Magnet Campagne Formulier.

Cronie, G., De Goeij, C., Wendriks, T., Van der Zee, J., & Van der Riet, A. (2024). *Order to Cash Monitor 2024-2025* (in Dutch). Retrieved January 6, 2025, from: Order 2 Cash Monitor 2024-2025 | Verbeter de liquiditeit van je onderneming.

De Goeij, C. (2021). *Logistieke sector factureert laat en krijgt laat betaald* (in Dutch). Logistiek.nl, Retrieved October 29, 2024, from: www.logistiek.nl/181167/logistieke-sector-factureert-laat-en-krijgt-laat-betaald

De Goeij, C., Onstein, A. T. C., & Steeman, M. A. (2016). Impediments to the adoption of reverse factoring for logistics service providers. In H. Zijm, M. Klumpp, U. Clausen, & M. Ten Hompel (Eds.), *Logistics and supply chain innovation* (pp. 261–277). Springer Int. Publishing.

De Goeij, C., Gelsomino, L. M., Caniato, F., Moretto, A. M., & Steeman, M. (2021). Understanding SME suppliers' response to supply chain finance: A transaction cost economics perspective. *International Journal of Physical Distribution & Logistics Management, 51*(8), 813–836.

Dekkers, R., de Boer, R., Gelsomino, L. M., de Goeij, C. A. J., Steeman, M., Zhou, Q., Sinclair, S., & Souter, V. (2020). Evaluating theoretical conceptualisations for supply chain and finance integration: A scottish focus group. *International Journal of Production Economics, 220*, Article 107451.

Dello Iacono, U., Reindorp, M., & Dellaert, N. (2015). Market adoption of reverse factoring. *International Journal of Physical Distribution & Logistics Management, 45*(3), 286–308.

Eisenhardt, K. M. (1989). Building theories from case study research. *Academy of Management Review, 14*(4), 532–550.

Elliot, V. H., De Goeij, C., Gelsomino, L. M., & Woxenius, J. (2020). Supply chain finance is not for everyone. *International Journal of Physical Distribution & Logistics Management, 50*(9/10), 775–807.

Fairchild, A. (2004). Using electronic invoicing to manage cash forecasting and working capital in the financial supply chain. In *ECIS 2004 Proceedings*, 29.

Fairchild, A. (2005). Intelligent matching: Integrating efficiencies in the financial supply chain. *Supply Chain Management: An International Journal, 10*(4), 244–248.

Grover, V., & Malhotra, M. K. (2003). Transaction cost framework in operations and supply chain management research: Theory and measurement. *Journal of Operations Management, 21*(4), 457–473.

Hobbs, J. E. (1996). A transaction cost approach to supply chain management. *Supply Chain Management: An International Journal, 1*(2), 15–27.

Hofmann, E., & Kotzab, H. (2010). A supply chain-oriented approach of working capital management. *Journal of Business Logistics, 31*(2), 305–330.

Hu, P. (2015). *Predicting and improving invoice-to-cash collection through machine learning*. Doctoral dissertation. Massachusetts Institute of Technology.

Intrum. (2024). *European Payment Report 2024*. Retrieved November 4, 2024, from: European Payment Report 2024 | Intrum.

Ireland, R. D., & Webb, J. W. (2007). A multi-theoretic perspective on trust and power in strategic supply chains. *Journal of Operations Management, 25*(2), 482–497.

Jimenez-Martinez, J., & Polo-Redondo, Y. (2004). The influence of EDI adoption over its perceived benefits. *Technovation, 24*(1), 73–79.

Koch, B. (2019). *The e-invoicing journey 2019-2025*. Billentis, July, 2019. Retrieved November 4, 2024, from: The E-Invoicing Journey 2019-2025.

Korotina, A., Müller, O., & Debortoli, S. (2015). Real-time business process intelligence. Comparison of different architectural approaches using the example of the order-to-cash process. *Wirtschaftsinformatik Proceedings, 2015*, 1710–1724.

Langlois, R. N. (2006). The secret life of mundane transaction costs. *Organization Studies, 27*(9), 1389–1410.

Nienhuis, J. J., Cortet, M., & Lycklama, D. (2013). Real-time financing: Extending e-invoicing to real-time SME financing. *Journal of Payments Strategy & Systems, 7*(3), 232–245.

Perego, A., & Salgaro, A. (2010). Assessing the benefits of B2B trade cycle integration: A model in the home appliances industry. *Benchmarking: An International Journal, 17*(4), 616–631.

Rindfleisch, A., & Heide, J. B. (1997). Transaction cost analysis: Past, present, and future applications. *Journal of Marketing, 61*(4), 30–54.

Silvestro, R., & Lustrato, P. (2014). Integrating financial and physical supply chains: The role of banks in enabling supply chain integration. *International Journal of Operations & Production Management, 34*(3), 298–324.

Van der Vliet, K., Reindorp, M. J., & Fransoo, J. C. (2015). The price of reverse factoring: Financing rates vs. payment delays. *European Journal of Operational Research., 242*(3), 842–853.

Williamson, O. E. (1975). *Markets and hierarchies*. The Free Press.

Williamson, O. E. (1979). Transaction-cost economics: The governance of contractual relations. *The Journal of Law and Economics, 22*(2), 233–261.

Williamson, O. E. (1985). *The economic institutions of capitalism: Firms, markets, relational contracting*. The Free Press.

Yin, R. K. (2009). *Case study research: Design and methods* (4th ed.). Sage Publications.

Chapter 11
Empowering Smallholder Farmers with Digital Supply Chain Management Tools

Ozgu Turgut

Abstract Smallholders are small-scale farmers who manage areas varying from less than 1 ha to 10 ha, and represent an important portion of the farming population, especially in developing countries. However, they are usually abandoned in a market dominated by big players or lack government support. Their poor conditions not only affect the well-being of these families, but also—as can be observed from the example of Turkey—the isolated position of these people causes the emergence of more intermediaries than necessary. This, in turn, results in hyperinflation, where neither the customer nor the producer is financially satisfied. By the same token, product waste due to poor logistical conditions increases significantly. Based on these observations, a digital market model together with an AI- and optimization-integrated decision support framework is proposed in this chapter. Analytical (AI algorithm-based predictive and mathematical modelling-based prescriptive) components of the digital tool are explained rigorously. Additionally, a case study is provided to elaborate on the potential contributions of the tool.

Keywords Digital · Decision Support · Optimization · Agriculture · Artificial Intelligence

11.1 Introduction

Agricultural enterprises drive worldwide growth and development. They provide expertise, technology, and finance while producing safe and nutritious food to especially growing populations (Elufioye et al., 2024). According to the report of FAO

O. Turgut (✉)
Department of Industrial and Systems Engineering, Turkish German University, Istanbul, Turkey

Y. Khojasteh, S.M. M. Jahan Abadi (eds.), *Digital Transformation of Supply Chains*, International Series in Operations Research & Management Science 372, https://doi.org/10.1007/978-3-032-01206-7_11

(2012), 80% of the farmland in sub-Saharan Africa and Asia is managed by smallholders (Smallholders are small-scale farmers who manage areas varying from less than 1 ha to 10 ha.). Out of the 2.5 billion people in poor countries living directly from the food and agriculture sector, 1.5 billion people live in smallholder households. Women comprise an average of 43% of the agricultural labor force of developing countries. For instance, this ratio is up to almost 50% in Eastern and Southeastern Asia and sub-Saharan Africa. This shows how important to support smallholders in terms of female-employment as well (World Food Program, 2024; INTELLIAS, 2024). Unlike widespread perceptions, sustainable smallholders can be productive. That is why agricultural enterprises are crucial from the Sustainable Development Goals (SDGs) point of view as well. A large study examining smallholder agriculture covering 286 projects, over 37 million hectares in 57 developing countries, found that when sustainable agriculture was adopted, average crop yields increased by 79%. Well-managed smallholder systems invest in building soil biomass and soil vegetative cover, which improves water filtration in case of floods and moisture retention in case of drought (Litvinova et al., 2020).

Although smallholders produce most of the world's food, they tend to be food insecure themselves: globally, they form the majority of people living in poverty. Helping to raise their incomes and improve their livelihoods means serving sustainable food systems, advancing food security and achieving Zero Hunger (Akkem et al., 2023). The agriculture supply chain (i.e. value chain) is defined as the process of moving food from farms to consumers (Hansson et al., 2023). Major global issues of agriculture supply chains are grouped under the headlines of weather dependence, perishability, variability, labor issues, and sustainability. Although the issues can be generalized under broad umbrella terms, the problems and solutions are usually unique to each region. As one of the food inflation champions of the World, Turkey for instance is notorious for the difference between the prices that farmer charges and the consumer embraces at the end. Which is a great indication of problems related to poor supply chain management. In this regard, this study proposes a baseline version of a digital tool that can remedy this problem by empowering the producer (i.e. smallholder) with data and at the same time easing the sky-rocketing food prices coming out of the consumers' pocket by avoiding extra brokers and logistics negotiators.

Two separate analytical modules are planned to be embedded into the tool: (1) Forecasting (2) Logistics planning. The proposed methodology for the backend of the first (i.e. Forecasting) module is an AI algorithm called "Transformer", which is the well-known algorithm behind many of the latest large language models (LLM). The output of this module will be the forecasted harvest quantity of each smallholder in a unit period. The proposed approach for the second (i.e. Logistics) module comprises mathematical modelling and optimization which support planning vehicle dispatching and pick-up routes to transport from the farmer. By functionality, the first module is predictive, the second module has a prescriptive nature. Implementation and algorithm details of each module are demonstrated using a sample case.

Overall, the proposed steps are parts of a business model which relies on digital supply chain management. The main idea behind the tool is promoting group action by attracting farmers to join logistics planning, and thus it acts as an intuitive catalyst for a "virtual union". Recall that empowering the farmer unions is one of the major remedies recommended by experts for smallholders' wellbeing.

Contribution of the study is an alternative digital market model to improve the management of a specific but significant type of agriculture supply chain which relies on an AI and optimization integrated decision support framework. The AI based predictive module of the tool -to the best of our knowledge- is a unique application of Transformer algorithm for time series data in agricultural context. On the other hand, the methodology integrated for the prescriptive module is a standard capacitated vehicle routing with heterogenous fleet with minor performance enhancements. The quality and the speed performance of the methodology is prioritized over novelty at this module for the sake of deployment practicality.

The manuscript progresses as follows: In the subsequent section the most recent noteworthy studies about farmer oriented digital supply chain management efforts are covered together with evident literature for the analytical methodologies being used. The succeeding section describes the proposed digital business model together with the components of the decision support schema behind the digital tool. Finally, a rigorous case study precedes the conclusion to reveal the details and contribution.

11.2 Literature Review

From three main aspects of sustainability, namely economic, social and environmental, economic perspective still has the utmost importance by placing the most urgent pressure on the producer especially in developing countries where the governmental support is unreliable. In recent years, the studies of Business Model Innovation (BMI) and Circular Economy (CE) issues have been much debated in the literature (Basile, 2021). Moreover, sustainable development is enhanced by the Green Economy (GE) industry descriptions or business analyses (Goyal et al., 2018; Lambert & Davidson, 2013). In this vein, Hansson et al. (2023) explores the interplay between managerial cognition and business decisions as revealed in semi-structured interviews from the sustainability point of view; and highlights the importance of knowledge transfer and government subsidies. Talerngsri-Teerasuwannajak and Pongkijvorasin (2021) show that end-market that values quality can significantly shape farmers' behaviors toward sustainable practice.

Vlachopoulou et al. (2021) investigate five indicative examples of "AgriFood-Tech" models, using the Business Model Canvas (BMC) framework to understand their building blocks and the required decisions that create, deliver, and capture value. Consequently, they highlight the importance of online sharing of information between the stakeholders and logistics services in the agribusiness sector. Shukla and Sengupta (2021) modify the "bottom of pyramid (BoP)" proposition which argues that firms can make a fortune by selling to poor people and at the same time help improve livelihood conditions in the region (Karnani, 2006). The major contribution in their work is recognizing the disadvantaged position of the farmer. Therefore, they modify the idea as "poor as supplier". Hence, they show how the design of incentives to stakeholders becomes a crucial factor toward successful implementation of BMI in subsistence markets. Mahdad et al. (2022) consider what it might take for agri-food firms to capitalize on interdependence by moving from traditional business models to business models based on collaboration and open innovation. They highlight once again farmer, and the consumer are the two key players of the agri-food.

Thus, according to the Brokerage business model, e-brokers develop a platform to enable the matching between potential buyers and sellers, and facilitate transactions by charging a fee or commission, acting in business-to-business (B2B), business-to-consumer (B2C), or consumer-to-consumer (C2C) markets (Strzebicki, 2015). Specifically, according to Rappa, different types of this model can be used (Rappa, 2000). It could be used as "Marketplace Exchange" which provides specific services during the whole process of the transaction, akin to a "Transaction Broker", undertaking a third-party payment system for trading parties, and as "Search Agent" or "robot", to find out specific information about the price and availability of a product. "Infomediaries (information intermediaries)" business models gather and share focused information on behalf of farmers, and supply chain partners for AgriFood products and their potential customers. Furthermore "The Data-Driven (DD)" model refers to organizations using data as a key resource for their business execution. Specifically, big data analysis drives cost-effective, innovative information processing, supporting enhanced analytics, insight, and decision-making. In this vein, "Data as a Service" refers to new ways of collecting, aggregating, and analyzing data, in order to organize a wide range of public and private data by the involved AgriFood parties and partners (Brownlow et al., 2015). Santhanam et al. (2018), explores drivers of business model innovation in food processing, and highlights that the increasing data availability leads companies to leverage advanced analytics to generate insights and learn how to run their businesses more efficiently (Gray et al., 2018). Nabradi and Kovács (2020) proposed e-Marketplace models and defined it markets that connect farmers, partners, and consumers based on a platform to enable accessing to information regarding trading markets and resources and easing exchanges between the involved parties. The proposed "Cloud systems and SaaS (Software as a Service)" infrastructure provides opportunity for farmers to pay a monthly fee instead of buying software, which provides access to it, while it is hosted on a cloud (Singh et al., 2020). Lezoche et al. (2020) emphasizes once again "Digital Transformation" builds business models that enable the customer and farmer win, leading towards "Everything-as-a-service" or "Capability-as-a-service" business model (called Xaas) in which customers only pay for the services or time that they actually use the product (Vlachopoulou et al., 2021).

As part of this digital information, AI technologies, including machine learning and big data analytics, have contributed to real-time data analysis, predictive maintenance, and resource optimization. Although there are many challenges waiting to be addressed such as data quality, infrastructure development, connectivity, cost, privacy, user adoption, and skill gaps among agricultural professionals, the integration of Artificial Intelligence (AI) in agriculture supply chains has revolutionized the approach towards demand forecasting and supply optimization, offering unmatched accuracy and efficiency. Elufioye et al. (2024) recommends that industry leaders and policymakers invest in infrastructure, promote AI research and development, and provide training to facilitate AI adoption for the following deployments areas:

- Y*ield forecasting:* Pal (2024) discusses the integration of AI into JIT systems, focusing on enhancing the accuracy of demand forecasting. Similarly, Padhan (2012) looked at the ARIMA model in India for forecasting agricultural productivity. Initially it was used by Badmus and Ariyo (2011) to anticipate the cultivated area and output of maize in Nigeria. The study relies on time series data from 1970 to 2005 (Akkem et al., 2023).
- P*rediction of market components*: It involves a comprehensive analysis of various factors influencing supply and demand, such as market trends, consumer behavior, commodity price prediction, and geopolitical events. Terada et al. (2023) emphasize the use of Auto-Regressive Integrated Moving Average (ARIMA) and Long Short-Term Memory (LSTM) models, utilizing historical transaction records to improve the performance of the demand forecasting system. Recently Amellal et al. (2024) use BERT transformer model for customer sentiment analysis, the Gated Recurrent Unit (GRU) model for demand forecasting, and the Bayesian Network for price prediction. Also, Ouaddi and Benhra (2024) use Transformer for demand forecasting.
- *Weather forecasting.*
- *Anomaly detection*: Disease and pest management are two crucial issues addressed by AI algorithms.

- In this study the whole framework that integrates an AI based forecasting technique with mathematical modelling-based optimization is proposed. The number of studies which present integrated frameworks have increased in recent years (Lu et al., 2024; Xu & Bo, 2024) due to the fruitful synergetic potential of innovation that stems from the power of data technologies as well as solid quality of solutions generated by well-known OR methodologies.

11.3 Digital Tool

The structure of the proposed digital unit and corresponding business model is summarized in Fig. 11.1 above. In its most generic form, the proposed tool is designed to organize and generate shipping advice to the smallholder which results in reduced logistics costs and increased "negotiation power". In this context, the type of

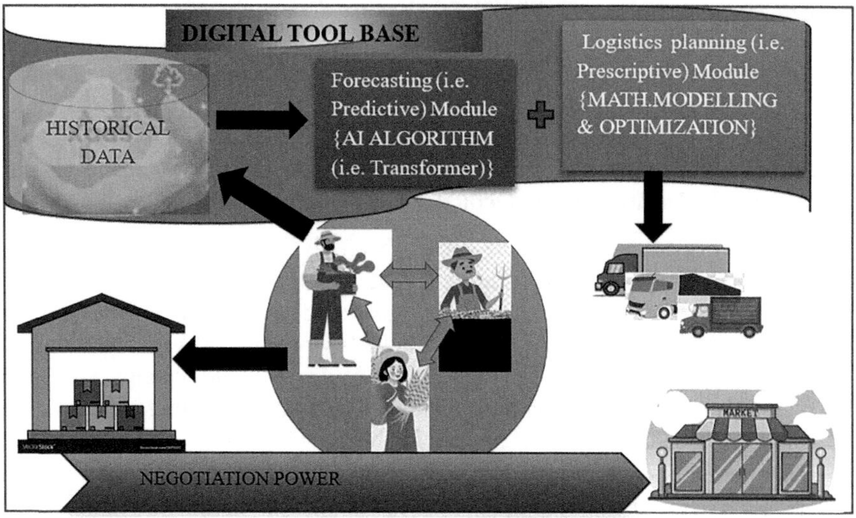

Fig. 11.1 Summary of the proposed (digital) business model

negotiation power[1] that is gained through the proposed tool is due to the knowledge of BATNA (the best alternative to a negotiated agreement) which improves according to the overall market insight provided to the smallholder. The necessary inputs of the analytical modules are historical data and information regarding nearby smallholders and logistics providers which can be obtained through www or mobile network. The only information that a farmer should enter is the type of product(s) she wants to transport and her address, all of which can be guaranteed to be kept confidential. The required historical data pertains to the last available harvest amounts of the product(s) which do not have to be available in precise or high volumes at the beginning. Indeed, this information can be scrapped from www or a relevant database for initial forecasts. With the help of the digital tool which can be deployed as a Xaas model, the farmer is able to establish its own "virtual union" which is endowed with data analytics. This does not only lead to benefitting from economies of scale which leads to reduced cost and easy access to "cold chain" transportation means, but also it helps to improve profit through empowering the smallholder with the information of alternative selling and supplier option(s).

11.3.1 Predictive Unit

AI algorithms such as Transformer and MAMBA are the most popular methods of recent years owing to the recent amelioration of LLMs (Large language models) and fusion models. In this study, the Transformer architecture is adopted to forecast

[1] https://www.pon.harvard.edu/daily/negotiation-skills-daily/types-of-power-in-negotiation/.

the next period's yield amount for a smallholder which acts similar fashion to the guessing of next word in the LLM context. This means, the "Transformer" code developed for this model process time-series-data, instead of meaningful sentences or word phrases. Accordingly, instead of embedding the words -so called tokens- as inputs as it would happen in an LLM setting, the time-series-data based number intervals are fed as input in the proposed structure. Although MAMBA, which is essentially a state-space model that is deployed using a smart matrix structure, became popular due to its time efficiency, attention is still competitive in terms of accuracy. When the decision does not have to be dynamic (i.e. the decision period allows us to wait for more than a couple of seconds) and the input size is not huge, time efficiency is not the priority, instead the accuracy of the estimates makes much difference. Therefore, in this study Transformer architecture is preferred in line with the general recommendations.

Transformer is an algorithm that learns mapping an input to the right output. In LLM context the right output is the next word/phrase; in image processing the missing pixels whereas in time series context, it is the value of the upcoming periods. Based on this, if the historical input is a series of nonnegative values (in our context, it is harvesting amount of each small holder), the available data set can also serve to deduce the possible range of outcomes. Then, this range of possible outcomes can be divided into equal intervals. Thus, the embedding maps the actual harvesting quantities to the indices of these intervals. Hence, input and output vectors are discretized through this embedding structure. The interval size used in discretization (i.e. unit embedding) is recommended to be selected based on logistical practicalities (Ex: unit carriage size).

The UNet architecture of the Transformer that is used in the case study is represented below (Fig. 11.2). Major building blocks of this structure are convolution and the "Attention" mechanisms. Convolution generates deep representations of the time series by increasing channels. Attention helps to detect extensive autoregressive behaviors in the series. In the baseline version of the proposed digital tool this component is used to forecast the weekly harvesting schemas of the smallholders.

11.3.2 Prescriptive Unit

After obtaining future values of the harvest for each farmer, it is fed into the next module which relies on a mathematical model and an optimization solver to determine optimum fleet to dispatch and corresponding routes. At this point, it is assumed that the farmers will make a deal with possible logistics providers will have already planned where to consolidate the products gathered from all the smallholders (e.g. a union depot, supermarket or factory storage) which is the destination. The tool draws all available marginal cost (per km) of collaborating logistics providers to the tool together with the location of the destination. Eventually, they will be ready to approve the order of the prescribed number and type of vehicle from the

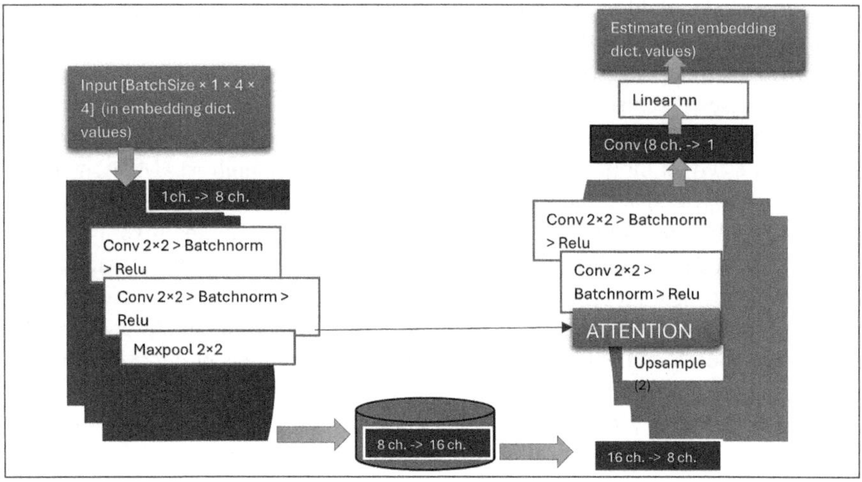

Fig. 11.2 The Transformer Architecture built in the tool that takes the embedding of historical harvest data in and generates the forecast for the next period

corresponding logistic firm. The following is the mathematical model that provides minimum cost vehicle type and route plan:

Sets:

F: Farmers set denoted by *f*.

N: Nodes set which is defined as $\{ F \cup destination\}$, denoted by *i* or *j*

M: Type of vehicles denoted by *m*

V_m: Vehicles of type *m* denoted by v_m

C: Connections (possible roads) set between nodes

Parameters:

$Cost_{ijm}$: Cost of travelling between node *i* and node *j* by vehicle type *m* (Marginal cost per km[2])

$Demand_f$: Harvest of farmer *f* that require a vehicle (i.e. transportation)

$VCap_m$: Maximum load capacity of vehicle type *m*

Decision Variables:

r_{ij} : Binary decision variable that takes value 1 if the road between node i and j is to be travelled

x_{ijv_m} : Binary decision variable that takes value 1 if the road between node i and j is to be travelled by vehicle v_m

z_{ijv_m} : Continuous (non-negative) decision variable that shows how much the vehicle v_m is carrying when it leaves node i to reach node j.

[2] https://www.scdigest.com/ONTARGET/24-06-26_ATRI_TRUCKING_COSTS_2024.PHP?cid=21472&ctype=content.

$$\min \sum_{ij\epsilon C; v_m \epsilon V_m; m\in M} Cost_{ijm} * x_{ijv_m}$$

s.t.

$$\sum_{i\in N} r_{ij} = 1 \quad \forall j \in N \tag{11.1}$$

$$\sum_{j\in N} r_{ij} = 1 \quad \forall i \in N \tag{11.2}$$

$$\sum_{m\in M; v_m \epsilon V_m} x_{ijv_m} = r_{ij} \quad \forall (i,j) \in C \tag{11.3}$$

$$\sum_{i\in N} x_{ijv_m} = \sum_{\forall i\in N} x_{jiv_m} \qquad \forall j \in N; m \in M; v_m \epsilon V_m \tag{11.4}$$

$$\sum_{i\in N, m\in M; v_m \epsilon V_m} z_{ijv_m} + Demand_j = \sum_{i\in N, m\in M; v_m \epsilon V_m} z_{jiv_m} \quad \forall j \in N; \tag{11.5}$$

$$z_{ijv_m} \geq Demand_i * x_{ijv_m} \qquad \forall (i,j) \in F; m \in M; v_m \epsilon V_m \tag{11.6}$$

$$z_{ijv_m} \leq \left(VCap_m - Demand_j\right) * x_{ijv_m} \qquad \forall (i,j) \in F; m \in M; v_m \epsilon V_m \tag{11.7}$$

In this model Eqs. (11.1) and (11.2) are the standard flow balance constraints of vehicle routing problem (VRP) without any resource restrictions which ensure all the farmers are visited and reaches to the destination. In addition to standard VRP, for our case, farmers generally need to plan for the type of vehicle they want to use as well, each of which is capacitated differently. Therefore, the formulation requires heterogenous fleet components. Accordingly, a binary variable (x_{ijv_m}) and a non-negative (z_{ijv_m}) keeps track of the vehicle is used and how much carried between the nodes of i and j respectively. Eq. (11.3) links the original road indicating binary, r_{ij} with the vehicle notified one, x_{ijv_m} while ensuring only one type of vehicle serves a farmer simultaneously. Equation (11.4) ensures that the entering vehicle leaves the node. Eq. (11.5) is the flow balance constraints in terms of the carried quantity, z_{ijv_m} considering the truck capacity and the predicted harvest of the smallholder. Finally, Eqs. (11.6) and (11.7) set the lower and upper bounds to carried quantity, z_{ijv_m} by ensuring the continuity of the path, demand quantity of the farmer as well as capacity of the vehicle. Although Eq. (11.6) does not change the result, it makes the formulation tighter and acts as valid cut which reduces the solving time.

The original r_{ij} variable of the vehicle routing formulation is kept in case the model requires customization with capacity cuts when the size of the virtual union gets bigger. Similarly, if the number of farmers increases the solving algorithm can be changed; for instance, it can be attacked using a branch and price schema as defined by Desrosiers et al. (2024). It should be noted that the model does not have to include time windows since farmers are generally flexible about dedicating time for loading. Hence there is no need to add time-window-related constraint in this context. The unit can be used for multiple product cases. In that case the predictive module will require historical data of all product portfolio and the resultant forecast of each product will be used to plan dedicated plans for each product assuming the destinations do not have to be same.

11.4 Case Description

To demonstrate how the proposed tool work in the digital business model, a pilot region in Antalya (Turkey) is selected. In particular, smallholders of the region who produce tomato are of interest for the case study. For the sake of improved precision, the smallholder classification is intensified further with two subgroups: smallholders who own less than 5 ha field, and the ones who own between 5–10 ha. The subgroups are called "mini farmer" and "mid-size farmer" respectively throughout the analysis. In this regard, the proposed digital market has capacity to gather all the mini and mid-size farmers who plant tomatoes as they download the application or register through the web site. Then by entering the type of their product, and if available weekly harvest amount of the last season, they will be able to get weekly harvest forecasts from the predictive module and proceed with prescriptive module to get their optimal logistics plan up until the market. The case is rich enough to be able to observe the common challenges of smallholders that stem from their individual production and transportation capacities.

Many smallholders own minivans or trucks. However, when the level of average yield is considered, these vehicles are not enough to transport all the production efficiently. In addition, economies of scale highlight using consolidated carriage when all the individual transportation expenses are considered. The transportation is handled generally through standard carriage boxes with 365 × 565 × 218 mm (inner) dimensions. On average this type of box can hold approximately 20 kgs of tomato. The calculations for the average weekly harvest of a smallholder are presented in Table 11.1 which is the historical data that is required be inputted into the prescriptive module. In this table, "Total kg of production" is calculated by multiplying the planted area from each smallholder type with the yield information of the region (150,000 kg/ha) and 0.5 (since there are generally 2 harvesting seasons). Then this quantity is converted into carriage units using the dimensions of the boxes mentioned above. This is followed by projections per smallholder (i.e. divided by 6 smallholders in each subgroup) and per week (i.e. divided by the number of production weeks within the harvest season). Based on the calculations, the mean transportation demand of a week can go as high as 6 vehicles of tilt truck trailer type which has approximately 2763 boxes of capacity (Fig. 11.3).

As mentioned above, the selected product of interest in this case study is tomato. Antalya region (Turkey) is the champion of last year in terms of production volume according to the reported numbers. Also, tomato is a very common and important

Table 11.1 Calculation of smallholder details for the case study

	Tomato Area planted in Antalya	Total kg of production	Total production in carriage units (for harvesting season long)	Production per smallholder in carriage units (for harvesting season long)	Average amount of carriage units delivered by a smallholder (weekly)
Mini farmer	14.1	1,060,804.7	53,040.2	9000.0	***529***
Mid-size farmer	38.2	2,864,172.8	143,208.6	24,300.0	***1429***

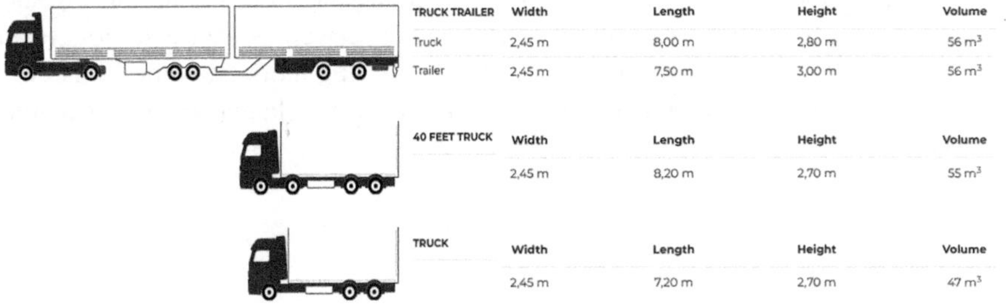

Fig. 11.3 Sample Vehicle Options from https://tirport.com/en/truck-dimensions/

product for most farmers on a global scale as well. However, it has its intricacies which affect the logistics operations which in turn change the profit amount significantly. The optimum carrying heat is between 10–13 °C for green tomatoes, 7–10 °C for pink tomatoes with 85–95% proportional moisture. Lower or higher storage temperature results in waste of the product. These facts motivate arranging dedicated logistics, i.e. "cold chain", even more.

The selected pilot region's last formally reported production numbers are 150 tons/ha from 20,000 ha dedicated farming field (TEPGE, 2023). Since the available information does not contain details for the individual smallholder of the region, the nationwide smallholder percentages are used to project the number of smallholders and their total production amount. The Government Statistic Institute[3] provided the information that 0.26% of total farming area belongs to smallholders. Among this 0.07 pertains to smallholders who own less than 5 ha field, i.e. "mini farmer", while the rest own between 5–10 ha which refer to the "mid-size farmer". When these ratios are multiplied with the dedicated field of the region (i.e. 20,000 ha), the mini-farmers plant around 14.1 ha, while the mid-size farmers have 38.2 ha of field for tomato. Then using the average farm size quantities, from the same source[3], approximate number of smallholders from each subgroup who harvest tomato in the Antalya region can be calculated. That is, considering a mini-farmer owns around 2.4 ha while a mid-size farmer owns 6.48 ha farm on average, the region might

[3] TurkStat, Agricultural Farm Structure Survey, 2016.

contain 6 mini and 6 mid-size farmers. This means the new business model must take care of the logistics planning of around (6 + 6)12 producers. Planning is based on the standard carriage units used by the farmers (i.e. box with 365 × 565 × 218 mm inner dimension). It is assumed that the tomato density of each box is around 20 kg. Based on this practical implication, harvest value embedding is prepared by dividing values in kgs by multiples of 20. Therefore, the output of the prediction is generated in terms of the number of carriage boxes instead of kgs; which means one can simply multiply back the prediction amount with the same scaling value in order to assess the output values in kg.

It is also mentioned above that farmers generally own minivans (6.8 m^3) or cargo truck (18 m^3) type of vehicles whose capacity cannot exceed 160 and 460 carriage units respectively. On the other hand, by connecting the proposed tool to logistics providers, bigger options that allow consolidation as well as access to cold-chain equipment (Fig. 11.3) can be arranged. That is, serving more than one farmer at a time with specially equipped storage conditions, i.e. better quality becomes plausible.

11.5 Results

As planned, the case study starts by running the predictive module. The module is used to estimate the time series for the weekly harvest amount of each smallholder. Figure 11.1 indicates that historical data is a key input of the predictive module. To start with, a real dataset of tomato harvest which belongs to another region of the World is adjusted to the pilot region. The raw time series represents is obtained from MIT's Garden data from Kaggle.[4] This set contains the harvesting amounts from a small vegetable garden in US. By filtering for the "big Beef" type of tomatoes within this data set, the weekly behavior of the harvesting volume for the focused product type can be obtained. Scaling this line in a way to match the reported total production values of the focused region gives an input (i.e. historical) time series for the prediction module of the tool.

When the range of the final weekly harvesting values are examined, 6174 number of carriage units appears as the maximum value. Considering that in a week no harvest is also possible, the data is assumed to be fluctuating within [0, 6174]. The size of the embedding vector (i.e. length of the token dictionary) is set as 100. Hence a grid size of 6174/100(=61.74) is needed for equally likely representation. All the historical values are converted to established embedding based on this grid dimension. Then the scaled time series fluctuated by picking a random coefficient between 0,90 and 1.10 to generate individual historical data for each of the 12 smallholders. Figures 11.4 and 11.5 represent the raw and the processed data used as input to the predictive module. In summary, these values are the weekly harvest amounts of 12

[4] https://www.kaggle.com/datasets/umerhaddii/lisas-vegetable-garden-data.

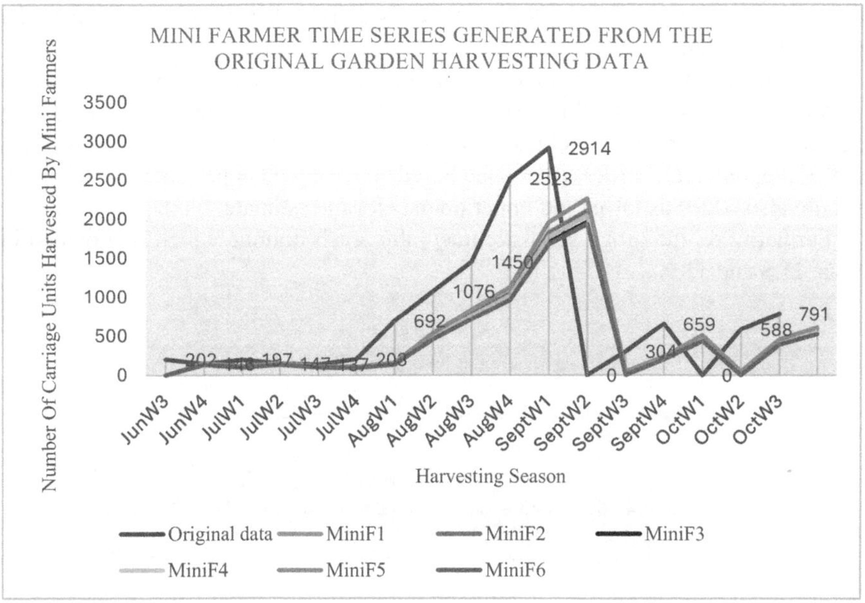

Fig. 11.4 Raw and adjusted (i.e. used) historical data for the weekly harvesting volume of the region (for Mini farmers)

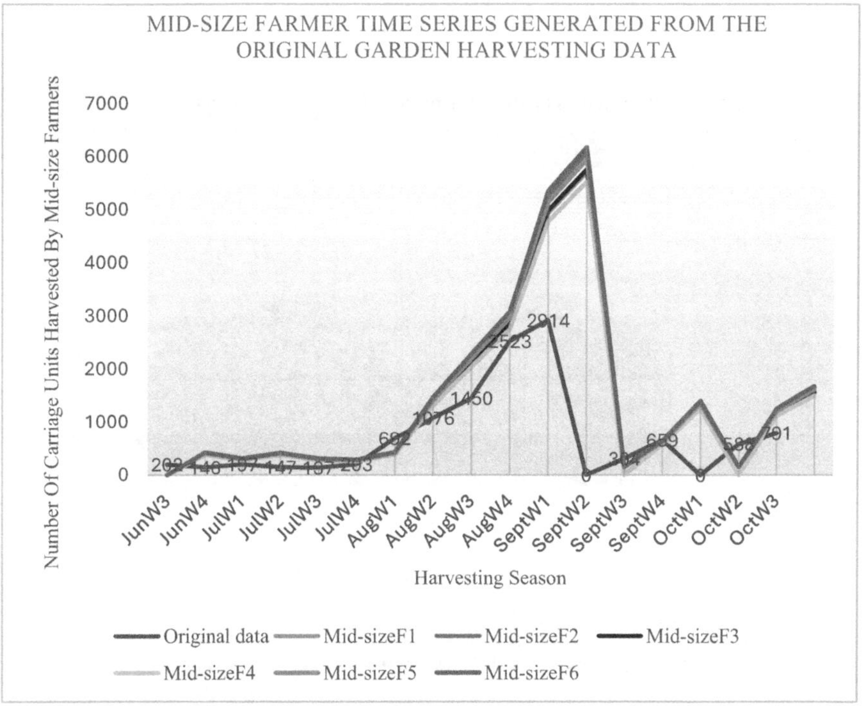

Fig. 11.5 Raw and adjusted (i.e. used) historical data for the weekly harvesting volume of the region (for Mid-size farmers)

different smallholders and converted into embedding values to be processed by the predictive module.

The Transformer algorithm with the architecture presented in Sect. 11.3.1 generated the forecasts for the third week of October. The forecasted value versus the actual value is represented in Figs. 11.6 and 11.7. Since the output is in terms of the embedding interval, borders calculated based on lower and upper limit of each interval are also added as lower and upper bounds for the estimate.

Furthermore, the progress of accuracy after each training epoch is provided in Figs. 11.8 and 11.9.

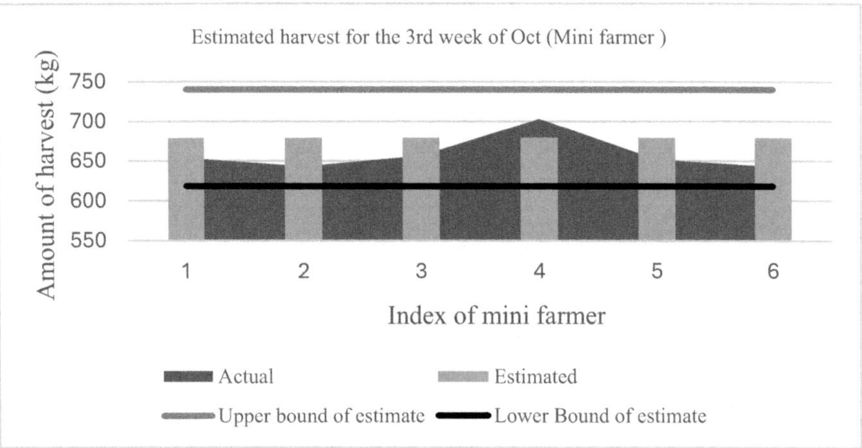

Fig. 11.6 Figure for forecasted vs historical(actual) delivery (Mini farmers)

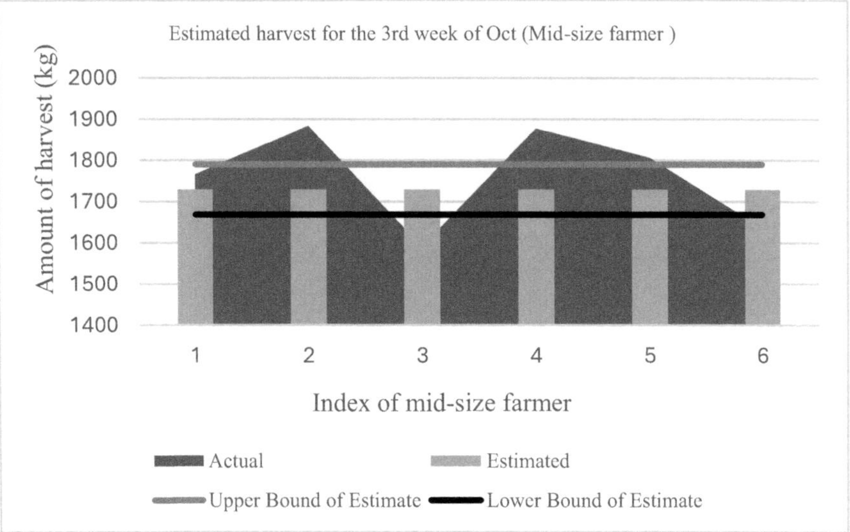

Fig. 11.7 Figure for forecasted vs historical (actual) delivery (Mid-size farmers)

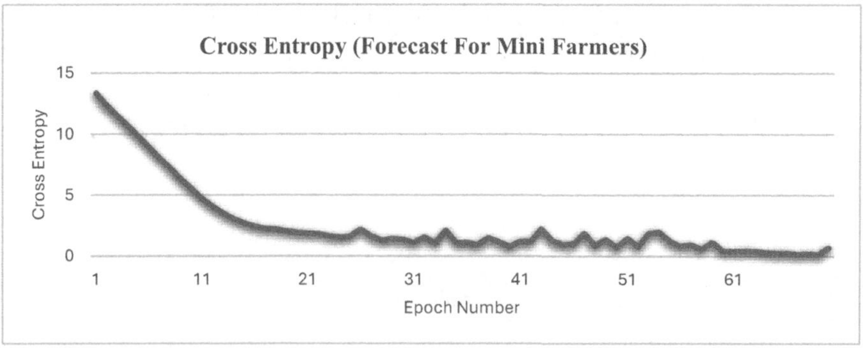

Fig. 11.8 Cross entropy progress of the transformer during training (on mini-farmer historical data)

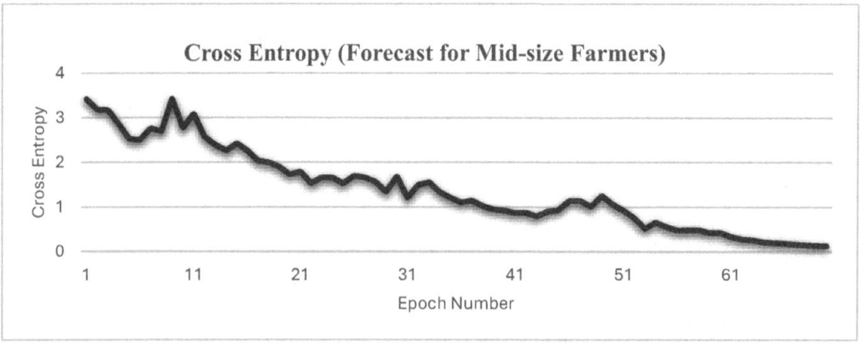

Fig. 11.9 Cross entropy progress of the transformer during training (on mid-size farmer historical data)

After predictive module generated weekly harvest forecasts, prescriptive module can be run based on the forecasts: The locations of the 12 farmers are generated randomly and presented in Fig. 11.10. This instance has been solved on a laptop of CPU model: Intel(R) Core(TM) i7-4558U CPU @ 2.80GHz, instruction set [SSE2|AVX|AVX2] with 2 physical cores, 4 logical processors. The Gurobi Optimizer (version 11.0.0 build v11.0.0rc2 (win64 - Windows 10.0 (19045.2))) is allowed to use up to 4 threads. The instance originally has 9516 columns and 39,684 nonzeros. After the pre-solve stage, Gurobi reduces the size to 420 continuous, 1368 integer components and solves it optimally after 948 iterations (0.02 sec). During the traverse of the branch and bound tree, Gomory (34), Lift-and-project (3), Cover (2), Implied bound (8), Clique(21), MIR (21), Flow cover (14), RLT(11) cuts are added. When the Eq. (11.6) is removed the number of iterations increases to 1288 (17.83 sec.) which shows the contribution of Eq. (11.6) as a valid cut.

After the optimization of prescriptive module, it is observed that 7 vehicles of tilt truck with trailer are needed to serve the region (Fig. 11.11). Based on this result, it

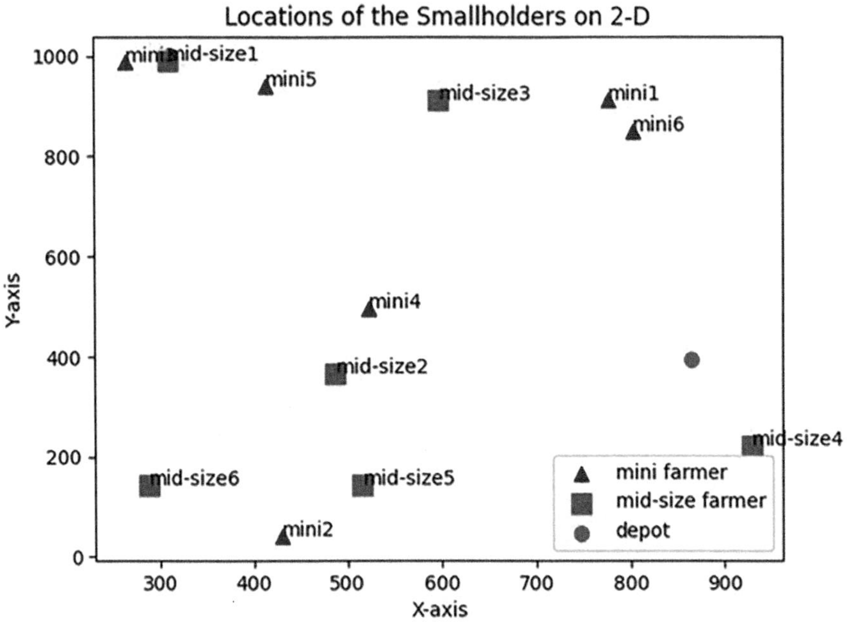

Fig. 11.10 Randomly generated coordinates in Antalya region

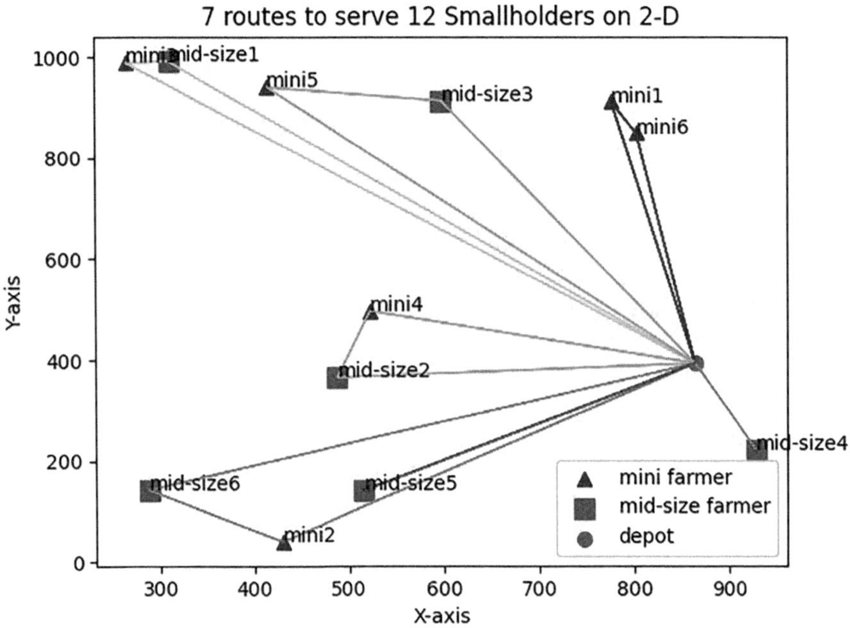

Fig. 11.11 Optimum vehicle dispatch plan generated by the prescriptive module based on the weekly harvest estimates

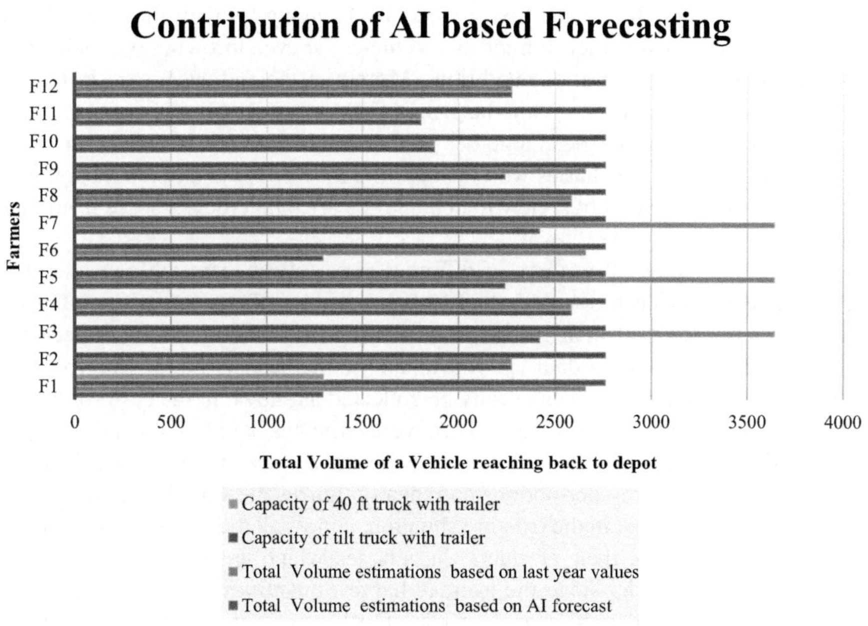

Fig. 11.12 Infeasible solutions generated when the prescriptive model runs based on last year's data instead of AI based forecasting

is possible to conclude that even this small group is benefiting from economies of scale.

In order to benchmark the contribution of the predictive part, another quick analysis is conducted. If the tool is equipped with only the prescriptive module but not the predictive part, in such cases the farmer needs to enter their own harvest estimates to come up with a logistics plan. It can be assumed that smallholders are wise enough to use naïve forecasting and use their historical value of the same period as estimate when they do not have a predictive tool. Figure 11.12 shows how much the forecasting-based plan deviates from the actual values compared with a plan generated solely based on naïve forecasting of last period's data. The optimum plan that is generated by the estimates of the predictive module stays feasible (i.e. the truck capacity is greater or equal to the total actual volume that would occur when prescriptive module prepares the dispatch plan based on the AI forecast). On the other hand, the plan generated by naïve forecasting would cause redundant capacity ordering for half of the farmers and therefore pay more than necessary.

11.6 Conclusions

In this chapter an agricultural business model relying on digital technologies and analytics is presented. It is subtle to understand the existing convention to recognize the contribution of such a model. Currently, many smallholders end up selling to underprices (on a first come first served basis), paying for their individual logistics (ex: transport using their limited size vehicles) or even throwing out their products due to poor transportation conditions. Moreover, the end-customers still suffer from hyper food inflation. With the help of the proposed digital infrastructure, smallholders will be able to get together (as a "virtual union") to benefit from avoiding unnecessary intermediaries who increase the food prices and obtain negotiation power relying on market overview to demand the price that they deserve from their customers in the most direct way.

In the case study, 12 farmers of different sizes (all of them are considered as smallholders since each own less than 10 ha) register to the proposed tool. For the generic use of the tool their historical harvesting information can be collected if available. If the historical data is not available immediately, similar data from the www or a relevant database can easily be collected as shown in the case study. The predictive module prepares the weekly harvest estimates based on this initial input. Then the tool transfers these forecasts to the prescriptive module, searches and reveals the logistic provider options and then optimizes the vehicle dispatch based on the selected options. In the existing situation, almost all the farmers need to drive multiple times to sell their products -in vehicles which are not compatible with "cold chain" standards- since the average harvest quantities generally exceed the capacities of standard mini-vans or trucks. Furthermore, if they want to wait for customers to arrive (ex: exporters, manufacturers, wholesalers etc.), they need to accept the price that is offered to them since they have almost no idea about their best alternative (i.e. negotiation power). On top of the uniting and planning power of the tool, the contribution of forecasting (i.e. predictive) module is demonstrated on the sample case (Fig. 11.12). As listed in Sect. 11.2, by empowering smallholders through such a digital tool, sustainability goals are being addressed from many aspects. Such a business model can be initiated by independent investors (as a Brokerage business model); or even more favored actuators, governments may come into play by providing training and incentives.

The proposed digital model with its analytical components is fertile enough to be customized for all sorts of smallholders around the World. As an example, the model can be extended for multi commodity by including other products which have compatible carrying conditions. For instance, avocado, banana, melon /watermelon, eggplant have togetherness properties with tomato. Additionally, the model can be extended to select between multiple selling options. Thus, the tool might evolve into an extensive market driver by connecting and including up-to-date information from the whole market such as customer portfolio, offered prices etc. Thus, it has potential to become an end-to-end supply chain planning tool which is centered around the farmer, i.e. producer. Finally, the forecasting algorithm can be improved to a fusion model which can utilize some intuitive additional qualitative data such as weather conditions or household demographics for quantitative time series deductions.

References

Akkem, Y., Biswas, S. K., & Varanasi, A. (2023). Smart farming using artificial intelligence: A review. *Engineering Applications of Artificial Intelligence, 120*, 105899.

Amellal, I., Amellal, A., Seghiouer, H., & Ech-Charrat, M. (2024). An integrated approach for modern supply chain management: Utilizing advanced machine learning models for sentiment analysis, demand forecasting, and probabilistic price prediction. *Decision Science Letters, 13*(1), 237–248.

Badmus, M. A., & Ariyo, O. S. (2011). Forecasting cultivated areas and production of maize in Nigerian using ARIMA model. *Asian Journal of Agricultural Sciences, 3*(3), 171–176.

Basile, V. (2021). The triple layered business model canvas in smart agriculture: The case of Evja startup. *Piccola Impresa/Small Business*, (2).

Brownlow, J., Zaki, M., Neely, A., & Urmetzer, F. (2015). *Data and analytics-data-driven business models: A blueprint for innovation* (Vol. 7, pp. 1–17). Camb. Serv. Alliance.

Desrosiers, J., Lubbecke, M., Desaulniers, G., & Gauthier, J. B. (2024). *Branch-and-Price, Technical report, Les Cahiers du GERAD G–2024–36*. GERAD, HEC.

Elufioye, O. A., Ike, C. U., Odeyemi, O., Usman, F. O., & Mhlongo, N. Z. (2024). Ai-driven predictive analytics in agriculture supply chains: A review: Assessing the benefits and challenges of ai in forecasting demand and optimizing supply in agriculture. *Computer Science & IT Research Journal, 5*(2), 473–497.

FAO. (2012). "*Smallholders and Family Farmers*" https://www.fao.org/fileadmin/templates/nr/sustainability_pathways/docs/Factsheet_SMALLHOLDERS.pdf

Goyal, S., Esposito, M., & Kapoor, A. (2018). Circular economy business models in developing economies: Lessons from India on reduce, recycle, and reuse paradigms. *Thunderbird International Business Review, 60*(5), 729–740.

Gray, B., Babcock, L., Tobias, L., McCord, M., Herrera, A., & Cadavid, R. (2018). *Digital farmer profiles: Reimagining smallholder agriculture*. Grameen Foundation.

Hansson, A. M., Pedersen, E., Karlsson, N. P., & Weisner, S. E. (2023). Barriers and drivers for sustainable business model innovation based on a radical farmland change scenario. *Environment, Development and Sustainability, 25*(8), 8083–8106.

INTELLIAS. (2024). "*How Technology Can Solve Agriculture Supply Chain Problems*". https://intellias.com/agriculture-supply-chain/

Karnani, A. G. (2006). *Fortune at the Bottom of the Pyramid: A Mirage*.

Lambert, S. C., & Davidson, R. A. (2013). Applications of the business model in studies of enterprise success, innovation and classification: An analysis of empirical research from 1996 to 2010. *European Management Journal, 31*(6), 668–681.

Lezoche, M., Hernandez, J. E., Alemany Díaz, M. D. M. E., Panetto, H., & Kacprzyk, J. (2020). Agri-food 4.0: A survey of the supply chains and technologies for the future agriculture. *Computers in Industry, 117*, 103187.

Litvinova, O. V., Abrosimova, M. S., Vasilyeva, O. G., Filippova, S. P., Gordeeva, L. G., & Nesterova, N. V. (2020). Digital platform as a liaison mechanism and a business model in agribusiness. In IOP conference series: Earth and environmental science, 604(1), p. 012033. IOP Publishing.

Lu, J., Ye, T., Chen, W., & Van Hentenryck, P. (2024). *Boosting Column Generation with Graph Neural Networks for Joint Rider Trip Planning and Crew Shift Scheduling*. arXiv preprint arXiv:2401.03692.

Mahdad, M., Hasanov, M., Isakhanyan, G., & Dolfsma, W. (2022). A smart web of firms, farms and internet of things (IOT): Enabling collaboration-based business models in the Agri-food industry. *British Food Journal, 124*(6), 1857–1874.

Nabradi, A., & Kovács, T. (2020). Types of platform based collaborative economy and its potential areas in agribusiness. *Western Balkan Journal Of Agricultural Economics And Rural Development, 2*, 9–19.

Ouaddi, A., & Benhra, J. (2024). Transformer-based model applied to customer demand forecasting: A comparative analysis. In *2024 sixth international conference on intelligent computing in data sciences (ICDS)* (pp. 1–8). IEEE.
Padhan, P. C. (2012). Application of ARIMA model for forecasting agricultural productivity in India. *Journal of Agriculture and Social Sciences, 8*(2), 50–56.
Pal, S. (2024). Optimizing just-in-time inventory management: A deep dive into AI-enhanced demand forecasting. *NOVYI MIR Research Journal, 9*, 62–70.
Rappa, M. (2000). *Business models on the web: Managing the digital Enterprise*. North Carolina State University.
Santhanam, N., Varanasi, S., Surana, K., Jacobson, Z., & Zegeye, A. (2018). *Food Processing & Handling; ripe for disruption?* McKinsey&Company.
Shukla, S., & Sengupta, T. (2021). Business model innovation in the agriculture supply chain at bottom of the pyramid: Evidence from India. *Strategic Change, 30*(5), 461–466.
Singh, S., Chana, I., & Buyya, R. (2020). Agri-info: Cloud based autonomic system for delivering agriculture as a service. *Internet of Things, 9*.
Strzebicki, D. (2015). The development of electronic commerce in agribusiness—The polish example. *Procedia Economics and Finance, 23*, 1314–1320.
Talerngsri-Teerasuwannajak, K., & Pongkijvorasin, S. (2021). Agricultural business model and upland sustainability: Evidence from northern Thailand. *Current Research in Environmental Sustainability, 3*, 100085.
TEPGE. (2023). (*Institute for Improving Agricultural Economics and Policy*). https://arastirma.tarimorman.gov.tr/tepge/Belgeler/PDF%20%C3%9Cr%C3%BCn%20Raporlar%C4%B1/2023%20%C3%9Cr%C3%BCn%20Raporlar%C4%B1/Domates%20%C3%9Cr%C3%BCn%20Raporu%202023-379%20TEPGE.pdf. Turkey.
Terada, L. Z., Cortez, J. C., López, J. C., Soares, J., Vale, Z., & Rider, M. J. (2023). Overload mitigation in electric vehicle smart charging algorithms using photovoltaic generation forecasting. In *2023 international conference on future energy solutions (FES)* (pp. 1–6). IEEE.
Vlachopoulou, M., Ziakis, C., Vergidis, K., & Madas, M. (2021). Analyzing agrifood-tech e-business models. *Sustainability, 13*(10), 5516.
World Food Program. (2024). "*Smallholder Market Support*". https://www.wfp.org/smallholder-market-support
Xu, J., & Bo, L. (2024). *Enhancing supply chain efficiency resilience using predictive analytics and computational intelligence techniques*. IEEE Access.

Part III
Control Towers in Supply Chains

Chapter 12
Supply Chain Control Towers

Jose M. Framinan and Paz Perez-Gonzalez

Abstract As a response to the need to quickly detect and act upon the performance deviations that characterise complex Supply Chains (SCs), the concept of the Supply Chain Control Towers (SCCTs) promises to make visible the SC operations by centralising the relevant real-time data in a single information hub, so these can be motorised and controlled, increasing the SC responsiveness and resilience, which in turn improves SC performance. As such, SCCTs have emerged as one of the key topics in SC management, and many companies have implemented them, in some cases reporting substantial advantages. In this chapter, we have extensively reviewed the increasing body of literature on SCCTs, and present a unified definition, components, and functionalities based on this analysis. Published study cases have been analysed in order to identify the potential advantages and implementation strategies. Finally, a number of research lines emerge from the analysis of the literature carried out.

Keywords Supply Chain Control Tower · Digital Transformation · Supply Chain Visibility · Supply Chain Resilience

12.1 Introduction

Extreme competition, short life-cycle products, and continuous change are among the most recognisable features of many industrial sectors nowadays, leading to the prominence of increasingly complex Supply Chains (SCs), often formed by a high number of partners that span across different locations and that constitute an intricate web where materials, information, and money exchanges must be carefully orchestrated. Although such complex interweaving of customers and suppliers in the SCs may be required to exploit the core competences of each partner and so to

J. M. Framinan (✉) · P. Perez-Gonzalez
Industrial Management, School of Engineering, University of Seville, Seville, Spain
e-mail: framinan@us.es

Y. Khojasteh, S.M. M. Jahan Abadi (eds.), *Digital Transformation of Supply Chains*, International Series in Operations Research & Management Science 372, https://doi.org/10.1007/978-3-032-01206-7_12

keep up with the fierce competition, it also makes the SCs extremely vulnerable to disruptions, thus requiring effective mechanisms to foresee, detect, and adapt to the new scenario as fast as possible. Therefore, it is not surprising that the concept of Supply Chain Control Towers (SCCTs) has emerged in recent years as one of the most attractive initiatives in SC management since —mimicking the air traffic management in an airport control tower— promises SC managers to have real-time visibility across End-to-End (E2E) SC operations by collecting, monitoring, and analysing the actual SC status according to a set of metrics, triggering alerts, and establishing corrective actions for deviations, thus allowing stakeholders to quickly react to unforeseen circumstances. As such, SCCT would act as an information hub that captures and uses SC data to support decision-making.

Some published experiences show that the expected benefits of SCCT include improved SC service and reduced operational costs, together with a significant reduction in the impact of disruptions in SC operation. By these means, it increases the efficiency, flexibility, and resilience of the SC, and, ultimately, allows the SC to gain competitive advantages. However, there is no clear agreement on the definition of the SCCT, their components and functionalities, or decision-making scope; therefore, very different systems with rather diverse functionalities are being labelled SCCT, making it difficult to assess what their true advantages and disadvantages are. Furthermore, their implementation poses a number of challenges, ranging from technological issues (such as e.g. the availability, connectivity and accuracy of real-time data across the SC), to process-related (such as the decisions to be encompassed in the SCCT) or organisational issues (such as e.g. the lack of trust among SC stakeholders to make the data visible, or the deep skills required in the organisation to properly exploit the SCCT).

In this chapter, we aim to provide a unified vision of what a SCCT is, its components and functionalities, advantages, and implementation issues. To do so, we review the growing —even if not large— body of literature on SCCTs, and connect it with other related concepts in the digital supply chain, such as Digital Twins (DTs), Robotic Process Automation (RPA), Artificial Intelligence (AI), or Internet of Things (IoT), among others, so the chapter may serve as a starting point for academics and practitioners interested in the topic. Furthermore, the main gaps in the current state-of-the-art and future research avenues /next-generation SCCTs are also discussed.

12.2 Concept and Evolution

Perhaps the most-often cited definition of SCCT is that of a decision support system merging different data streams from various subordinate levels in the SC and displaying the consolidated information at a higher level for the purpose of monitoring and control of processes while pursuing the goal of optimal process operation (Alias et al., 2014b; Brintrup, 2020), which essentially aligns with the definitions of several standardisation bodies, see e.g. (DIN IEC 60050-351, 2009; VDI 4493, 2009). The centre of the SCCT concept is to increase the resilience of the SC and to manage its disruptions. Supply chain resilience is the result of deliberate actions which increase flexibility and adaptability to prepare for unexpected events, that is risks, affecting business performance, so that the SC can respond to disruptions and, ideally, maintain continuity of operations (Ponomarov & Holcomb, 2009). Indeed, different empirical works link SC responsiveness with SC visibility (Williams et al., 2013), so, according to this scheme, the ability to quickly spot the potential problems that can be offered by a SCCT can naturally help to improve SC resilience (DeVoe et al., 2023).

As it has been mentioned in Sect. 12.1, the notion of a SCCT originates from aviation, where a controller coordinates aeroplane traffic in sky. The concept behind a SCCT is similar in that it monitors orders, inventory, shipments, forecasts and their status (Cooke, 2014) and, indeed, initial experiences on SCCTs focused on the orchestration and visibility of the transportation operations as a response to the increased complexity of the globalised SCs (Vlachos, 2021). In this regard, several initial experiences labelled such system as a *transportation* or *logistic* control tower (see e.g. Alias et al. 2014a, b), and even some references refer to different types of SCCTs depending on the specific part of the SC that is monitored. However, such function-oriented vision of the SCCT has been abandoned in favour of a cross-functional, E2E vision of the scope of the SCCT.

The concept of a SCCT is mentioned in the literature as earlier as in 2006 (Aronsson & Huge Brodin, 2006), and the first implementations described include that by Scania in 2003 to monitor its part suppliers and inbound transportation (Cooke, 2014), or that carried out in 2009 by Unilever (Sheffi, 2015b) to centralise its transportation procurement and operations. Indeed, the consolidation of the companies' contract logistic operations seems to have been a main driver of the first implementations of SCCTs (Cooke, 2014).

Since then, the interest on SCCT has been continuously increasing, so building a SCCT is considered one of the five key steps to manage SCs (Mena et al., 2014), being considered one of the main four topics in the development in SC management over the next years (Schiffer & Doerr, 2020). Indeed, by 2015, 10 of the Gartner Supply Chain top 25 have a SCCT in place (Bhosle et al., 2015). Although many companies have developed ad hoc SCCTs, there are also different players in the market offering global SCCT solutions (see e.g. Trzuskawska-Grzesińska, 2017, and the reports on the merits of these different *off-the-shelf* solutions by Gould & Brennan, 2022 or Lund & Payne, n.d.). However, several authors (such as e.g. DeVoe et al., 2023) warn about considering the SCCT just a tool that can be potentially purchased, as their processes and their staff's skills may be more important than the technological system.

Figure 12.1 summarises the evolution of the SCCT concept since it first appearance, becoming more and more important for the efficiency of the companies. The milestones of the evolution presented in Fig. 12.1 have been deducted from cases study in the academic literature that will be presented in detail in Sect. 12.6. Nowadays SCCT is seen as a key factor in SC performance, and it is widely accepted that a trade-off between the solutions provided in the market and a customized tool should be achieved in order to take into account important components for its

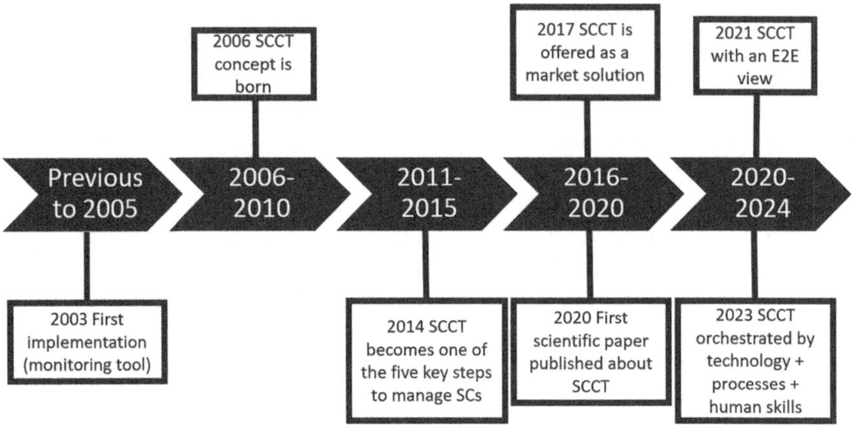

Fig. 12.1 Evolution of the SCCT concept

implementation beyond technological aspects, such as the processes involved and the human resources interacting with the technology.

The specific components of a SCCT will be discussed in Sect. 12.4, once the main functionalities of SCCTs are presented in the next section.

12.3 Definition and Functionalities

As with similar initiatives in SC management, there is no clear-cut definition of what the functionalities of the SCCT should be. In some earliest descriptions, five functions are usually mentioned, i.e.: planning & routing, auditing & reporting, forecasting, event management, and decision making (Cap Gemini, 2011; Yan et al., 2012). However, these descriptions, albeit practical, present some overlapping, since e.g. planning is usually considered a decision-making activity, and clearly event management must depend on some reporting system. Other functional classification also overlaps some capabilities, e.g. in Wertheimer (2011) it is mentioned that the functions include: quality and integrity of the information; operational execution; real-type dynamic management of materials and transportation flows; and the use of analytical tools and metrics.

Note also that another distinction could be made depending on the nature of the decisions carried out by the SCCT. However, while some authors state that the SCCT operates at the strategic, tactical and operational level of the SC (Bennett et al., 2023), most authors consider it to be an operational/tactical tool (see e.g. Sheffi, 2015a or DeVoe et al., 2023), which seems to be more in tune with the need to require real-time data (Gerrits et al., 2022).

Adopting a more abstract view in line with the definition of SCCT given in Sect. 12.2 (i.e. that of consolidating data for monitoring and control in order to achieve optimal operation of the SC), the following high-level functionalities, described in detail in the following subsections, can be identified:

- Data integration.
- Analytics.
- Decision-Making.

The aforementioned functionalities essentially agree with the four *levels* of a SCCT described by other researchers (see e.g. Baumgraß Anneand Dijkman, 2015, Becker et al., 2019, or Verma et al., 2020): Level 1 refers to data availability and transparency, whereas levels 2 and 3 refer to the alert-generation and decision support, respectively. Level 4 refers to autonomous or automated decision-making without human intervention. Note that these levels can be seen as incremental in the sense that each level is a prerequisite for the next ones, which can also serve as a blueprint for the implementation of a SCCT. Interestingly, these three first levels essentially coincide with the phases of implementation of the SCCT in one of the first reports on the topic (Cap Gemini, 2011).

Note that, although these three functionalities are mentioned by most of the literature, they are not universal and some authors (see e.g. Brintrup, 2020 consider that the analytics and decision-making functionalities are left to the humans. Other authors distinguish between *traditional* and *intelligent* SCCTs (see e.g. Zhang & Xiao, 2023, Vlachos, 2021, or Naik & Kumar, 2021), being the latter the ones that implement analytics and decision-making. For other authors (see e.g. Rai et al., 2012, Duarte et al., 2023), a system providing both real-time, integrated data, and their analysis in terms of an alert system is what technologically supports SC *visibility*, whereas, for most scholars, SC visibility is simply an enabler of the decision-making functions in the SCCT (Wycislak, 2023).

12.3.1 Data Integration

Ideally, the SCCT allows a single data access point across the SC, thus effectively integrating all the relevant data required for the operation of the SC. Note that, in order to achieve such integration, the *relevance* of the data has to be established, i.e. it has to be decided which data are required, and which are not, what in turns requires to have identified one or more SC processes that are going to be monitored and controlled. In other words, not all data in the SC are integrated, but only those belonging to the processes that are to be managed, and with the degree of detail required for their management. Such identification of the relevant data is far from being trivial and, in general, it requires a map of the SC (MacCarthy et al., 2022). Other alternative would be to integrate all data in the SC, but this does not seem realistic given the sheer amount of data and the current status of the technology.

Once the relevant data to be integrated have been identified, at least three different aspects have to be considered regarding its integration: (i) scope, (ii) granularity, and (iii) recency. The scope refers to the number of SC processes that are encompassed by the SCCT, whereas the granularity refers to the level of detail/integration of these data. Finally, the recency refers to the frequency used to update these data in the SCCT. Regarding the scope, although some definitions of SCCT stress that its scope is the entire SC (end-to-end or E2E), most implementations described in the literature focus instead on some part of the SC. With respect to the granularity, it is believed that different levels of granularity of these integrated data should be provided (Patsavellas et al., 2021), so the individuals can drill-down to identify specific metrics (Handfield et al., 2020).

Finally, regarding its recency, most literature claim that a critical feature of the SCCT is to have real-time data, but again this seems to represent an idealised goal of the SCCT rather than the reality described in most case studies. With respect to the update of the data, most references do not mention the frequency of the updates, although some mention that the data are refreshed in the order of minutes (see Cooke, 2014 or Handfield et al., 2020).

Clearly, one prerequisite for the integration of the data is their digitalisation, as well as to ensure their quality (Wycislak, 2023). Despite the increase in IT (Information Technologies) investments, SC digitalisation often extends up to first tier suppliers (Bienhaus & Haddud, 2018) and logistics service providers (Cichosz et al., 2020), or limited to specific applications (Vlachos, 2021).

In most cases, the data required in the SCCT are already captured in some of the information systems of the companies, but their typically require at least a *middleware* to translate and consolidate these data and possibly their storage in a *data warehouse* or *data lake* (see Sect. 12.4.1). In addition to the technical problems arising from integrating and maintaining such disparage data coming from different operational systems, there may exist problems related to the lack of trust among partners to make their data available to the others.

12.3.2 Analytics

The data integrated in the SCCT are analysed in a continuous manner. In general, it is usual to distinguish three layers of analytics (Dietrich et al., 2010):

- Descriptive analytics, a set of technologies and processes that use data to understand and analyse business performance (what happened? why did it happen?). Some authors (see e.g. Bennett et al., 2023; Delen & Zolbanin, 2018) decompose this layer into descriptive analytics (what?) and diagnostic analytics (why?), although, for our purposes, we will not make such distinction. The descriptive analytics techniques mentioned include data visualization, reporting, dashboards, and query/drill-down techniques, or root-cause analysis.
- Predictive analytics, a set of mathematical techniques to uncover explanatory and predictive models of business performance that connect data inputs and outputs (what will happen?). These include data mining, pattern recognition and alert, simulation, forecasting, or predictive models (including regression, time series, or Machine Learning).

- Prescriptive analytics, a set of mathematical techniques that computationally determine a set of alternative action or decisions given a complex set of objectives, requirements and constraints to improve business performance. Basically, entails optimisation, either in the sense of exact solutions, approximate solutions, or obtaining solutions taking into account data uncertainty (stochastic optimization).

From the analysis of the literature, the degree of analytics present in the case studies of the SCCT varies greatly, although it is considered that at least certain degree of descriptive analytics beyond mere dashboards should be employed in the SCCT (Kulkarni, 2023) to benefit from its implementation (Gupta, 2022). Predictive analytics are frequently also included as a SCCT functionality, as it could use the real-time data to identify the strengths and weaknesses of the SC (Mukherjee, 2023). Regarding prescriptive analytics, we believe that this aspect cannot be separated from the manner in which decision-making is conducted, so we will consider it an option within the next feature to be discussed in Sect. 12.3.3.

The typical manner in which analytics are implemented in a SCCT is to establish performance indicators that are monitored to identify deviations in the actual SC performance with respect to the expected one. Given the sheer amount of data to be monitored, some mechanisms to prioritise the data according to their impact and urgency must be put in place in order to prevent overwhelming the planner by vast amounts of unsorted data (Alias et al., 2015). Naturally, this leads to an exception management process, where parameters are set for each relevant indicator in the SCCT so the system can automatically identify when these indicators exceed their acceptable values and generates the corresponding alerts/exception messages. In some cases, the alert generation is probabilistic-based (see e.g. Gerrits et al., 2022 where an alert is generated if the probability of stockout at a warehouse exceeds a certain threshold), which is naturally linked to the more general concept of SC risk management. Finally, note that the alert generation mechanisms can be implemented both using descriptive analytics, i.e. simply collecting the values of the relevant indicators and comparing them with the pre-defined expected ones, or using sophisticated prescriptive analytics. It is clear that the sophistication of the techniques and their added value are correlated (Delen & Zolbanin, 2018), as they provide non-trivial alerts and behaviour patterns that otherwise may pass unnoticed.

12.3.3 Decision-Making

The information/alerts elaborated by the SCCT analytics upon the integrated data can be presented to the (human) planners to make the corresponding decisions (Brintrup, 2020), or they can be aided by a decision support system that offers different intervention options together with their estimated impact (Gerrits et al., 2022). As mentioned before, different techniques can be used to support the process, including simulation models, or SC Digital Twins (SCDT) (see e.g. the

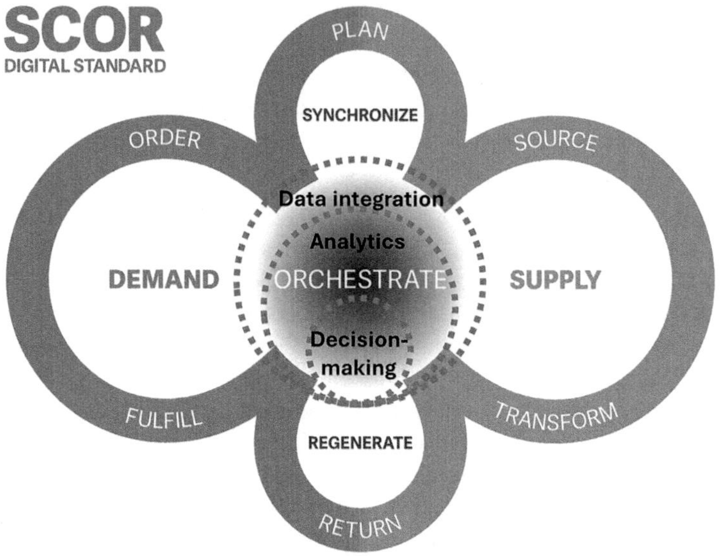

Fig. 12.2 Contribution of SCCT functionalities in the Orchestration of the SC. Adapted from ASCM (2025)

application by Guidani et al., 2024). The SCDT is an effective living model of the SC that continuously adapts to the changes in current operations and forecasting the coinciding physical twin (Liu et al., 2021). However, some authors (Bennett et al., 2023; MacCarthy & Ivanov, 2022) consider the development of a DT as an add-on to the functionalities of the SCCT, or an *iteration* of the SCCT (Brintrup, 2020) (note that, instead, other authors rather consider than the SCCT is part of the DT (Ivanov et al., 2019).

One extreme case of the decision-support could be the fully automation of the decision-making process —particularly for some well-established decisions—, so humans can concentrate on the management of the exceptions. It has been noted that, depending on the process, typically 50–70% of rule-based decisions can be automated (Smith, 2022).

It is clear that the SCCT functionalities discussed (data integration, analytics, and decision-making) are related to the orchestration of the SC. In this regard these functionalities can be mapped to the Orchestrate process in the SCOR digital model (ASCM, 2025). The Orchestrate process is the only level-0 process, related to the integration activities and enablement of SC strategies. Therefore, the SCCT features (highly related among them) can be seen as a contribution to the Orchestrate process as they affect all SC functionalities: Demand, Synchronize, Supply, Regenerate (see Fig. 12.2), as well as the processes (see Sect. 12.4.2).

12.4 Components of a Supply Chain Control Tower

As it has been mentioned before, rather than a specific piece of software/infrastructure (Maheshwari et al., 2023), a SCCT is composed of technology, organisation and processes. Indeed, to achieve the advantages sought by SCCTs, they need to be considered more than a technological tool, thus stressing the need to focus on processes and people (DeVoe et al., 2023), aspects that sometimes are overlooked in favour of the technological issues. It is considered that, despite their increasing importance among practitioners, there are relatively few papers dealing with SCCT in the academic literature (Bennett et al., 2023), and particularly there is a need to explore the managerial aspects of SCCT implementation (Maheshwari et al., 2023). In the next subsections, these three components of a SCCT (technology, processes and people) are discussed.

12.4.1 Technology

From a technology viewpoint, SCCT is a (usually cloud-based) digital platform or information repository that acts as an integrator of existing information systems across the SC, consolidating data from internal and external data sources. Internal data sources include business software internal to the company, such as Transport Management Systems (TMS), Warehouse Management Systems (WMS), Enterprise Resource Planning (ERP), Manufacturing Execution Systems (MES), or APS (Advanced Planning and Scheduling) systems, among others. External data sources include tracking information from external partners, or live traffic feeds, among others. Since it is unlikely that all partners and systems use the same software from the same vendor, a middleware is needed to translate and consolidate these data (Cooke, 2014). A data warehouse or *data lake* may be required to consume the critical data from the different applications (Banker, 2011) and to create a harmonised version of these business objects (such as e.g. purchase order, stock-keeping-units, etc.) that may be differing in the various information systems. Finally, the already mentioned analytics and decision support capabilities should be embedded in some IT infrastructure (Cap Gemini, 2011).

Therefore, a SCCT typically entails setting up a rather complex IT, where different elements (such as data storage systems, algorithms, visualization tools) should be deployed to be used in a multi-user (and probably multi-location) environment, thus also needing to deal with the issues of security and accessibility to the different parts of the system. Indeed, the SCCT is considered to use all three IT technological types, i.e. long-linked technologies such as IoT and big data collected at various points of the SC, mediating technologies such as EDI, cloud platforms etc. to connect the different partners/systems, and intensive technologies such as Machine Learning models and optimisation algorithms (Vlachos, 2021). These technological challenges must be tackled also keeping in mind the technological flexibility and scalability required so the SCCT can accommodate to the changes in the needs of the planners (Naik & Kumar, 2021).

As it can be seen from the above discussion, technology is an enabler of the SCCT functionalities presented in Sect. 12.3, and a careful analysis of the technologies involved in the SCCT should be carried out, in order to facilitate the interoperability and minimize unnecessary data.

12.4.2 Processes

In order to successfully deploying the SCCT, the as-is processes in the SC have to be assessed, capturing the visibility needs and evaluating the benefits (Cap Gemini, 2011). Note that it is recommended that the SC processes within the SCCT should be highly mature (Klesman & Cherici, 2017). A high-level to-be concept for the processes has to be designed. Since the entire SC processes are not fully integrated, the development of a SCCT may require a reengineering across the organisation (Trzuskawska-Grzesińska, 2017). Typical tools from business process management, such as 5-Whys, can be used (Vlachos, 2021). Also, the corresponding KPIs have to be developed, perhaps extending the visibility in order to include financial measures to gain the support of the financial staff (Cap Gemini, 2011).

As a summary, the impact of the SCCT in the different processes should be analysed in depth before being implemented, also taking into account the overall effect that SCCT can have. In this regard, the SCOR digital standard (ASCM, 2025) presents SCCT as a best practice and provides a list of the processes more suitable to be controlled by the SCCT, including the related metrics and human resources involved.

12.4.3 People

From the people viewpoint, there is usually a dedicated team in charge of the SCCT (Cooke, 2014). It is worth noting that the SCCT breaks from the traditional SC where each planner is in charge of a specific function within the SC (i.e. transportation manager, manufacturing manager, inventory manager, etc.). However, the SCCT requires the role of *network manager* that spans functional silos and focuses on the customer, product, and market (Naik & Kumar, 2021), a role that is sometimes defined as *T-shaped people* (Sean Culey, 2020). In addition, new skills are required to transform the work of these network managers into an exception-driven management using information integration, analytics and rule-based decisions, thus focusing on creating more strategic value (Liotine, 2019). Note that, for these roles, IT-skills and business and supply chain management skills should be blended (see e.g. Dalporto & Venn, 2020 or Jha et al., 2022).

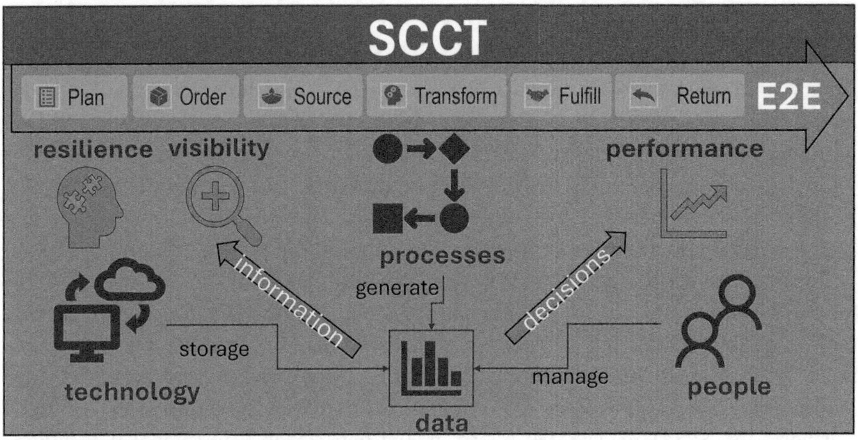

Fig. 12.3 Relation between components and objectives in a SCCT

Finally, as conclusion of this section, it should be noted that the connection among the presented components is key for a functional SCCT. Figure 12.3 shows how the components of the SCCT (technology, people and processes) provide the data required by the SCCT in order to impact on the visibility, performance, and resilience of the processes across the SC, identified in the next section as the main objectives of the SCCT. Note that the processes included in the E2E vision have been extracted from the first level provided by the SCOR digital standard (ASCM, 2025).

12.5 Objectives

There are different contributions reporting the (theoretical) objectives of SCCTs. These objectives can be grouped into three layers:

- SC visibility and transparency
- SC responsiveness and resilience.
- SC performance

These three layers serve to group most of the advantages reported in the literature. Nevertheless, some authors state that one objective of SCCT is process orchestration (Mukherjee, 2023), but we see process orchestration more as a result of the combination of E2E data integration and E2E decision-making. The layers are discussed in the next subsections.

12.5.1 SC Visibility and Transparency

A survey among global SC executives stated in 2018 that their worst challenge they face was the lack of visibility of the SC (Mukherjee, 2023). These results are in line with previous surveys (Patsavellas et al., 2021). This need of SC visibility increases with their complexity in terms of sophisticated products with shorter life cycles, shorter customer expected lead times and increased replenishment lead times, as well as with complex organisations and information systems (Christopher, 2011).

In the aforementioned context of lack of visibility across the SC, it is clear that having a single point where all relevant data are centralised and updated in (near) real-time constitutes a big advantage for decision making, Indeed, it is the increase in SC visibility provided by the SCCT what it is often referred as one of the main causes for their adoption (see e.g. Alias et al., 2014a and Sharabati et al., 2022). Furthermore, SCCT ensure the unicity of the data, establishing what it is called a *single-point-of-truth* (Bennett et al., 2023), which can produce improved organisational models (Patsavellas et al., 2021). Establishing such single version of the otherwise disperse (and potentially different) data across the SC serves to make a single diagnostic of the problems in the SC for all stakeholders in the SC, which naturally can lead to better decision making.

In order to achieve these advantages in terms of visibility, it is critical to ensure the quality of the data (Wycislak, 2023). Furthermore, it is to note that several studies (such as e.g. Malmstedt & Bäckstrand, 2022) link the SC visibility to the scope of the SCCT, i.e. to the integration of E2E data. Since most reported applications of SCCT only integrate part of the SC processes, this objective should not be overestimated.

12.5.2 SC Responsiveness and Resilience

Once (and if) relevant data for SC operations are available in real-time, it is clear that an earlier detection of the deviation of KPIs from their intended targets can be accomplished, and the corresponding actions or modifications of the plans can be activated, thus increasing the responsiveness and resilience of the system, including the fast identification of SC disruptions (Gluttig, 2022). Indeed, this seems to be the case, as a survey conducting among 85 companies in Indonesia outline the positive impact of SCCT in increasing SC resilience (Syahchari et al., 2022).

12.5.3 SC Performance

Finally, as a result of the two aforementioned layers, an improvement on SC performance can be expected from the implementation of the SCCT.

Some of these improvements come from the reduction of costs associated to the digitalisation of the SC processes (Smith, 2022). Other improvements come from the acceleration of some processes induced by the integration of the information (Vlachos, 2021), and from the fact that, in a SCCT, the time of the employees devoted to routine, low-value activities, is saved so the employees are involved into more value-added activities (Bennett et al., 2023).

Malmstedt & Bäckstrand (2022) link the SC performance to the addition of intelligence (i.e. decision-making) on top of the data integration and analytics, as otherwise the decisions would be simply reactive and not pro-active. Another study on a SCCT for the incoming and outgoing processes of a company reports costs saving of up to 20% also attributes the increase in performance to the usage of a decision-making system to optimally book the docks (Wycislak, 2023). These findings may be in synch with the findings of some cases studies where the performance of the SC does not seem to have significantly increased with a SCCT (Dwiyana et al., 2022).

Operational advantages in term of quality and reliability have been reported in the usage of a SCCT in a pharmaceutical company (Sharabati et al., 2022). A reduction of the forecast variability in more than one third has been reported by Merck's Group, with the overall inventory levels reduced between 5 and 10% with no negative impact on service (Castellanos, 2018), in line with other implementations reporting similar savings (see e.g. Yan et al., 2012, Cooke, 2014). This is perhaps the most evidenced translation into KPIs of the SCCT objectives.

Finally, in a multi-case study conducted by Trzuskawska-Grzesińska (2017), it is stated that companies also perceive that their SCCTs added value to their customers, and other authors state that it can also play a pivotal role to implement sustainability principles by providing more sustainable plans and insights into sustainable supply chain practices (Din et al., 2023), also derived from the necessary digitalisation of the SC by implementing paperless processes with electronic proofs and tracking (Wycislak, 2023).

From Sects. 12.3, 12.4 and 12.5, it is possible to have a general idea about the capabilities of a SCCT (see Fig. 12.4), taking into account its main components, the different functionalities included, and the objectives of the system. As it can be seen, these aspects are highly interrelated, and the selection of the functionalities should be carried out carefully to control the impact of the SCCT on the SC.

12.6 Application and Success Factors

Although several contributions describe generic implementation, very few include some details to gain some insights into the application of SCCTs. In general, there is a lack of empirical evidence on what is the best way to implement a SCCT (Vlachos, 2021), as well as to understand how the potential advantages materialise (Patsavellas et al., 2021). In the latter reference, the authors provide a tool to determine the suitability of the implementation of a SCCT by a company.

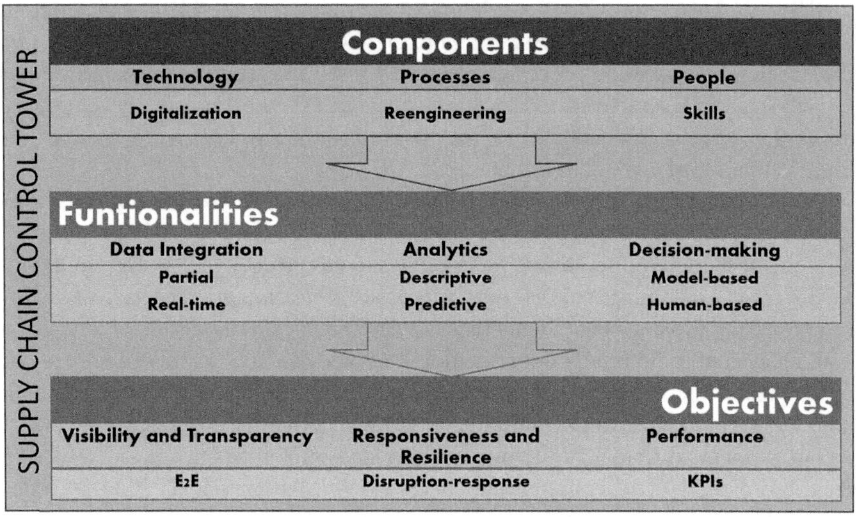

Fig. 12.4 Components, functionalities and objectives of the SCCT

As explained in Sect. 12.2, the concept of SCCT attracted the attention in the scientific community very recently, being the first contribution from 2020. The research carried out about the topic has been oriented to the implementation of SCCTs in SC of different sectors, such as manufacturing, Fast-Moving Consumer Goods (FMCG) or pharmaceutical, among others. In this section, a brief overview of these contributions is carried out, in order to provide some insights about the features of the SCCTs described in the literature. Table 12.1 summarizes these features, classified for each of the functionalities given in Sect. 12.3 (Data integration, Analytics and Decision Making).

In Verma et al. (2020), the application of decision support in a SCCT of a FMCG company is presented. Vlachos (2021) describes the implementation of a SCCT in a company in the sector of health technologies. Interestingly, the company decided to focus on its core competences and outsourced its logistics to a leading 3PL. After developing a SCCT in close collaboration, the SCCT was expanded to cover inbound logistics, warehouse and, ultimately, E2E SC operations. Finally, the SCCT was internalised by the company, who now owns the SCCT. Gerrits et al. (2022) describe a SCCT for IT hardware spare parts, focusing on the problem of automatic alert generation, a similar case than in Duarte et al. (2023). In Wycislak (2023), the case of a SCCT in a FMCG company is discussed, whereas Kulkarni (2023) presents a SCCT for the container terminal operations. Guidani et al. (2024) describe a SCCT in the agri-food industry that encompasses the agriculture, processing and packaging, logistics and distribution, and the consumption. Environmental and performance indicators are presented in real-time, and decision-making is supported by simulating a range of scenarios. Finally, Kumar and Barros Gomez (2024) describe the implementation of a SCCT in a pharmaceutical industry for the procurement

Table 12.1 Summary of the features of the applications of SCCTs

Reference	Sector	Data Integration	Analytics	Decision Making
Verma et al. (2020)	Fast-moving consumer goods	Partial (logistics)/ Real-time	Descriptive	Model-based (Fuzzy neuro system)
Vlachos (2021)	Manufacturing	E2E/Real-time	Descriptive	Human-based
Duarte et al. (2023)	Spare parts (OEM)	Partial (spare parts/ Real-time	Predictive (automatic alert generation)	Human-based
Gerrits et al. (2022)	IT hardware spare parts	Partial (spare parts/ Real-time	Predictive (automatic alert generation)	Model-based
Kulkarni (2023)	Maritime terminal operations	Partial (transportation)/ Real time	Descriptive	Human-based
Wycislak (2023)	Fast-moving consumer goods	Partial (transportation)/real time	Descriptive	Human-based
Guidani et al. (2024)	Food supply chain	E2E/Real-time	Predictive	Model-based (Digital twin)
Kumar and Barros Gomez (2024)	Pharmaceutical company	Partial (procurement) /n.d.a.	Predictive (Machine learning)	Human-based

process. They use different Machine Learning algorithms in the analytics part to provide a consistent classification of the materials.

Regarding the implementation process of SCCTs, Cap Gemini (2011) is an original paper describing how to set up a SCCT. They divide their approach in two phases. In the first one (denoted *strategic stage*), essentially an as-is analysis of the relevant SC process and an effort to establish the corresponding to-be processes is carried out, being in this aspect similar to any other business management process. In the second one, the system is deployed. These two phases roughly coincide with the first two described in Vlachos (2021), who examines the implementation of a SCCT in a large manufacturing company, or in Yan et al., (2012), who describes the implementation of a SCCT for an SME. In this case the implementation is divided into three stages, i.e. (i) initiation, (ii) live, (iii) continuous improvement, being the third one an important element to ensure that the SCCT is capable to capture the dynamic nature of the business. These similar three steps are mentioned by Smith (2022). Mukherjee (2023) describes the steps for the successful implementation of a SCCT, using equivalent steps, even if in their case the need of a detailed deployment plan is stressed. There are tools to assess the convenience of implementing a SCCT that can be used (Milenkovic et al., 2019 apply the Value Case Methodology to assess the feasibility of the implementation of a SCCT in the rail-freight sector whereas in Patsavellas et al. (2021) the authors provide a tool to determine the suitability of the implementation of a SCCT). Other option is to choose an incremental approach, and most organisations start with a proof of concept to validate the idea of the SCCT and then move throughout the SC for a broader SCCT (Cooke, 2014).

It Is to be noted that, in any case, a SCCT is not a product that can be purchased and installed (Woodward, 2012), as SCCT rely on existing software (ERP, TMS— Transport Management Systems, WMS— Warehouse Management Systems etc.) that enable the centralisation of operations (Wycislak, 2023), and practitioners consider that having processes for data management and information is a prerequisite for SCCT (Bennett et al., 2023). Indeed, the typical architecture of a SCCT is that of a system orchestrating individual business information systems (see e.g. Smith, 2022). Finally, it is mentioned that it is important to properly establish and manage the expectation, as, in some cases, overambitious goals were set (Wycislak, 2023).

Taking into account the experience of the cases of study analysed, the following success factors in the implementation of SCCT identified in the reviewed literature can be summarised as follows:

- Careful and exhaustive planning & analysis: The steps to carry out the implementation of the SCCT must be carefully designed. These steps involve an exhaustive analysis of the SCCT components (technology, processes and people), in order to estimate the impact of the implementation. The impact of the technological solution in the processes should be also checked. Finally, continuous improvement should be conducted by analysing the quantity and quality of the information provided by the SCCT.
- Iterative integration of SCCT functionalities: The features discussed above (data integration, analytics and decision making) should be introduced iteratively due to the difficulty to obtain the information needed, and the complexity of the methodologies involved (from data visualization to optimization).
- Scalable implementation and scope: A gradual implementation starting with the core processes and then expanding the scope with the objective of covering the E2E SC is recommended.

12.7 Research Lines

Analysing the literature presented in the previous section, some research gaps have been identified, and consequently a number of research lines arise in order to enhance the SCCT capabilities. First, some research lines related to the functionalities are presented:

- *Research on Data Integration*: Most experts coincide that today's SCCT provide rather simplistic, limited visibility despite claiming E2E visibility (Liotine, 2019). Note that the factors that best characterise SC visibility are the accessibility, quality and usefulness of the data (Somapa et al., 2018), aspects that may require further research. Most of the literature assume that these data are readily available, with sufficient quality and granularity for the different decisions, which we believe it is hardly the usual case. Another overlooked aspect is the recency of the data: while real-time data would be ideally required, it is clear that this is not neither realistic nor required for some decisions, or it may be that even such continuous update of all data could introduce nervousness in the system, leading to poor decision-making. In this regard, additional research could be conducted to develop a framework for the decisions covered by the SCCT, from which the data requirements could be derived, including their intended granularity and frequency of update depending on the decisions for which they will be used. This would lead to addressing other data-related issues such as the storage of the same data at different aggregation levels (and possibility in a distributed manner to facilitate its access) while keeping a single point of truth.

- *Research on SCCT Analytics:* Probably, the main open research question regarding the analytics functionality in SCCT is related to the automated alert generation. As discussed before, the alert system is seen as a very important part of the SCCT. However, there is not a common solution for alert generations and the existing literature has not addressed so far how these alerts and trends can be identified and generated (Duarte et al., 2023). A critical issue related to alert generation is to establish non-trivial mechanisms or thresholds to generate the alert since, in some cases, it has been reported that too many alerts are generated, and the planners in the SCCT do not have time to check all these alerts (Gerrits et al., 2022), thus limiting the potential benefits of the alert generation system.

In Ye et al. (2022), a two-stage system is proposed to identify and to assess the risks in a SCCT: In a first stage, alerts (risks) are identified whenever a pre-defined threshold of an indicator is exceeded. In a second stage, a knowledge base management system is employed to determine the probability and impact of the risk, together with other relevant aspects, such as the alternative strategies and decision makers involved. So far, the system is purely theoretical, and it has not been tested either in real-life or in a controlled experiment. A similar system is proposed in Duarte et al. (2023) to identify critical spare parts. First, Machine Learning models – based in the data gathered by the SCCT– are used to predict the consumption of each item in a spare parts warehouse. Then, the probability of stockouts for each part according to the corresponding estimation is computed, and the k-th parts with the highest probability are generated as an alert. Although these works represent some preliminary research in this area, a number of aspects are still missing, including the ability of the alert generation system to distinguish between patterns that reveal an uncommon behaviour of the indicators (being thus the cause of a true alert in the SCCT) from variations in the values of these indicators that simply obey to changes in the business dynamics. Even if all these aspects are addressed, it is also to note that the SCCT framework for analytics essentially mimics that of the basic decision support systems in the sense that they make the decision maker aware of some non-conformant behaviour of the system, but do not aim to provide them actionable insights or plausible courses of action to correct this behaviour. In this regard, a tighter integration with SCCT Decision-Making (see next bullet) would be required.

- *Research on SCCT Decision-Making:* This is, undoubtedly, the less mature research area in SCCT, which seems natural as a result of the evolution of the SCCT discussed previously. However, the area of decision support systems for supply chain management is not new and, in this regard, perhaps a preliminary research work would consist of aligning the existing literature on these decision support systems (including APS—Advanced Planning and Scheduling, DT—Digital Twins) with the incipient experiences on SCCT decision-making to obtain a more coherent picture of what advances are required in this area, and what can be learnt from past initiatives. In this regard, it would be also interesting to discriminate between what it could be perceived as a re-branding or re-labelling of existing commercial solutions, and what it is truly the addition of functionalities/solutions beyond the current state of the art. Among the latter, autonomous self-decisions and self-learning is an area where enhancements are expected, although in the current SCCT remain somehow futuristic (Becker et al., 2019). From a technical point of view, such step would require a much tighter integration between SCCT's functionalities, which in turn would require an implementation approach is at odds with the gradual, scalable implementations that are suggested in the current literature. Furthermore, from a human point of view, additional studies would be required to assess the degree of acceptance of these automated solutions by the different decision makers in the process.

In addition to the previous lines, additional research avenues related to conceptual aspects of the SCCT have been identified:

- *Architecture of the SCCT*. Whereas most literature describes the SCCT as a monolithic architecture where all data and decisions across the SC is centralised, this vision may not be feasible/suitable in some cases. Indeed, the earliest designs of SCCT architecture are decentralised (see e.g. Hofman, 2014). In Hopman et al. (2022), different SCCT architectures are discussed, along with the main factors favouring the adopting of centralised/decentralised SCCTs. Indeed, what are the components and the most suitable SCCT architecture constitutes an area where more research is required, with just a few references (see e.g. Shou-Wen et al., 2013 or Baumgraß and Dijkman, 2015) presenting specific architectures for the SCCT. Increasing the research in this area may also help to delineate the similarities and differences between the SCCT concept and other related decision-support systems, such as the AIC (Acquire, Interpret, and Contextualise) (Fertier et al., 2021).
- *Extensions of the SCCT concept*. Whereas the SCCT concept has mainly focused on the control of material and information flows in the SC, the financial flows can be also included in the concept, as discussed in the paper by Wu et al. (2021). Another extension/variant of the SCCT is that of *Service Control Towers* (SCTs), employed for the centralisation of data and their analysis and optimisation in after-sales services supply chains (Topan et al., 2020). Furthermore, the SCCT can be seen as part of an *ecosystem supply network of networks* that may need to be orchestrated (Sherman, 2019), a similar idea to that of *Cross Chain Control*

Center (Trzuskawska-Grzesińska, 2017) to manage federated supply chain networks (Handfiel & Linton, 2017). Finally, the concept of the CT can help to implement new business models, such as Planning Services, where companies do not buy the software but the planning information (forecasts, re-order points and quantities, etc.) that is produced by a service company using a CT to gather the relevant data from the company (Rustenburg, 2016), being one of the services provided by the 4PL (Giusti et al., 2019). In Alias et al. (2015), a business model canvas for a SCCT (albeit mostly restricted to logistics) is presented.

Another possible extension is the application of the SCCT concepts into other management domains, such as hospitals (Meekings & Briault, 2013).

12.8 Conclusions

As a response to the need to quickly detect and act upon the performance deviations that characterise complex SCs, the concept of the SCCT promises to make visible the SC operations by centralising the relevant real-time data in a single information hub, so these can be motorised and controlled, increasing the SC responsiveness and resilience, which in turn improves SC performance. As such, SCCTs have emerged as one of the key topics in SC management, and many companies have implemented them, in some cases reporting substantial advantages. However, there is a lack of understanding of what an SCCT is and how it relates to other initiatives, what their main functionalities and components are, and what are their advantages and critical success factors for their application.

In this research, we have extensively reviewed the increasing body of literature on SCCTs, and presented a unified definition and functionalities based on this analysis. The main components are identified: technology, processes and people. These components are highly related among them, and they should be deeply analysed previous to the implementation of a SCCT. Therefore, the specific requirements for the company in one of these three areas become a key issue for a successful implementation. Additionally, the main functionalities of the SCCT have been identified and classified around three topis, i.e.: Data Integration, Analytics and Decision-Making. These features also serve to define the scope of the SCCT and to determine the role of the human resources after the implementation: from decision making based on the visualization of the different data provided by the SCCT, or promptly reacting to unforeseen events due to automated alert generation mechanisms embedded in the SCCT, to following recommendations provided by sophisticated optimization/machine learning methods (or, ideally, simply supervising the implementation of the plans provided by these algorithms).

A number of case studies have been analysed in order to study the potential advantages of these systems, as well as some critical success factors related to their implementation. The picture that emerges from this analysis shows that, notwithstanding their enormous potential for improving supply chain operations under the premise of higher visibility, SCCT is still an incipient area where more documented experiences are required to properly assess its contribution and to provide guidelines for its implementation. Up to now, it seems that, in practice, the visibility of E2E operations provided by the SCCTs is limited by the scope of the implementations, and that the decision-making capabilities are scarce, particularly with respect to the automation of these decisions. In this regard, several research lines are proposed to enhance SCCT capabilities in the previously identified functionalities, in some cases by aligning existing initiatives on decision support systems for supply chain management with the SCCT framework. At the same time, the analysis carried out shows that the SCCT concept itself is evolving to encompass the visibility and monitoring of operations beyond the SC scope.

Acknowledgement This research has been partially supported by the CTA project CONTENT (CONtrol Tower for ENd-To-end operations).

References

Alias, C., Özgür, C., Jawale, M., & Noche, B. (2014a). Analyzing the potential of future-internet-based logistics control tower solutions in warehouses. In *Proceedings of 2014 IEEE international conference on service operations and logistics, and informatics, SOLI 2014* (pp. 452–457). https://doi.org/10.1109/SOLI.2014.6960767

Alias, C., Özgür, C., Jawale, M., & Noche, B. (2014b). Evaluating the potential of future-internet-based transportation control tower solutions using the example of a logistics service provider. *CSME International Congress*. https://www.researchgate.net/publication/284173626_Evaluating_the_potential_of_Future-Internet-based_transportation_control_tower_solutions_using_the_example_of_a_logistics_service_provider

Alias, C., Goudz, A., Jawale, M., & Noche, B. (2015). Generating a business model canvas for future-internet-based logistics control towers. In *2015 4th IEEE international conference on advanced logistics and transport, IEEE ICALT 2015* (pp. 257–262). https://doi.org/10.1109/ICADLT.2015.7136592

Aronsson, H., & Huge Brodin, M. (2006). The environmental impact of changing logistics structures. *The International Journal of Logistics Management, 17*(3), 394–415. https://doi.org/10.1108/09574090610717545

ASCM. (2025). *Association for Supply Chain Management*, available at https://www.ascm.org/corporate-solutions/standards-tools/scor-ds/

Banker, S. (2011, February). Manufacturing Supply Chain Agility: Simple Concepts from Complex Technology. *InTech*, 20–23.

Baumgraß, A., Dijkman, R., Grefen, P. W. P. J., Pourmirza, S., Völzer, H., & Weske, M. H. (2015). A software architecture for transportation planning and monitoring in a collaborative network. In L. M. Camarinha-Matos, F. Bénaben, & W. Picard (Eds.), *Risks and resilience of collaborative networks* (pp. 277–284). Springer International Publishing.

Becker, J., Wayne, J., & Shaikh, S. (2019). The automotive industry's journey toward the autonomous supply chain. *Supply Chain Quaterly, Q4*. https://digital.thescxchange.com/supplychain/library/item/q4_2019_mobile/3784616/

Bennett, G., Radke, A. M., Moradlou, H., & Reefke, H. (2023). Improving supply chain resilience with a control tower approach beyond Covid-19. In *Advanced technologies and the Management of Disruptive Supply Chains* (Vol. 8, pp. 181–209). Springer Science and Business Media Deutschland GmbH. https://doi.org/10.1007/978-3-031-45229-1_8/FIGURES/5

Bhosle, G., Kumar, P., Griffin-Cryan, B., van Doesburg, R., Sparks, M., & Paton, A. (2015). *Supply Chain Control Towers: Is the Hype over?*

Bienhaus, F., & Haddud, A. (2018). Procurement 4.0: Factors influencing the digitisation of procurement and supply chains. *Business Process Management Journal, 24*(4), 965–984.

Brintrup, A. (2020). Artificial intelligence in the supply chain: A classification framework and critical analysis of the current state. *The Oxford Handbook of Supply Chain Management*, 209–236. https://doi.org/10.1093/OXFORDHB/9780190066727.013.24

Cap Gemini. (2011). *Global Supply Chain Control Towers Achieving end-to-end Supply Chain Visibility*.

Castellanos Sara. (2018). Germany's Merck Introduces Automation to Supply Chain. *Wall Street Journal, Sept 18*.

Christopher, M. (2011). *Logistics & Supply Chain Management*. Financial Times Prentice Hall.

Cichosz, M., Wallenburg, C. M., & Knemeyer, A. M. (2020). Digital transformation at logistics service providers: Barriers, success factors and leading practices. *International Journal of Logistics Management, 31*(2), 209–238.

Cooke, J. (2014). Control towers. In *Protean supply chains: Ten dynamics of supply and demand alignment* (pp. 113–131). John Wiley and Sons, Ltd. https://doi.org/10.1002/9781118759721.CH7

Dalporto, A., & Venn, R. (2020). Supply chain leadership, transparency, workforce development and collaboration through control tower implementation. *Journal of Supply Chain Management, Logistics and Procurement*. https://hstalks.com/article/5849/supply-chain-leadership-transparency-workforce-dev/

Delen, D., & Zolbanin, H. M. (2018). The analytics paradigm in business research. *Journal of Business Research, 90*, 186–195. https://doi.org/10.1016/j.jbusres.2018.05.013

DeVoe, S., Iampieri, M., & Mattson, E. (2023, May). Supply chain foundation: Building the elusive control tower. *Supply Chain Management Review*. https://www.scmr.com/article/supply_chain_foundation_building_the_elusive_control_tower

Dietrich, B., Johnson, C., & Dziekan, C. (2010). *in Business Analytics The Analytics Journey An IBM view of the structured data analysis landscape: descriptive, predictive and prescriptive analytics*. https://doi.org/10.1287/LYTX.2010.06.01.

DIN IEC 60050-351. (2009). *Leittechnik - Begriffe (Internationales Elektrotechnisches Worterbuch: Grundbegriffe der Leittechnik*.

Din, M. S. E., Masengu, R., & Bentley, Y. (2023). The role of supply chain control towers in facilitating sustainable supply chain management practices. In *Sustainable marketing, branding, and reputation management: Strategies for a greener future* (pp. 310–327). https://doi.org/10.4018/979-8-3693-0019-0.ch016

Duarte, E., de Haro Moraes, D., & Padula, L. L. (2023). An empirical study of learning-to-rank for spare parts consumption in the repair process. *Expert Systems, 40*(10). https://doi.org/10.1111/EXSY.13441

Dwiyana, M., Dyah Kusumastuti, R., & Bustaman, Y. (2022). Enhancement of supply chain control tower to reduce inventory parts of heavy equipment at PT XYZ. *Journal of World Science, 1*(12), 1105–1117. https://doi.org/10.58344/jws.v1i12.115

Fertier, A., Martin, G., Barthe-Delanoë, A. M., Lesbegueries, J., Montarnal, A., Truptil, S., Bénaben, F., & Salatgé, N. (2021). Managing events to improve situation awareness and resilience in a supply chain. *Computers in Industry, 132*. https://doi.org/10.1016/j.compind.2021.103488

Gerrits, B., Topan, E., & van der Heijden, M. C. (2022). Operational planning in service control towers—Heuristics and case study. *European Journal of Operational Research, 302*(3), 983–998. https://doi.org/10.1016/j.ejor.2022.01.025

Giusti, R., Manerba, D., Bruno, G., & Tadei, R. (2019). Synchromodal logistics: An overview of critical success factors, enabling technologies, and open research issues. *Transportation Research Part E: Logistics and Transportation Review, 129*, 92–110. https://doi.org/10.1016/J.TRE.2019.07.009

Gluttig, G. (2022). Resilient supply chains: A practical guide for successful implementation. In *Supply chain resilience: Insights from theory and practice* (Vol. 17, pp. 183–190). Springer Nature. https://doi.org/10.1007/978-3-030-95401-7_15/FIGURES/2

Gould, I., & Brennan, C. A. (2022). *Control Tower Technology Value Matrix 2022.*

Guidani, B., Ronzoni, M., & Accorsi, R. (2024). Virtual Agri-food supply chains: A holistic digital twin for sustainable food ecosystem design, control and transparency. *Sustainable Production and Consumption, 46*, 161–179. https://doi.org/10.1016/J.SPC.2024.01.016

Gupta, A. (2022, September). What Is a Supply Chain Control Tower and What's Needed to Deploy One? *Supply Chain 24/7.* https://www.supplychain247.com/article/what_is_a_supply_chain_control_tower_and_whats_needed_to_deploy_one

Handfiel, R., & Linton, T. (2017). INTERACTIVE! The emergence of federated supply chain networks. In *The LIVING supply chain : The evolving imperative of operating in real time* (pp. 73–100). Wiley.

Handfield, R., Finkenstadt, D. J., Schneller, E. S., Godfrey, A. B., & Guinto, P. (2020). A commons for a supply chain in the post-COVID-19 era: The case for a reformed strategic national stockpile. *Milbank Quarterly, 98*(4), 1058–1090. https://doi.org/10.1111/1468-0009.12485

Hofman, W. (2014). Control tower architecture for multi - and synchromodal logistics with real time data. *ILS 2014 - 5th International Conference on Information Systems, Logistics and Supply Chain.*

Hopman, M., Fransen, R., van Meijeren, J., & Zubin, I. (2022). A Framework on Centralised to Decentralised Logistics Control Structures Applied in Two Case Studies. *Lecture Notes in Computer Science (Including Subseries Lecture Notes in Artificial Intelligence and Lecture Notes in Bioinformatics), 13557 LNCS*, 369–381. https://doi.org/10.1007/978-3-031-16579-5_25/FIGURES/3.

Ivanov, D., Dolgui, A., Das, A., & Sokolov, B. (2019). Digital supply chain twins: Managing the ripple effect, resilience, and disruption risks by data-driven optimization, simulation, and visibility. *International Series in Operations Research and Management Science, 276*, 309–332. https://doi.org/10.1007/978-3-030-14302-2_15/FIGURES/6

Jha, A. K., Verma, N. K., & Bose, I. (2022). Measuring and managing digital supply chain performance. In *The digital supply chain* (pp. 199–214). Elsevier. https://doi.org/10.1016/B978-0-323-91614-1.00012-5

Klesman, S., & Cherici, E. (2017). *The Features of an Operative Control Tower.*

Kulkarni, M. (2023). The role of the control tower in increasing the productivity in container terminal operations. *Journal of Maritime Research, 20*(1), 30–33. www.jmr.unican.es

Kumar, B., & Barros Gomez, P. (2024). Procurement control tower: Proof of concept through machine learning and natural language processing. *Supply chain. Management Review.* https://www.scmr.com/article/procurement-control-tower-proof-of-concept-through-machine-learning-and-natural-language-processing/Control_Towers

Liotine, M. (2019). Shaping the next generation pharmaceutical supply chain control tower with autonomous intelligence. *Journal of Autonomous Intelligence, 2*(1), 56–71. https://doi.org/10.32629/jai.v2i1.34

Liu, M., Fang, S., Dong, H., & Xu, C. (2021). Review of digital twin about concepts, technologies, and industrial applications. *Journal of Manufacturing Systems, 58*, 346–361. https://doi.org/10.1016/j.jmsy.2020.06.017

Lund, P. O., & Payne, T. (n.d.). *Magic Quadrant for Supply Chain Planning Solutions.*

MacCarthy, B. L., & Ivanov, D. (2022). The digital supply chain—Emergence, concepts, definitions, and technologies. In *The digital supply chain* (pp. 3–24). Elsevier. https://doi.org/10.1016/B978-0-323-91614-1.00001-0

MacCarthy, B. L., Ahmed, W. A. H., & Demirel, G. (2022). Mapping the supply chain: Why, what and how? *International Journal of Production Economics, 250*, 108688. https://doi.org/10.1016/J.IJPE.2022.108688

Maheshwari, P., Kamble, S., Kumar, S., Belhadi, A., & Gupta, S. (2023). Digital twin-based warehouse management system: A theoretical toolbox for future research and applications.

International Journal of Logistics Management. https://doi.org/10.1108/IJLM-01-2023-0030/FULL/PDF

Malmstedt, A., & Bäckstrand, J. (2022). How to predict disruptions in the inbound supply chain in a volatile environment. *Advances in Transdisciplinary Engineering, 21*, 638–649. https://doi.org/10.3233/ATDE220182

Meekings, A., & Briault, S. (2013). The "control tower" approach to optimising complex service delivery performance. *Measuring Business Excellence, 17*(3), 15–27. https://doi.org/10.1108/MBE-11-2011-0031/FULL/PDF

Mena, C., Christopher, M., & van Hoek, R. (2014). *Leading procurement strategy: Driving value through the supply chain*. Kogan Page Ltd..

Milenkovic, M., Bojovis, N., & Val, S. (2019). A value case approach for improving the quality of rail freight services: Control tower concept. In M. Ioannis (Ed.), *8th international symposium and 30th National Conference on operational research* (pp. 140–146) http://eeee2019.teiwest.gr

Mukherjee, A. (2023, June). The critical process of supply chain control tower integration orchestration. *Supply Chain Management Review*. https://www.scmr.com/article/the_process_of_supply_chain_control_tower_integration_orchestration

Naik, S., & Kumar, P. (2021, September). The New Essential for Supply Chains: Intelligent Execution Control Tower Part I. *Supply Chain Management Review*. https://www.scmr.com/article/the_new_essential_for_supply_chains_intelligent_execution_control_towe1

Patsavellas, J., Kaur, R., & Salonitis, K. (2021). Supply chain control towers: Technology push or market pull—An assessment tool. *IET Collaborative Intelligent Manufacturing, 3*(3), 290–302. https://doi.org/10.1049/CIM2.12040

Ponomarov, S. Y., & Holcomb, M. C. (2009). Understanding the concept of supply chain resilience. *The International Journal of Logistics Management, 20*(1), 124–143. https://doi.org/10.1108/09574090910954873

Rai, A., Pavlou, A., Im, G., & Du, S. (2012). Interfirm IT capability profiles and Communications for Cocreating Relational Value: Evidence from the logistics industry. *MIS Quarterly, 36*(1), 233. https://doi.org/10.2307/41410416

Rustenburg, J. W. (2016). Planning Services: A Control Tower Solution for Managing Spare Parts. *Lecture Notes in Logistics*, 239–259. https://doi.org/10.1007/978-3-319-22288-2_14/TABLES/3

Schiffer, M., & Doerr, D. M. (2020). Development of the supply chain management 2040 - opportunities and challenges. *Proceedings of the Conference on Production Systems and Logistics*, 29–37. https://doi.org/10.15488/9644

Sean Culey. (2020, October). Gaining Control in a Turbulent World: The Case for Supply Chain Control Towers. *Supply Chain Brain*. https://www.supplychainbrain.com/articles/32098-gaining-control-in-a-turbulent-world-the-case-for-supply-chain-control-towers?v=preview

Sharabati, A. A. A., Al-Atrash, S. A., & Dalbah, I. Y. (2022). The use of supply chain control tower in pharmaceutical industry to create a competitive advantage. *International Journal of Pharmaceutical and Healthcare Marketing, 16*(3), 354–375. https://doi.org/10.1108/IJPHM-08-2020-0064/FULL/PDF

Sheffi, Y. (2015a). Detecting disruption. In *The power of resilience: How the best companies manage the unexpected* (pp. 189–214). MIT Press. https://ebookcentral.proquest.com/lib/uses/detail.action?pq-origsite=primo&docID=4093101

Sheffi, Y. (2015b, September 15). Preparing for Disruptions Through Early Detection. *MIT Sloan Management Review*. https://sloanreview.mit.edu/article/preparing-for-disruptions-through-early-detection/

Sherman, R. J. (2019, June). I can see clearly now…the constraints are gone. *Supply Chain Management Review*. https://www.scmr.com/article/i_can_see_clearly_nowthe_constraints_are_gone

Shou-Wen, J., Ying, T., & Yang-Hua, G. (2013). Study on supply chain information control tower system. *Information Technology Journal, 12*(24), 8488–8493.

Smith, G. A. (2022). Supply Chain Control Towers-Benefits & Framework for Implementation. *Journal of Business Forecasting, 30*. www.supplychainmavens.net

Somapa, S., Cools, M., & Dullaert, W. (2018). Characterizing supply chain visibility—A literature review. *International Journal of Logistics Management, 29*(1), 308–339. https://doi.org/10.1108/IJLM-06-2016-0150/FULL/PDF

Syahchari, D. H., Sudrajat, D., Lasmy, L., Herlina, M. G., Kiatama, C., & Jordaan, H. K. W. (2022). Achieving Supply Chain Resilience Through Supply Chain Control Tower And Supply Chain Agility. *ACM International Conference Proceeding Series*, 195–198. https://doi.org/10.1145/3512676.3512709.

Topan, E., Eruguz, A. S., Ma, W., van der Heijden, M. C., & Dekker, R. (2020). A review of operational spare parts service logistics in service control towers. *European Journal of Operational Research, 282*(2), 401–414. https://doi.org/10.1016/j.ejor.2019.03.026

Trzuskawska-Grzesińska, A. (2017). Control towers in supply chain management– Past and future. *Journal of Economics and Management, 27*, 114–133. 10.22367/jem.2017.27.07.

VDI 4493. (2009). *Leitstand für Die Intralogistik.*

Verma, R., Koul, S., & Singh, G. (2020). Intelligent decision-making: Using control tower at a logistics company. In *2020 IEEE international conference on computing, power and communication technologies, GUCON 2020* (pp. 550–554). https://doi.org/10.1109/GUCON48875.2020.9231108

Vlachos, I. (2021). Implementation of an intelligent supply chain control tower: A socio-technical systems case study. *Production Planning and Control, 34*(15), 1415–1431. https://doi.org/10.1080/09537287.2021.2015805

Wertheimer, K. E. (2011, December). Taking a Control Tower Approach. *Supply Chain Management Review*. https://www.scmr.com/article/taking_a_control_tower_approach

Williams, B. D., Roh, J., Tokar, T., & Swink, M. (2013). Leveraging supply chain visibility for responsiveness: The moderating role of internal integration. *Journal of Operations Management, 31*(7), 543–554. https://doi.org/10.1016/j.jom.2013.09.003

Woodward, M. (2012, December). 5 Supply Chain Predictions For 2013. *Supply Chain Management Review*. https://www.scmr.com/article/5_supply_chain_predictions_for_2013

Wu, C., Liu, J., & Zhang, H. (2021). Data ecology and accurate portrait: Optimization of credit risk system for SMEs in supply chain finance based on big data technology. *Journal of Risk Analysis and Crisis Response, 11*(4), 163–175. https://doi.org/10.54560/jracr.v11i4.310

Wycislak, S. (2023). From real-time visibility to operational benefits—Tensions on unfinished paths. *International Journal of Logistics Management, 34*(5), 1446–1474. https://doi.org/10.1108/IJLM-03-2022-0126

Yan, W. J., Tan, P. S., Koh, N. W., Tan, Y. Q., & Zhang, A. N. (2012). Towards better supply chain visibility - The design and implementation of a supply chain system S-ConTrol to support an operational HQ in Singapore. *IEEE International Conference on Industrial Engineering and Engineering Management*, 971–975. https://doi.org/10.1109/IEEM.2012.6837885

Ye, C., Zaraté, P., & Kamissoko, D. (2022). A DSS Based on a Control Tower for Supply Chain Risks Management. *Lecture Notes in Business Information Processing*, *447 LNBIP*, 124–136. https://doi.org/10.1007/978-3-031-06530-9_10

Zhang, X., & Xiao, G. (2023). Digital control tower empowers supply chain resilience improvement from a dynamic capability perspective. In *Digitalization and management innovation II* (Vol. 376, pp. 186–194). https://doi.org/10.3233/FAIA230730

Chapter 13
The Role of Control Towers in Improving Supply Chain Agility

Natalia Szozda

Abstract The evolution of supply chains has been profoundly shaped by technological advancements, shifting consumer expectations, and global disruptions such as the COVID-19 pandemic. These factors have driven a transition toward digital, customer-centric supply chain management, where Supply Chain Control Tower (SCCT) technology plays a pivotal role. Implementing new technologies, including control tower solutions, is crucial to integrating agility, responsiveness, visibility, and transparency into the supply chain structure. This chapter aims to introduce the concept of SCCT in the context of a digital, customer-centric supply chain. This will be achieved by reviewing theoretical foundations and illustrating the functionality of control towers through case studies of the Electrolux and Procter & Gamble supply chains. The research methodology includes a case study analysis supported by a survey study conducted among 228 supply chains, incorporating elements of qualitative analysis.

Keywords Control tower · Customer-centric supply chain · Digital supply chain · Responsiveness · Visibility · Agility · Transparency

13.1 Introduction

The COVID-19 pandemic acted as a catalyst for rapid digital transformation in supply chains (Choi et al., 2020). It revealed significant global logistics and manufacturing weaknesses, causing widespread disruptions in material sourcing, production, and inventory management (SAP Insights n.d.). These disruptions forced companies to rethink their supply chain strategies, seek alternative suppliers, and adopt digital tools to improve real-time visibility (Choi et al., 2020). The growing trend of

---

N. Szozda (✉)
Department of Logistics, Wroclaw University of Economics and Business, Wrocław, Poland
e-mail: natalia.szozda@ue.wroc.pl

Y. Khojasteh, S.M. M. Jahan Abadi (eds.), *Digital Transformation of Supply Chains*, International Series in Operations Research & Management Science 372, https://doi.org/10.1007/978-3-032-01206-7_13

short-term, high-volume online consumer purchases has driven significant structural changes in supply chain management. To accommodate the growing demand for e-commerce and direct-to-consumer deliveries, companies have had to reconfigure and modernize their traditional supply chain frameworks (Chatterjee et al., 2021). This included implementing a more rigorous evaluation process for partners, suppliers, and distributors to ensure reliability and resilience (Chaffin et al., 2024; Haren & Simchi-Levi, 2020; Linton & Vakil, 2020). The pandemic also underscored the importance of flexibility and agility in supply chains (Ivanov, 2022; Kazancoglu et al., 2022; Syahchari et al., 2022). In response, companies have invested in advanced digital technologies such as Artificial Intelligence (AI), Machine Learning (ML), and Blockchain to enhance their forecasting accuracy, optimize inventory levels, and improve overall supply chain efficiency (Cheng, 2024). These investments are designed to create more responsive and agile supply chains to withstand future disruptions and meet changing customer demands.

Simultaneously, the customer-centric approach has become a dominant trend in supply chain management. According to Potter et al. (2015), customers are playing an increasingly influential role, driven by their evolving expectations, fluctuating requirements, and fast-paced lifestyles. The supply chain's objective is to adapt to these changing market needs effectively and actively engage customers to ensure seamless integration between entities. A customer-centric supply chain prioritizes the customer throughout the entire process, from product design to delivery (Christopher, 2016). Customers expect greater control and visibility over their purchasing experience, with a strong emphasis on transparency and proactive supply chain communication (Esper et al., 2020). The focus has shifted from the product to ensuring a personalized and seamless customer service. In this customer-centric supply chain, customers not only have the option to choose from available products but also have the ability to influence the supply chain through new technologies (Chen et al., 2024; Esper et al., 2020). A Gartner study in 2021 revealed that around 83% of supply chains aim to enhance the customer experience, which is defined as the customer's shopping experience, with a focus on digital transformation and distribution system improvements. New technologies enable customers to determine when, how, and where they access their service or product (Tih et al., 2016). Technological advancements in distribution systems empower customers to access products or services on their terms, removing traditional barriers such as geographical or informational constraints (Brynjolfsson et al., 2013). Consequently, organizations are integrating omnichannel structures and real-time data analytics to improve their service offerings and meet fluctuating market demands (Brynjolfsson et al., 2013; Chatterjee et al., 2021).

The integration of the supply chain is crucial not only in the downstream, but also in the upstream supply chain. Adopting an agile sourcing approach makes delivering the expected products to the customer possible. Thus, integration and closer connections within the supply chain are necessary to respond rapidly to changing conditions on both the distribution and sourcing sides (Christopher & Lee, 2004). The need for visibility, responsiveness, and agility in dynamic alignment requires that real-time information be accessible to all decision-makers, with operational synchronization mechanisms and feedback loops. This supply chain management approach can be achieved through advanced analytics to enable supply chain integration, with a control tower as the supporting technology (Vlachos, 2021).

This chapter explores the role of control tower technology in enabling digital customer-centric supply chains. It aims to bridge the gap between theoretical concepts and practical applications, offering a deeper understanding of this evolving research area. The analysis is based on a literature review, case studies, and qualitative research conducted among 228 supply chains. The chapter highlights how control towers enhance visibility, responsiveness, transparency, and agility in modern supply chains.

The chapter is structured as follows: Sect. 13.2 presents a comprehensive literature review on control tower technology, outlining its evolution, key functionalities, and impact on supply chain management. Section 13.3 provides case studies from Electrolux and P&G, illustrating the implementation of control towers in customer-centric supply chains and offering a comparative analysis of their approaches. Section 13.4 synthesizes insights from the literature review and case study findings to develop a conceptual model of the digital customer-centric supply chain leveraging control tower technology. Section 13.5 discusses the validation of the proposed model, assessing its applicability and effectiveness. Section 13.6 concludes with key findings, managerial implications, and future research directions, emphasizing the strategic role of the control tower in shaping the next generation of supply chains.

13.2 Literature Review

The development of computer technology and Enterprise Resource Planning (ERP) systems has prompted companies to integrate various operational aspects, particularly in supply chain management (Bozarth & Handfield, 2019; Christopher, 2021). This was the beginning of centralizing and automating supply chain management, *coinciding with control tower technology's emergence* in the 1980s. The Supply Chain Control Tower (SCCT) is an advanced software solution that acts as a central hub for overseeing, monitoring, and coordinating supply chain processes (Christopher, 2021). The goal is to improve supply chain performance (Chen et al., 2024) and enhance efficiency (Din et al., 2023). This technology is crucial for ensuring smooth operations, addressing issues as they arise, and optimizing processes. The primary objective of SCCT is to determine the most cost-efficient solutions while maintaining superior customer service standards. Real-time monitoring of customer service levels, including timing, efficiency, and overall service level, is essential for proactive and reactive supply chain risk management (Wyciślak & Pourhejazy, 2023; Ye et al., 2022). According to Capgemini (Bhosle et al., 2011), SCCT is a cross-divisional organization with integrated information hubs that provide visibility into the supply chain. The SCCT is a comprehensive planning and execution system designed to efficiently address resource limitations, process deviations, and operational challenges through real-time corrective and preventive measures (Trzuskawska-Grzesińska, 2017). Zhang and Xiao (2023) emphasize the role of digitalization in the control tower, describing it as a hub for visibility, decision-making, and real-time action based on comprehensive information and analytics. The control tower is designed to manage and control cross-functional and cross-company decision-making and execution, optimizing the entire network in alignment with strategic objectives and short- and long-term decisions.

Control tower is crucial for managing supply chains at three primary levels (Patsavellas et al., 2021): strategic, tactical, and operational, as identified by Bhosle et al. (2011). At the strategic level, the control tower oversees the entire network in the supply chain, ensuring comprehensive control. The tactical level involves proactive planning of procurement, distribution, and operational activities to align with market demands and customer-centric strategies. For example, this level includes organizing transportation to enhance sustainability, as discussed by Klesman and Cherici (2017). On the other hand, the operational level focuses on executing strategies through day-to-day operations, such as transportation management, inventory tracking, and adverse event management. The SCCT system is designed to adapt to a company's constantly changing environment by defining new parameters, such as inventory levels or ordering plans, and making adjustments when issues arise (Patsavellas et al., 2021). This solution considers the various participants in the supply chain and the incentives that can disrupt it. SCCT facilitates collaboration across trading partners, manufacturers, logistics providers, and customers, streamlining coordination at various levels.

SCCT is based on four key pillars: the integration of people, processes, knowledge (data), and technology infrastructure (Accenture, 2022; Bhosle et al., 2011; Din et al., 2023), as shown in Fig. 13.1. The first pillar emphasizes the essential role of people in gathering information, monitoring performance, and using insights from the control tower to enhance overall performance. The second pillar, processes, requires effective procedures for managing and integrating supply chain activities as a prerequisite for utilizing the software. These processes encompass the management and integration of delivery, procurement, and transportation. The third pillar emphasizes the importance of users' comprehensive understanding of supply chain management processes and proficiency in leveraging SCCT technology for optimization purposes. Additionally, the advancement of digital technology has led to an increased demand for robust data management capabilities (DeVoe et al., 2023). Organizations that utilize SCCT must effectively manage current and accurate internal and external data.

The foundational pillar of SCCT incorporates cloud-based solutions that leverage cutting-edge technologies, including the Internet of Things (IoT), cloud computing, big data and data analytics, along with AI and ML. According to SAP

Supply Chain Control Tower (SCCT)			
people	processes	knowledge / data	technological infrastructure

Fig. 13.1 Pillars of SCCT

Insights (n.d.), cloud computing enables extensive data storage and processing, providing global access to information. IoT technology allows object connectivity, enabling data collection and analysis relevant to supply chain operations. This integration of devices and sensors across the supply chain facilitates instant data gathering and examination. Big data and data analytics are crucial for utilizing large volumes of data from various sources, such as sensors, ERP systems, and external data, to monitor and improve process efficiencies. AI and ML are crucial in predicting potential issues, automating decision-making, and optimizing logistics operations.

Additionally, SCCT incorporates digital twin technology, a pivotal simulation tool for finding optimal solutions and enhancing supply chain procedures (Maheshwari et al., 2024; Naik & Kumar, 2021). Regarding IT infrastructure support, the control tower is equipped with Key Performance Indicators (KPIs) and dashboards (SAP Insights, n.d.), which primarily focus on process optimization, time assessment, cost analysis, and ensuring data quality control. This sophisticated integration of technologies and analytical tools is the foundation of SCCT, driving supply chain visibility, efficiency, and performance enhancement.

SCCT has a wide range of applications. The primary functionalities of this solution include the following (Patsavellas et al., 2021; SAP Insights, n.d.; Bhosle et al., 2011):

1. *Real-time monitoring and predictive alerts*: These features utilize advanced algorithms to detect potential risks by analyzing critical data sources related to supply and demand fluctuations. A real-time view of key component and material inventory, supplier capacity, and anticipated demand enables manufacturing and assembly plants to identify and address potential risks proactively.
2. *Route and resource optimization:* This entails using algorithms to plan the most efficient delivery routes, manage vehicle fleets, and optimize the use of warehouse resources.
3. *Inventory management:* This encompasses optimizing stock levels, streamlining product rotation, ensuring adequate inventory availability at local and regional levels, and identifying shortages or ageing inventory before they become critical issues.
4. *Operational management:* Insights into production performance trends, company capacity utilization, and schedule adherence provide an early warning of impending operational problems.
5. *Scenario planning and risk management:* Integrated demand and supply scenario modelling and trade-off analysis enable more effective navigation through uncertainty by considering growth potential and risk. It also facilitates identifying and managing potential supply chain disruptions, including delays, quality issues, and changes in demand.

SCCT solution providers classify control towers based on their functional characteristics. For example, SAP (SAP Insights, n.d.) identifies four main categories: logistics, operational transportation, analytical, and inventory control tower.

The key advantages of utilizing a SCCT include the following (Patsavellas et al., 2021; SAP Insights, n.d.; Bhosle et al., 2011; Syahchari et al., 2022; Chaffin et al., 2024):

1. *Supply chain visibility and transparency:* The control tower gathers data from various sources to provide comprehensive visibility, allowing companies to address potential disruptions promptly. Visibility refers to observing and understanding the entire supply chain, including tracking, monitoring, reviewing, and detecting the status of products, materials, services, and information at every stage and process level (Somapa et al., 2018). On the other hand, transparency indicates an organization's willingness to grant all supply chain trading partners immediate and unrestricted access to information (Sodhi & Tang, 2019). This enables users to take corrective measures in response to adverse events and adjust planned activities as needed, demonstrating the ability to react quickly when challenges arise.
2. *Supply chain resilience and agility: Resilience refers to recovering from unexpected disruptions or market anomalies without significantly impacting supply, production, and outputs* (Ponomarov & Holcomb, 2009). On the other hand, agility is the organization's flexibility in adapting to changing business conditions and its ability to quickly adjust supply and production accordingly (Swafford et al., 2008). These factors enable continuous supply chain transformation within a dynamic business model. In a rapidly changing environment, different supply chain segments require different approaches to ensure management success. Managers can make informed decisions about possible management changes based on automatically calculated KPIs and their forecasts.
3. *Responsiveness and proactive risk management*: This refers to an approach that focuses on anticipating and preventing problems before they occur rather than reacting to them afterwards. It involves gathering information and adjusting to changes in the environment, including shifts in demand, productivity, regulations, or any other factors that could impact business value (Can Saglam et al., 2021). In this process, predictive analytics and scenario planning are instrumental in mitigating risks and optimizing decision-making.
4. *Optimization of decision-making processes and improving collaboration within the supply chain*: This term refers to organizations using advanced analytical tools and information technology to make better-informed and more productive supply chain management decisions. SCCT involves collecting data from all supply chain links, analyzing it, and customizing the resulting information. The control tower promotes cooperation between departments and partnerships. Collaborative work platforms and dashboards are being developed to address potential conflicts of interest and facilitate faster decision-making. Recommendations for operational optimization, demand and risk management are made based on the collected data, and improvements are implemented to minimize costs and promote sustainable supply chain practices (Din et al., 2023).

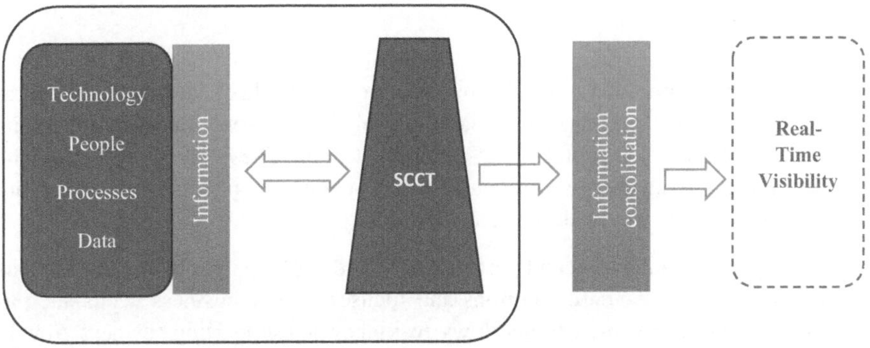

Fig. 13.2 Flow of information in SCCT (Klesman & Cherici, 2017)

The main goal of implementing SCCT is to provide added value, which can be categorized into two primary areas: improving decision-making processes and enhancing business performance (Klesman & Cherici, 2017), as shown in Fig. 13.2.

The implementation of decision-making support tools, in conjunction with KPIs, comprehensive reports, and real-time information from the supply chain, can significantly improve precise decision-making processes. Integrating these tools is projected to decrease operational costs by approximately 10–15% (Kumar, 2022). This is accompanied by increased customer satisfaction, reduced operational anomalies, and the effective use of predictive analytics, all contributing to improved business performance. Additionally, SCCT can support companies in achieving their sustainability objectives, such as reducing GHG emissions by optimizing vehicle and modality usage, improving last-mile delivery, and enhancing fuel efficiency (Kumar, 2022).

13.3 Case Studies

The case studies in this chapter section examine supply chains that have recently undergone significant transformations, particularly in adopting a customer-centric approach. One of the key technologies driving these changes is the control tower, which facilitates a demand-driven approach, agility, responsiveness, and transparency in the supply chain. These technologies and other solutions pave the way for implementing a demand-centric supply chain. This will be explained in the case studies to provide a comprehensive understanding of the evolving nature of supply chains. The case studies focus on Electrolux and P&G, both of which have undergone digital transformation in recent years and utilize control tower technology for this purpose.

13.3.1 Electrolux

The global pandemic had a significant impact on Electrolux's supply chain operations. As a result, the company initiated a global transformation program for its strategic ambition, Electrolux 2025+ Supply Chain. The goal of digitizing the Electrolux supply chain is to create a consumer-centric supply chain. The program is executed in three waves (Jorgensen, 2022):

1. Supply chain foundation and performance: In this phase, Electrolux implements and accelerates standard solutions and focuses on key business needs such as OEM flows, direct-to-customer flows, resiliency and supporting business growth.
2. Simplify and connect with suppliers and customers: In this phase, Electrolux successfully tackles complexity. Seamless integration of customers and suppliers enables efficient management of demand and supply chain changes.
3. Digital and prescriptive supply chain network: In this phase, the ultimate goal is to create a Consumer-Centric Network (CCN), a fully digitalized and orchestrated supply chain network that is aligned with consumer preferences and expectations.

Electrolux's CCN supply chain transformation encompasses several key areas, including efficiency and agility, the control tower concept, advanced and digital analytics, process excellence and optimization, omnichannel structures and sustainable logistics. The company's main goal is to improve its agility, which can be achieved by implementing digital transformation and introducing a control tower into its supply chain structure. Electrolux's supply chain agility requires collaboration between downstream and upstream teams, as shown in Fig. 13.3.

Electrolux was among the first organizations worldwide to implement Blue Yonder's Luminate Control Tower solution. The goal was to integrate downstream and upstream planning. The control tower technology facilitates integration across various key areas, including supply and primary transportation planning, production

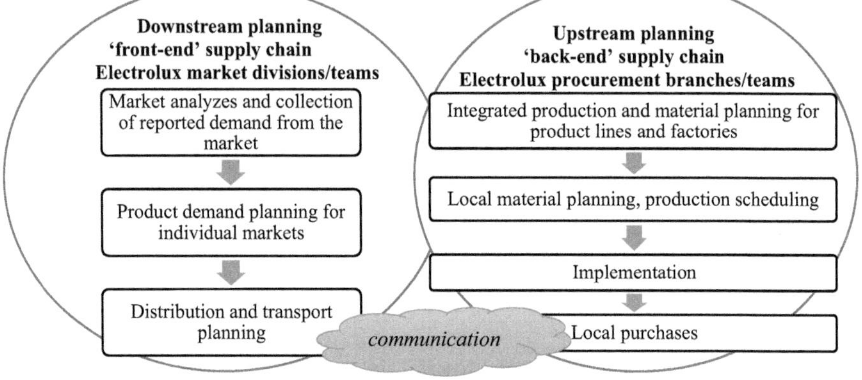

Fig. 13.3 Agile management in the Electrolux supply chain (Szozda, 2022)

and materials planning, OEM supply chain, spare parts planning, supply chain excellence, and commissioning and analytics. Supply chain management at Electrolux is carried out in Management Centers, which are dedicated business units that bring together logistics and supply chain management specialists in a single location and are dedicated to specific regions of the global market. For example, a team of approximately 150 specialists and managers from a single location in Poland is responsible for planning and coordinating the flow of products and information in Electrolux subsidiaries, factories, and distribution centers in Europe, Africa, and Asia. The control tower supports procurement and transportation planning, production scheduling, and material management within the production process. Additionally, it oversees the planning of transportation and distribution of finished products to customers, as well as planning spare parts inventory, analyzing and controlling supply chain performance. Each planner is provided with a real-time view of stock levels and plans for individual departments and divisions. This is made possible by the Blue Yonder platform, which provides planners with an ecosystem that enables synchronization of planning and scheduling. The solution includes planning spare parts inventories and analyzing and controlling supply chain efficiency using advanced analytics and dashboards.

One of the most crucial aspects of integrating a supply chain is fostering collaboration between Electrolux and its clients. The company's objective is to optimize its processes and establish omnichannel structures to better serve its customers while maintaining sustainable logistics practices. These factors are anticipated to contribute to the optimization of supply chain response time, including the reduction of lead time, time to awareness, time to survive, and time to recover. An efficient distribution system is achieved by synchronizing the activities of various branches and departments. The system is divided into two categories of deliveries: direct deliveries to customers, which account for 30% of all deliveries, and deliveries from regional distribution centers, which account for the remaining 70%. The lead time for direct deliveries of customized products is 4 weeks, and customers can monitor the status of their orders in real time. The regional distribution centers for other products deliver products to the warehouse within 1–10 days, thus ensuring that they are available in time to meet customer deadlines. To provide a seamless shopping experience for customers, Electrolux's supply chain partners with DHL for delivery and product returns, and Hermes and Geodis for product connection and installation. The system continuously analyzes real-time user interactions to anticipate, mitigate, and swiftly address potential disruptions within the supply chain. Communication between supply chain entities is facilitated through the use of management cockpits and client portals, commonly known as dashboards. These dashboards enable the generation of reports on inventory, distribution, production, or labor utilization, as well as tracking transportation and monitoring manufactured and distributed products.

Blue Yonder's Luminate Control Tower has improved Electrolux's responsiveness and proactivity throughout its supply chain, strengthening customer service and end-to-end visibility. This solution allows the company to react rapidly to market changes and apply agile methodology in managing supply chain flows. The implementation of the Luminate Control Tower has enabled Electrolux to better serve clients by providing early information, resolving issues in real time, and adopting a more agile approach. The visibility of the supply chain has significantly improved, with every link in the supply chain having access to relevant information. It is also worth noting that data quality has been improved, and data gathering has become more efficient. The introduction of the control tower has increased added value, ensuring tasks are prioritized according to set KPIs and utilizing AI for more accurate forecasts. This has enabled supply chain managers to spend less time searching for information and more time leveraging data effectively. Overall, this transformation has enabled the organization to efficiently monitor, analyze, and manage critical supply chain data in real time.

13.3.2 Procter & Gamble

Procter & Gamble's (P&G) supply chain is considered one of the most innovative in the world. P&G was among the first organizations to establish a supply chain management center, which functions similarly to the Electrolux supply chain solution. One such center is located in Warsaw, in Poland and manages end-to-end supply chain operations in over 100 countries, from planning raw materials to delivering the finished products to the customer. This ensures faster and more efficient product delivery to retailers and consumers. It's important to note that P&G has been recognized for the ninth year in a row as one of four supply chain Masters in Gartner's Supply Chain Top 25 ranking. P&G's digital supply chain transformation, which helps address customer issues, was a key factor in this recognition. The technologies P&G invests in to support this transformation include cloud computing, control tower operation incorporating AI, and automation. These technologies are aimed at facilitating supply chain integration, enhancing supply chain flexibility and resilience, and introducing new approaches to product category management, product monitoring, and home usage, among other things.

P&G uses an end-to-end approach, enabled by digital capabilities, to speed up product delivery to the market. They have built a synchronized network based on real-time data to meet customers' changing needs and created a customer-centric approach. P&G has made its supply chain more flexible and efficient by using flow centers that operate on the principle of cross-docking. About 60% of the inventory is cross-docked, with products turning over in just 4 days. P&G implemented a Consumer Driven Supply Network (CDSN), leveraging real-time sales data to enhance responsiveness to consumer demands (Gupta, 2020). To achieve this transformation, P&G has implemented structural changes in its supply chain, including flow centers, control tower operations, Supply Network Operation (SNO), Global Data Synchronization Network (GDSN), and distributor connect.

Control tower operations provide a platform for P&G's supply chain integration, supporting P&G's supply chain management centers. Cloud computing gathers information from suppliers, warehouses, logistics providers, stores, and customers.

Subsequently, this information is analyzed using AI, among other tools. The structure of P&G's SCCT is illustrated in Fig. 13.4. The control tower helps monitor all transportation activities, including inbound, outbound, raw materials, and finished goods. It has been shown to reduce the 'deadhead' movement by approximately 15%, which in turn lowers costs and carbon monoxide emissions. This improvement leads to better customer service and reduced inventory across the supply chain.

Another solution implemented at P&G to integrate the supply chain through planning, distribution logistics and customer service is called SNO. It includes the following functionalities: product demand forecasting, production planning and monitoring. This tool monitors the flow of information along the supply chain, from suppliers, production facilities and distribution centers to the customer's shelf.

P&G has implemented GDSN to improve communication with retailers. This standardized data warehouse allows the company to automate commerce with its retail partners, eliminating the need for human intervention. When all parties use a common data warehouse, where data is maintained in a dynamic and accurate manner, it significantly reduces the likelihood of errors in orders.

A digital interface, named 'distributor connect', facilitates direct communication between P&G and distributors, reducing inventory and improving service. This initiative benefits both the distributor and P&G by streamlining their supply chain.

P&G has strengthened its supply chain by incorporating these solutions into its CDSN, emphasizing customer-centricity as a core principle. It has been instrumental in product design and production, allowing for the customization of products to meet customer requirements and enabling faster adjustments to production lines. For instance, during the 2021 Texas storm, which impacted the supply of raw materials and disrupted 60% of the US liquid laundry industry, P&G utilized new technologies to respond and uphold supply chain resilience swiftly. Within 28 days, P&G developed new products from available raw materials, ensuring product

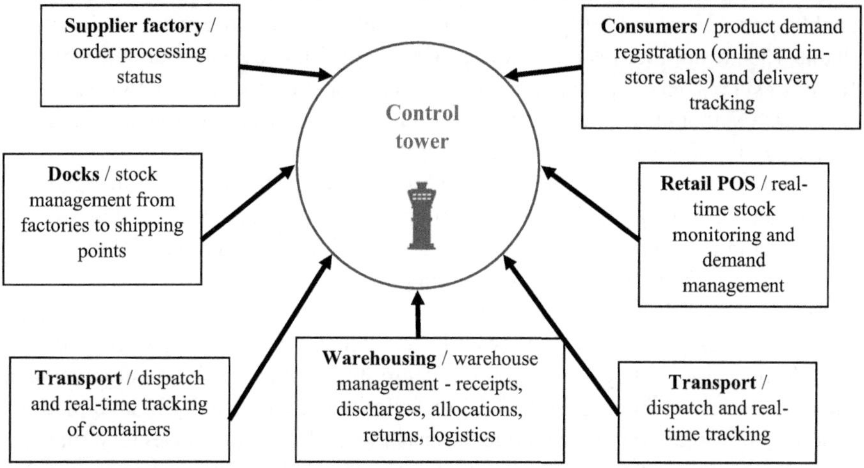

Fig. 13.4 Control tower operations in P&G (Szozda, 2022)

availability in the market. Additionally, P&G has utilized ML algorithms to revolutionize the product category management process, significantly reducing its duration and streamlining operations. Previously, it took several weeks to plan product placements. Now, the new algorithms dynamically optimize product availability in physical stores and online, adjusting in real time based on reported demand, 24 h a day, 7 days a week, with updates every 15 min.

P&G's supply chain digitization efforts will soon encompass the manufacturing and operational processes, including the introduction of robotics for packaging and autonomous vehicles for transportation. In addition, the company is enhancing its capabilities in big data and business analytics to efficiently collect and manage data across the supply chain. These tools will improve the understanding and coordination of operations within the supply chain. Currently, P&G utilizes various planning and analytics solutions like Blue Prism, Databricks, MS Azure, Power Platform, Rulex, and Knime. Moreover, the company is working on developing additional solutions to enable a comprehensive analysis of the entire supply chain, which will help enhance responsiveness to customer needs.

13.3.3 Comparative Analysis: Electrolux vs. Procter & Gamble

A comparison of the two presented case studies reveals that their goals are very similar. Both aim to achieve integration with the customer by implementing new technologies into their structures and transforming their operations into digital supply chains. Table 13.1 examines the digital transformation efforts in supply chain management at Electrolux and P&G, highlighting key similarities and differences and showcasing their unique approaches to creating flexible, efficient, and responsive supply chains in the digital era.

Electrolux and P&G have deployed cutting-edge control tower technologies to establish real-time visibility and seamless integration throughout their supply chains. Electrolux focuses on building CCN using Blue Yonder's Luminate Control Tower to synchronize upstream and downstream operations, while P&G has developed CDSN that incorporates an AI-driven control tower integrated with cloud computing to align supply with demand dynamically.

Electrolux emphasizes seamless collaboration through centralized management centers and the integration of upstream (procurement) and downstream (market teams) planning, whereas P&G employs tools such as SNO and GDSN to connect all supply chain entities for end-to-end integration.

In addressing disruptions, Electrolux leverages predictive analytics to manage real-time issues, while P&G demonstrates resilience by swiftly innovating product lines during crises. A key focus for both Electrolux and P&G is the development and adoption of new technologies. Electrolux relies heavily on Blue Yonder's platform which incorporates AI for accurate forecasting and real-time inventory management. Meanwhile, P&G utilizes a mix of tools such as Blue Prism, Databricks, MS

Table 13.1 Comparison—Electrolux vs. P&G

Aspect	Electrolux	Procter & Gamble (P&G)
Corporate context	Global household appliance leader with a strong focus on sustainability and customer satisfaction	Leading consumer goods company with a vast brand portfolio and recognized for supply chain excellence
Digital transformation goal	CCN focused on agility, efficiency, and sustainability	CDSN for dynamic supply-demand alignment and enhanced resilience
Control tower	Luminate control tower by blue yonder, enabling synchronized upstream and downstream planning	AI-driven control tower integrated with cloud computing for real-time monitoring and analytics
Integration approach	Collaboration through centralized management centers and dashboards for real-time data visibility	Tools like SNO and GDSN for end-to-end supply chain integration
Disruption management	Predictive analytics for real-time issue resolution and agile response to supply chain challenges	Swift product innovation during disruptions, as demonstrated in the 2021 Texas storm
Technologies	Blue Yonder's platform, AI for accurate forecasting, and dashboards for supply chain visibility	Various planning and analytics solutions like blue prism, Databricks, MS azure, power platform, Rulex, and Knime, robotics, and ML for real-time optimization and process automation
Sustainability	Focus on sustainable logistics, efficient deliveries, and reduced lead times	Reduction of 'deadhead' transport, waste minimization, and adoption of robotics in packaging
Customer interaction	Omnichannel delivery strategies with real-time order tracking and seamless customer experience	Automated retailer communication, distributor connect, and dynamic product availability optimization

Azure, Power Platform, Rulex, Knime, robotics, and ML algorithms for dynamic product availability and operational insights.

Both companies also prioritize sustainability, with Electrolux focusing on sustainable logistics and P&G minimizing waste through reduced 'deadhead' transportation.

Customer-centricity is a shared goal, with Electrolux enhancing omnichannel experiences and P&G optimizing customer service through automated retailer communication and distributor connect tools.

13.4 A Model of a Customer-Centric Supply Chain Leveraging the Control Tower Technology

When discussing the digital customer-centric supply chain model supported by control tower technology, it is essential to reference the digital supply chain models proposed in the literature. Two particular approaches warrant special attention: the models proposed by Calatayud et al. (2019) and Garay-Rondero et al. (2020).

Calatayud et al. (2019) highlight three groups of technologies in supply chain structures: IoT, cloud computing, and AI. They refer to a supply chain that uses these technologies as the 'self-thinking' supply chain, where all systems and facilities are linked into one coherent system through the use of IoT technologies. The authors emphasize the importance of these technologies in creating a more efficient and effective supply chain. Vast amounts of data are generated, stored, and analyzed via IoT and AI in real time. This enables the continuous monitoring of supply chain performance and early identification and management of potential risks. Transparency of supply chain processes and their resilience to disruption are essential. These are supply chains where the integration of management and automation of decision-making is observed. The digital supply chain has greater flexibility, responsiveness and adaptability.

In the Digital Supply Chain Model within Industry 4.0, Garay-Rondero et al. (2020) identified six dimensions. The first dimension is the core of the model that utilizes cloud computing and cloud robotics (Calatayud et al., 2019). It functions like a brain powered by ML or AI, enabling the digitalized supply chains to be more flexible and to respond rapidly as conditions change. The second dimension describes the physical network structure, interconnected with Industry 4.0 technology enablers and features through Cyber-Physical Systems (CPS). The third dimension focuses on implementing Industry 4.0 concepts, enablers, and features. The fourth dimension is based on significant flows, which include inbound material flow, outbound finished products/goods flow, services flow, information flow, knowledge flow, financial resources flow (money/profit), risk and return flows of goods/services. The fifth dimension is the virtual value chain, which offers a view that encompasses the entire chain and uses the Industry 4.0 technology facilitators. The goal of the virtual value chain is to maintain transparency, communication, collaboration, flexibility, responsiveness and accuracy throughout the entire structure. The last dimension is the convergence of the digital and physical worlds, creating a new system where the real and digital worlds meet and interact symbiotically.

Based on the research by Calatayud et al. (2019) and Garay-Rondero et al. (2020), the digital customer-centric supply chain, utilizing the control tower solution, should be an expansion of their suggested frameworks to encompass the customer's role, as depicted in Fig. 13.5.

In the model presented in Fig. 13.5, the customer-centric supply chain is controlled and managed by integrating multiple layers of technology and decision-making tools. The control tower acts as the central hub for monitoring, analyzing, and managing the entire supply chain ecosystem. It collects real-time data from

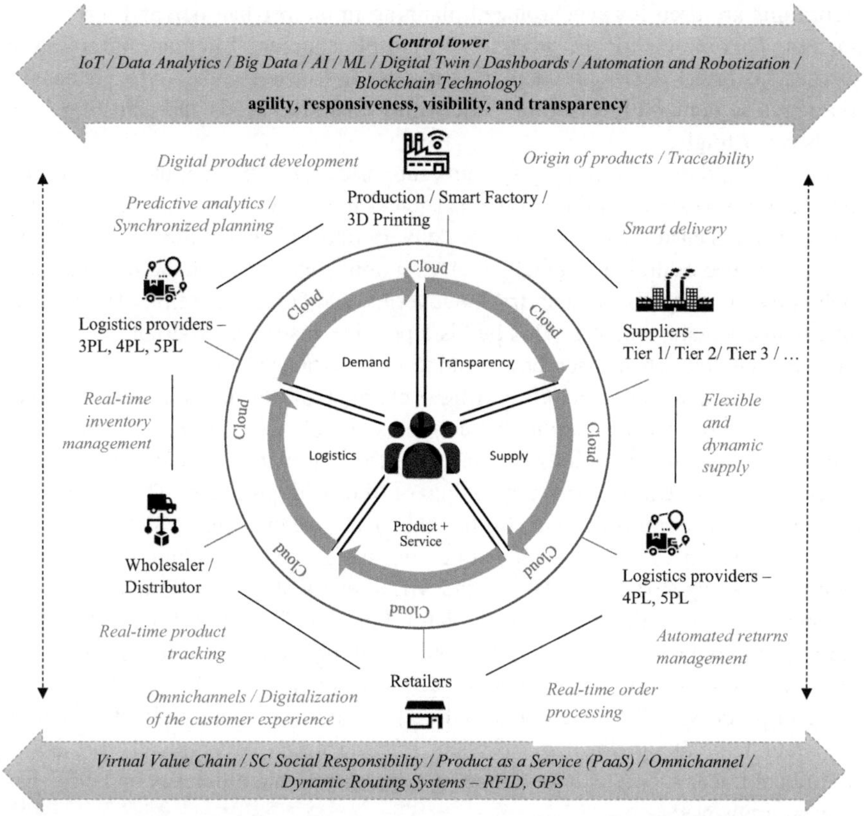

Fig. 13.5 Digital customer-centric supply chain model using control tower technology

various sources, including suppliers, logistics providers, manufacturers, wholesalers, distributors, retailers and customers, creating a seamless information flow across the entire supply chain. This ensures that every link in the supply chain is aligned and responsive to customer needs.

In the proposed model, the customer has the power to decide which product to purchase, and full product customization is possible. Customers have access to comprehensive product information, including origin, materials, and supply chain details, empowering them to make informed and socially responsible purchasing decisions. If the customer decides to make a purchase, their order will be processed in real time. The sales platform integrates the order management system, which immediately registers and processes customer orders, reducing the risk of errors and delays. The system utilizes various sales channels that are integrated into one cohesive omnichannel system, ensuring a consistent customer experience at every stage and every point (Brynjolfsson et al., 2013). Stock levels are verified in real time using RFID or barcode technology. Stock alerts are automatically set to notify the supply chain team when stocks are running low. Predictive analytics and ML algorithms are used for synchronized planning in the product demand forecasting process. This allows for the analysis of multiple sources of demand information, such as historical data, current market trends, and customer behavior. It enables more precise demand forecasting, anticipating customer needs and adjusting stock levels accordingly.

Flexible and dynamic supply systems supported by GPS technology are implemented to respond in real time to events such as traffic, weather, and order urgency. These systems ensure a higher on-time delivery rate, enhancing the customer experience. The use of digital channels, such as mobile apps, SMS, and email, provides the ability to confirm shipment, track deliveries, and confirm receipt, offering real-time delivery status updates. This builds a positive customer experience, increases customer satisfaction and purchases, and improves transparency.

Returns management, another critical element of the model, is streamlined through the control tower's ability to track and automate return processes. Customers can initiate and monitor returns online, and the automated system ensures faster acceptance, verification, and processing of returned products. This guarantees quicker refunds or replacements, further enhancing the customer experience.

The model in Fig. 13.5 demonstrates the integration of technologies such as cloud computing, IoT, big data, AI, and ML at every stage of the supply chain to enhance decision-making and operational efficiency. The control tower leverages these technologies to create a 'digital core', enabling constant communication and interaction among companies, devices (such as machines, vehicles, and tools), people, and processes. This interconnected network allows for multidirectional communication, breaking down the barriers of time and space typically present in traditional linear supply chains. Consequently, operational efficiency and effectiveness are improved.

As indicated by the model, managers of digitized supply chains rely on communication platforms that connect suppliers and customers, and control tower systems that streamline and automate both input and output processes. These systems support downstream and upstream processes, ensuring the alignment of supply chain activities with customer demands. By monitoring performance in real time and leveraging predictive analytics, the control tower facilitates proactive decision-making, such as demand forecasting and inventory adjustments.

13.5 Findings and Evaluation

The initial validation studies of the model presented in Fig. 13.5 were conducted as part of a comprehensive supply chain study by the Global Manufacturing Research Group (GMRG). The studies on the implementation and utilization of digitalization in supply chains were part of a dedicated questionnaire module. Given the focus on the specific control tower technology, the following analysis is based on responses from the open-ended questions section of the survey.

The study, conducted between 2023 and 2024, encompassed 228 supply chains worldwide. Among the surveyed supply chains, 180 declared the use of digitalization in their structures. In many cases, digitalization was partial, encompassing selected processes or the use of specific systems, such as ERP, MRP, MES, or WMS. Only 12 out of the 228 supply chains implemented management centers enabling real-time monitoring, coordination, and optimization of logistics processes, supported by control tower technology.

The analysis of open-ended responses indicates that control tower technology plays a crucial role in integrating the supply chain with the customer. Managers emphasized that integration enhances planning and decision-making, reduces costs, increases transparency, improves customer service, and shortens response times. The number of mentions related to efficiency improvements in various areas is presented in Fig. 13.6.

Planning and decision-making support in the analyzed supply chains using control tower technology is manifested through the use of digital tools for demand forecasting and synchronization of supply with demand, including production and inventory planning. An essential element is the integration of systems such as ERP, WMS, MRP, and MES, as well as the use of the integrated Sales and Operational Planning (S&OP) method. Additionally, cloud technologies and big data analysis enable the integration of data from production and supplier orders to the end customer. The benefits of these solutions include reduced risk of overproduction or shortages and better resource utilization, which significantly enhances supply chain flexibility and agility.

Control tower technology contributes to cost reduction primarily through process automation, better planning, and waste reduction. Real-time reporting ensures data visibility and enables cost-effective decision-making. Key benefits include lowering storage costs and minimizing losses resulting from process inconsistencies.

Control tower technology enhances transparency by providing comprehensive real-time monitoring across all levels of the supply chain. It allows stakeholders to

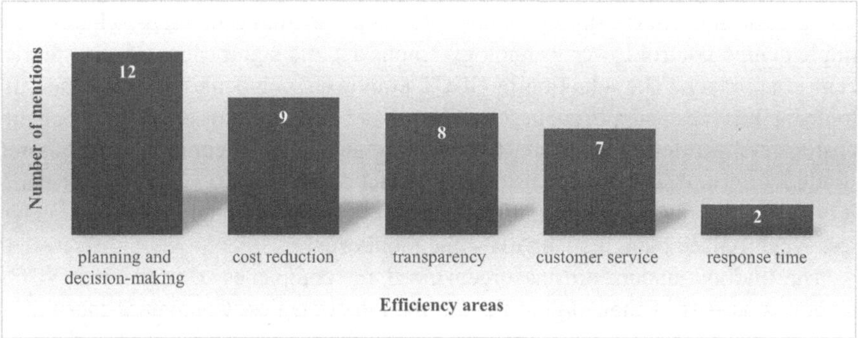

Fig. 13.6 Mentions of efficiency areas in the supply chain using a control tower to integrate the supply chain with the customer

share critical information by integrating data from production to the end customer and effectively managing inventory. Technologies such as cloud computing, digital twins, and big data support these processes. Transparency facilitates faster problem identification and resolution while improving coordination between departments and partners in the supply chain.

One significant area mentioned by the surveyed supply chains is improved customer service. Managers highlighted the importance of synchronizing actions with customer expectations, which translates into higher satisfaction levels measured by the OTIF (On Time In Full) metric. Faster order fulfillment and product availability lead to fewer complaints and returns.

The analysis also highlights the importance of response time in the context of control tower technology. Responses indicate the ability to quickly identify problems, such as delivery delays, enabling real-time corrective actions. Faster responses to changes in product demand were also emphasized, ensuring better customer service. An essential factor influencing efficiency is the integration of systems, allowing for improved product flow and timely reactions to changes. Process automation also plays a key role in shortening task completion times and responding to unexpected events.

13.6 Conclusions and Further Research

The integration of a control tower into supply chain management represents a transformative approach that enhances visibility, responsiveness, transparency, and agility. This technology model underscores its ability to improve the efficiency of supply chain operations and the quality of customer service, ultimately elevating the overall shopping experience. As demonstrated through the case studies of Electrolux and P&G, incorporating advanced digital technologies improves decision-making, enhances resource allocation, expands sustainability efforts, and increases the capacity to respond to disruptions and evolving customer demands. Research conducted among 228 supply chains confirms that SCCT plays a crucial role in developing customer-centric supply chains. Managers operating in supply chains with implemented control tower technology emphasize the significance of a customer-centric approach. The adoption of SCCT empowers companies to place the customer at the core of supply chain activities. On the one hand, this technology benefits customers by offering real-time tracking, omnichannel integration, and tailored product offerings, enhancing customer satisfaction and loyalty. On the other hand, it enhances supply chain adaptability to market changes through predictive analytics, synchronized planning, and real-time monitoring.

The findings underscore the urgent need for companies to implement SCCT solutions as part of their digital transformation strategies. Companies that fail to adopt real-time data analytics, AI-driven decision-making, and end-to-end supply chain integration risk falling behind in today's increasingly competitive global market. Case study findings and survey data from 228 supply chains validate that SCCT is critical for seamless integration in modern supply chains. They demonstrate that the implementation of SCCT enables a transition to the most advanced supply chain structures, characterized by a customer-centric approach in the digital environment. Furthermore, they highlight that the primary areas of efficiency improvement achieved through these solutions are visibility, responsiveness, transparency, and agility. SCCT provides real-time visibility, enhances coordination among stakeholders, supports agility and ensures that customers remain informed throughout the process. A digital customer-centric supply chain powered by control tower technology enhances service optimization, boosts customer satisfaction, and ensures product alignment with evolving consumer expectations. Together, these elements underscore the transformative potential of SCCT in shaping the future of supply chain management.

Although the presented case studies and research findings do not exhaustively address the topic or fully validate the digital customer-centric supply chain model using control tower technology, they identify key areas and research directions for future exploration, including:

1. *Broader industry applications*: Future research should explore the implementation and effectiveness of SCCT in various sectors, including healthcare, energy, and agriculture, to uncover industry-specific best practices and cross-sector insights.
2. *Technology synergies*: Investigating the potential of emerging technologies like quantum computing, Blockchain, and 5G to augment the capabilities of SCCT could yield valuable insights.
3. *Ethical and privacy concerns*: With increasing reliance on data-driven systems, research into safeguarding customer and partner data while ensuring ethical compliance and adherence to regulatory standards is critical.
4. *Scalability for SMEs*: Studies on cost-effective and scalable SCCT solutions tailored for small and medium-sized enterprises could expand accessibility and adoption.
5. *Long-term impacts and resilience*: Longitudinal studies examining the sustained effects of SCCT on supply chain resilience and business performance will provide a deeper understanding of their value.
6. *Customer empowerment and engagement*: The role of SCCT in fostering deeper customer engagement through transparency and customization deserves further exploration, particularly in light of changing consumer preferences.
7. *Integration with circular economy models*: Research into how SCCT can support circular supply chains by enabling efficient product lifecycle management, recycling, and waste reduction is essential for advancing sustainability goals.

In summary, the findings highlight that SCCT is not merely a tool for operational improvement but a supply chain management strategy that ensures customer integration drives innovation and prepares supply chains for the challenges of an increasingly complex and digitalized world. The future of SCCT lies in the integration of AI-powered predictive analytics, quantum computing, and autonomous supply chain networks. As supply chains become more complex, organizations must remain agile, leveraging emerging technologies to stay resilient and competitive.

Funding The study was financed by the National Science Centre, Poland, as a research project no. 2019/35/B/HS4/00056.

References

Accenture. (2022, May). *Supply chain Control tower - from visibility to value.* https://www.accenture.com/pl-pl/insights/consulting/supply-chain-control-tower#accordion-7ad3695c02-item-f13f17b86e

Bhosle, G., Kumar, P., Griffin-Cryan, B., van Doesburg, R., Sparks, M. A., & Paton, A. (2011). Global Supply Chain Control tower. Achieving end-to-end Supply Chain Visibility. *Capgemini Consulting.* https://www.capgemini.com/wp-content/uploads/2017/07/Global_Supply_Chain_Control_Towers.pdf

Bozarth, C., & Handfield, R. (2019). *Introduction to operations and supply chain management.* Pearson Education.

Brynjolfsson, E., Hu, Y. J., & Rahman, M. S. (2013). Competing in the age of omnichannel retailing. *MIT Sloan Management Review, 54*(4), 23–29.

Calatayud, A., Mangan, J., & Christopher, M. (2019). The self-thinking supply chain. *Supply Chain Management, 24*(1), 22–38. https://doi.org/10.1108/SCM-03-2018-0136

Can Saglam, Y., Yildiz Çankaya, S., & Sezen, B. (2021). Proactive risk mitigation strategies and supply chain risk management performance: An empirical analysis for manufacturing firms in Turkey. *Journal of Manufacturing Technology Management, 32*(6), 1224–1244. https://doi.org/10.1108/JMTM-08-2019-0299

Chaffin, C., Maywald, J., Reiman, A., & Glassburner, A. (2024). Systematic review of supply chain control tower critical success factors and resilience effects. *Operations and Supply Chain Management: An International Journal, 17*(3), 236–251. https://doi.org/10.31387/oscm0580439

Chatterjee, S., Chaudhuri, R., & Vrontis, D. (2021). Examining the global retail apocalypse during the COVID-19 pandemic using strategic omnichannel management: A consumers' data privacy and data security perspective. *Journal of Strategic Marketing, 29*(7), 617–632. https://doi.org/10.1080/0965254X.2021.1936132

Chen, S., Cohen, M. A., & Lee, H. (2024). Enhancing customer-supplier coordination through customer-managed inventory. *Management Science, 70*(12), 9073–9088. https://doi.org/10.1287/mnsc.2021.03658

Cheng, Y. (2024). Enabling technologies in supply chain operations: An examination of smart technologies on supply chain performance with moderation of virtual reality. *International Journal of Electronic Business, 19*(3), 296–310. https://doi.org/10.1504/IJEB.2024.139792

Choi, T. Y., Rogers, D., & Vakil, B. (2020, March 27). Coronavirus is a wake-up call for supply chain management. *Harvard Business Review.* https://hbr.org/2020/03/coronavirus-is-a-wake-up-call-for-supply-chain-management

Christopher, M. (2016). *Logistics and supply chain management.* Pearson Higher Education.

Christopher, M. (2021). Re-thinking supply chain strategy. In E. Sweeney & D. Waters (Eds.), *Global logistics. New directions in supply chain management.* Kogan Page.

Christopher, M., & Lee, H. (2004). Mitigating supply chain risk through improved confidence. *International Journal of Physical Distribution & Logistics Management, 34*(5), 388–396. https://doi.org/10.1108/09600030410545436

DeVoe, S., Iampieri, M., & Mattson, E. (2023). Supply chain foundation: Building the elusive control tower. *Supply Chain Management Review*. https://www.scmr.com/article/supply_chain_foundation_building_the_elusive_control_tower

Din, M. S. E., Masengu, R., & Bentley, Y. (2023). The role of supply chain control towers in facilitating sustainable supply chain management practices. In R. Masengu, O. T. Chiwaridzo, S. Bigitimana, & B. Christina (Eds.), *Sustainable marketing, branding, and reputation management: Strategies for a greener future* (pp. 310–327). https://doi.org/10.4018/979-8-3693-0019-0.ch016

Esper, T. L., Castillo, V. E., Ren, K., Sodero, A., Wan, X., Croxton, K. L., & Goldsby, T. J. (2020). Everything old is new again: The age of consumer-centric supply chain management. *Journal of Business Logistics, 41*(4), 286–293. https://onlinelibrary.wiley.com/doi/pdfdirect/10.1111/jbl.12267

Garay-Rondero, C. L., Martinez-Flores, J. L., Smith, N. R., Caballero Morales, S. O., & Aldrette-Malacara, A. (2020). Digital supply chain model in industry 4.0. *Journal of Manufacturing Technology Management, 31*(5), 887–933. https://doi.org/10.1108/JMTM-08-2018-0280

Gupta, R. (2020). *Supply Chain to Supply Network: P&G*. https://www.linkedin.com/pulse/supply-chain-network-pg-ragini-gupta/

Haren, P., & Simchi-Levi, D. (2020, February 28). How coronavirus could impact the global supply chain by mid-march. *Harvard Business Review*.

Ivanov, D. (2022). Viable supply chain model: Integrating agility, resilience and sustainability perspectives—Lessons from and thinking beyond the COVID-19 pandemic. *Annals of Operations Research, 319*(1), 1411–1431. https://doi.org/10.1007/s10479-020-03640-6

Jorgensen, H. (2022). *Electrolux—Supply Chain, Perform & Transform*. https://www.procurescape.com/electrolux-supply-chain-perform-transformation/

Kazancoglu, I., Ozbiltekin-Pala, M., Mangla, S. K., Kazancoglu, Y., & Jabeen, F. (2022). Role of flexibility, agility and responsiveness for sustainable supply chain resilience during COVID-19. *Journal of Cleaner Production, 362*, 132431. https://doi.org/10.1016/j.jclepro.2022.132431

Klesman, S., & Cherici, E. (2017). *The features of an operative control tower*. Maastricht Univeristy & Caroz B.V. https://www.supplychainmagazine.nl/wp-content/uploads/2017/04/Whitepaper-Control-Towers.pdf

Kumar, P. (2022). Control tower Delivers Visibility and Action. *Accenture*. https://www.accenture.com/us-en/blogs/business-functions-blog/supply-chain-control-tower

Linton, T., & Vakil, B. (2020, March 5). Coronavirus is proving we need more resilient supply chains. *Harvard Business Review*.

Maheshwari, P., Kamble, S., Kumar, S., Belhadi, A., & Gupta, S. (2024). Digital twin-based warehouse management system: A theoretical toolbox for future research and applications. *International Journal of Logistics Management, 35*(4), 1073–1106. https://doi.org/10.1108/IJLM-01-2023-0030

Naik, S., & Kumar, P. (2021). The new essential for supply chains: Intelligent execution control tower. Part I. *Supply Chain Management Review*.

Patsavellas, J., Kaur, R., & Salonitis, K. (2021). Supply chain control towers: Technology push or market pull—An assessment tool. *IET Collaborative Intelligence Manufacturing, 3*(3), 290–302. https://doi.org/10.1049/cim2.12040

Ponomarov, S. Y., & Holcomb, M. C. (2009). Understanding the concept of supply chain resilience. *The International Journal of Logistics Management, 20*(1), 124–143. https://doi.org/10.1108/09574090910954873

Potter, A., Towill, D. R., & Christopher, M. (2015). Evolution of the migratory supply chain model. *Supply Chain Management: An international Journal, 20*(6), 603–612. https://doi.org/10.1108/SCM-06-2015-0231

SAP Insights. (n.d.). *SAP Insights: Supply chain Control towers: Providing end-to-end visibility*. https://www.sap.com/uk/insights/supply-chain-control-tower.html

Sodhi, M. S., & Tang, C. S. (2019). Research opportunities in supply chain transparency. *Production and Operations Management, 28*(12), 2946–2959. https://doi.org/10.1111/poms.13115

Somapa, S., Cools, M., & Dullaert, W. (2018). Characterizing supply chain visibility–a literature review. *The International Journal of Logistics Management, 29*(1), 308–339. https://doi.org/10.1108/IJLM-06-2016-0150

Swafford, P. M., Ghosh, S., & Murthy, N. (2008). Achieving supply chain agility through IT integration and flexibility. *International Journal of Production Economics, 116*(2), 288–297. https://doi.org/10.1016/j.ijpe.2008.09.002

Syahchari, A. H., Sudrajat, D., Lasmy, L., Herlina, M. G., Kiatama, C., & Jordaan, H. K. W. (2022). Achieving supply chain resilience through supply chain control tower and supply chain agility. In *ICCMB '22: Proceedings of the 2022 5th international conference on computers in management and business* (pp. 195–198). https://doi.org/10.1145/3512676.3512709

Szozda, N. (2022). *Zarządzanie produktem w łańcuchu dostaw w dobie transformacji cyfrowej.* Publisher of the Wroclaw University of Business and Economics., No. 310, ISBN 9788376959948 (in Polish). https://doi.org/10.15611/2022.995.5

Tih, S., Wong, K.-K., Lynn, G. S., & Reilly, R. R. (2016). Prototyping, customer involvement, and speed of information dissemination in new product success. *Journal of Business & Industrial Marketing, 31*(4), 437–448. https://doi.org/10.1108/JBIM-09-2014-0182

Trzuskawska-Grzesińska, A. (2017). Control towers in supply chain management—Past and future. *Journal of Economics and Management, 27*(1), 114–133. https://doi.org/10.22367/jem.2017.27.07

Vlachos, I. (2021). Implementation of an intelligent supply chain control tower: A socio-technical systems case study. *Production Planning & Control, 34*(15), 1415–1431. https://doi.org/10.1080/09537287.2021.2015805

Wyciślak, S., & Pourhejazy, P. (2023). Supply chain control tower and the adoption of intelligent dock booking for improving efficiency. *Frontiers in Energy Research Systems, 11*–2023. https://doi.org/10.3389/fenrg.2023.1275070

Ye, C., Zaraté, P., & Kamissoko, D. (2022). A DSS based on a control tower for supply chain risks management. In A. P. Cabral Seixas Costa, J. Papathanasiou, U. Jayawickrama, & D. Kamissoko (Eds.), *Decision support systems XII: Decision support addressing modern industry, business, and societal needs. ICDSST 2022. Lecture notes in business information processing, 447.* Springer. https://doi.org/10.1007/978-3-031-06530-9_10

Zhang, X., & Xiao, G. (2023). Control tower empowers supply chain resilience improvement from a dynamic capability perspective. *Digitalization and Management Innovation II, 396*, 186–194. https://ebooks.iospress.nl/doi/10.3233/FAIA230730

Chapter 14
Digital Control Towers in Automotive Supply Chains

Emre İpekçi and Arda Toygar

Abstract In the automotive supply chain, service points designated for after-sales vehicle maintenance inherently lack adequate storage capacity. Therefore, inventories at these points must be minimized, necessitating efficient management of material flows from a centralized source. This chapter examines the potential application of digital control towers within automotive supply chain management. A SWOT analysis evaluates digital control tower implementation, and a subsequent TOWS analysis develops aligned strategic approaches. Digital control towers enhance operational efficiency by enabling optimal stock levels, just-in-time production, accurate and rapid data flow, effective supplier management and communication, streamlined document processes, and reliable data provision for production planning. However, weaknesses including challenges in managing inaccurate information, high start-up implementation costs, and difficulties related to technological adaptation. A digital control tower affords opportunities for demand-driven production planning, increased competitiveness, enhanced adaptability to market changes, improved management of special operations, and support for digital twin technologies, but also poses threats confusion in stock tracking, supplier resistance and bias, failure to monitor updates, confidentiality demands, and persistent data security challenges.

Keywords Supply chain management · Control tower · Digital technologies · Automotive industry · SWOT analysis · TOWS analysis

E. İpekçi
Faculty of Political Sciences, Department of Maritime Business Administration, Samsun University, Samsun, Türkiye

A. Toygar (✉)
Fatsa Faculty of Marine Sciences, Department of Maritime Business Administration, Ordu University, Ordu, Türkiye
e-mail: ardatoygar@odu.edu.tr

Y. Khojasteh, S.M. M. Jahan Abadi (eds.), *Digital Transformation of Supply Chains*, International Series in Operations Research & Management Science 372, https://doi.org/10.1007/978-3-032-01206-7_14

14.1 Introduction

Ripple effects in supply chains can significantly impact the functioning and management of the entire chain (Dolgui & Ivanov, 2021; Ivanov & Dolgui, 2020). Major economic crises and shocks in the 20th and 21st centuries have generated ripple effects, especially in global supply chains. For example, the economic crisis in 2008 and the COVID-19 pandemic caused disruptions in global supply chain operations and destabilized global trade volumes (Guan et al., 2020; Ivanov, 2020; Sarkis et al., 2020). A significant increase in freight rates was observed during the COVID-19 pandemic and political disruptions such as the Russia-Ukraine conflict and the Israel-Hamas conflict has led to cost increases in raw material procurement and transportation processes. Such disruptions can cause volumetric losses in supply chain operations and limit trade mobility (Handfield et al., 2020; Toygar & Yildirim, 2023; Toygar & Yıldırım, 2022).

Digital technologies enable current supply chain management approaches to differ significantly from traditional methods. Current supply chains provide transparent and up-to-date information that businesses need on the supplier side. While simultaneous communication on a thread-by-thread basis is one of the main reasons for this, changes in end-customer demand can also be responded to quickly. The dynamic nature of the supply chain allows for a just-in-time response to planning and execution actions across all chains, rather than the disparate planning cycles that caused delays and out-of-sync responses across multiple chains in the past. Digital supply chain solutions are a necessity for enabling synchronized and transparent interaction between multiple nodes in the chain regardless of how robust the collaboration between businesses and their suppliers is. In this context, control tower applications in supply chains are a method used by businesses to manage their supply chains. A digital control tower has functions such as instant visibility of each customer order and the date range of products needed, selection of carriers and management of transportation organizations with Standard Operating Procedures (SOPs) (Ojasalo et al., 2024). This digital technology has the potential to reduce costs in supply chain management while operational efficiency is increased. An integrated information-sharing infrastructure with carriers enables digital control tower applications to communicate changes in estimated delivery times of shipments in transit to customers. This paves the way for dynamic changes in the production plan, making businesses more agile. Control tower applications in supply chains are planned to manage the flow of materials and products in the chain without disruption and error. These operations aim to manage the flow of goods with minimal errors by using digital technologies to prevent potential disruptions, especially during the supplier management stage. Furthermore, thanks to the visibility provided throughout the supply chain, businesses do not need a separate data entry for systems such as Enterprise resource planning (ERP).

The automotive industry is considered to have one of the most efficient and best organized supply chains to date (Sinha et al., 2020). One of the main reasons for the acknowledgment of this idea is the optimal management of supply chain costs to adapt to the competitive conditions in the globalized automobile manufacturing market. In addition, the concept of minimum inventory at the center of the lean production concept, which was implemented by Toyota years ago and been accepted and implemented by almost the entire automotive industry, further emphasizes the importance of effective supply chain management in this industry (Yamamoto et al., 2019). The automotive industry is scattered in many different countries as a result of its inherent structure. Especially large-scale enterprises prefer to be close to the market with production centers established in different continents. While the adoption of just in time production methods has enabled automotive suppliers to place their production centers as close as possible, the dissemination of customers all over the world has made businesses global.

The ripple effects caused by economic and political crises have further increased the importance of operations such as control towers, especially in recent years. The control tower, developed to provide control and visibility over the supply chain, enables businesses to make faster and more effective decisions during ripple effect periods. In addition to providing supply chain services, a control tower enables logistics service providers to offer specialized logistics services in which they specialize to their customers. The automotive supply chain management is different from others in the fact that the service points where after-sales vehicle repair and maintenance is conducted are not structured to accommodate a storage function at the same time. In addition, spare parts availability at service points for after-sales services is also an important issue (Khajavi et al., 2014). Therefore, automotive manufacturers conduct their supply chains in two separate stages, namely pre-production and post-production (Panwar et al., 2022). The stock at service points should be kept at a minimum and the required material flow should be managed from a single source, taking future needs into account. These differences necessitate a focus on the examination of automotive supply chain management. However, there is a lack of sufficient studies that address the effectiveness and efficiency of the digital control tower for the automotive industry. In this context, the chapter explores the implementation and impact of digital control towers in automotive supply chain management, with a particular focus on their role in addressing industry-specific challenges. It also aims to contribute to the existing literature by examining and developing strategies for the impact of a digital control tower in automotive supply chain management. This is achieved through a SWOT (Strengths, Weaknesses, Opportunities, Threats) and a TOWS (Threats, Opportunities, Weaknesses, Strengths) analyses. The SWOT analysis serves as a diagnostic tool to classify the internal and external factors that influence the success of digital control towers in automotive supply chain management. Strengths and weaknesses are internal to the system, while opportunities and threats stem from the external environment. The findings of the SWOT analysis highlight several factors influencing the efficiency and effectiveness of control towers in supply chain management. Building on the SWOT analysis, the TOWS framework provides a structured approach to translating these findings into actionable strategies. Unlike SWOT, which primarily categorizes factors, TOWS emphasizes the interconnections between these factors to guide strategic decision-making. By combining SWOT's diagnostic capacity with TOWS's prescriptive potential, this chapter provides a holistic framework for analyzing and addressing the challenges faced in digital control tower implementation within the automotive industry. Therefore, this chapter is important to address the supply chain challenges that arise due to the complex and global nature of the automotive industry and provide digital solutions to address these challenges. The theoretical framework of the chapter is explained in the following section.

14.2 Theoretical Framework

14.2.1 Control Towers in Supply Chain Management

A supply chain is a structure composed of interconnected logistics links. This structure is planned to manage the flow of materials and products within the chain without disruption and error. The supply chain generates a lot of data with these overarching components. The key challenges facing businesses nowadays are how to collect, classify and, most importantly, make sense of this data to manage operations better (Seyedan & Mafakheri, 2020). Disruptions in the chain can have negative consequences that can lead to bottlenecks in production and a loss of customers in distribution. Therefore, the design of the supply chain should be based on the conditions that suit the needs of the businesses best. Many factors can cause disruptions in the functioning of supply chains even if they are perfectly designed from start to finish.

The control tower technologies are one of the best examples of collaborative technologies in the supply chain that utilize powerful problem-solving technologies as well as integrating multiple workers, all types of data generated in the chain, and communication methods. It enables the management of variations in supplier shipments and efficient outcomes in changing conditions through predictive insights generated by data collected on the network (Sinha et al., 2020). A control tower has two functions. The first is the management of inbound and outbound logistics operations (Ojasalo et al., 2024). The main function of the control tower in the logistics pillar is the selection of a transportation mode or modes for routes and load types, and who will be the carriers and freight forwarders to work these modes. The other function is supplier management. At this stage, the control tower keeps track of the orders established in the system for suppliers and directs the suppliers about shipments that are due. Logistics planning of loading requests from suppliers and the exception management of any information discrepancies on an order basis also takes place at this stage. Control tower applications in the supply chain aim to manage the flow of goods with minimal errors by using digital technologies to prevent potential disruptions, especially at the supplier management stage. The task of these applications is to aggregate data and information on flows and inventories along the chain, monitor the status of each node and link in real time and highlight deviations on planned activities and events (Christopher, 2023).

While managing supplier affairs seems to be at the heart of control tower applications, this application provides real-time visibility of information on material and production flows, as well as key data on distributors, customers, and products in a centralized location. This feature of the application prevents reacting to events in the market and allows for a pre-emptive reaction to potential disruptions (Taschner & Charifzadeh, 2020). This also minimizes transportation and storage costs by enabling optimum stock levels while supply problems in the chain are avoided. The following section describes the digital technologies incorporated by the control tower.

14.2.2 Digital Technologies Incorporated in the Control Tower

Control tower application is needed to make sense of the technologies available for supply chain management within the industry 4.0 concept. Therefore, a control tower that can interpret and control the signals collected from sensors, Radio-frequency identification tags and the like is needed to take advantage of the benefits that can potentially be derived from the implementation of Internet of Things (IoT) technology and help maintain the fast flow of information through the chain in an error-free manner (Christopher, 2023). Each technology used throughout the system must be managed correctly for the smooth functioning of the control tower. In the first stage, the visibility of all kinds of data generated within the chain in the system must be ensured to manage the control tower efficiently (Ojasalo et al., 2024). The correct functioning of the cyber-physical systems used to collect this data is a requirement for the smooth operation of the system. Digital technologies in the control tower are visualized in Fig. 14.1.

A control tower application is usually web-based or controlled by interfaces from software vendors such as System Analysis Program Development (SAP). The data needed by the application is usually generated in the system by the relevant actors, but thanks to IoT, it can also be collected from sensors and obtained from actors in the chain (Kepczynski et al., 2018). Control tower technologies can provide artificial intelligence-supported recommendations using the information in the system to businesses on issues such as dynamic inventory control, guidance on when stocks should be sold or shipped at the latest, demand forecasting and the actual batch inventory level.

While traditional control tower management relied on the service provider entering data into the application after it was received from suppliers, currently collected data is automatically transferred to cloud-based platforms and then processed by sophisticated algorithms that make sense of it and provide information and data for predictive and prescriptive analytics. For example, the control tower module developed by SAP enables data collected from sensors to be transferred to a cloud and be visible across the chain just in time (Kepczynski et al., 2018). Control tower

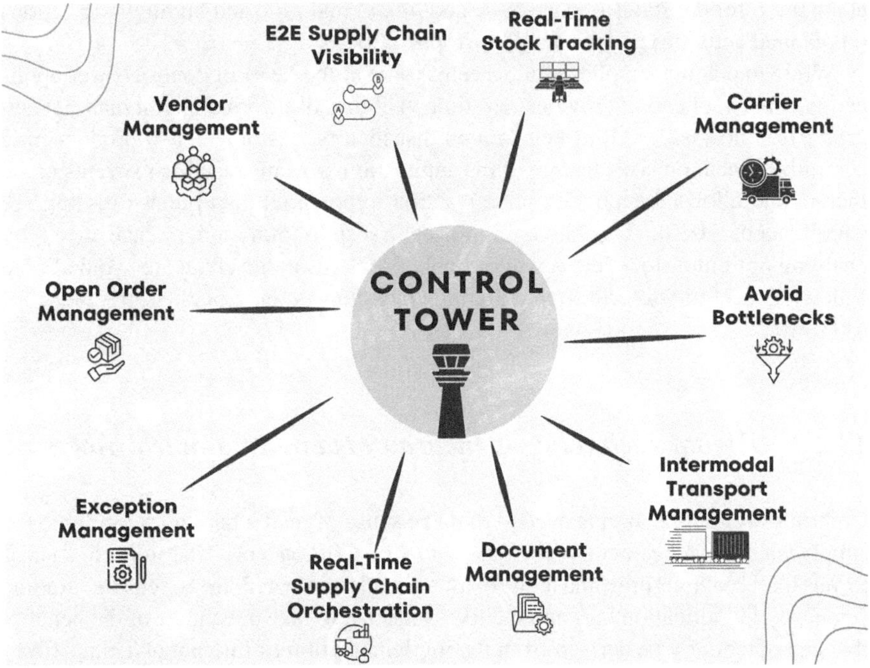

Fig. 14.1 Digital technologies in control towers

management can consist of information provided by suppliers and accessed through computers or mobile devices as well as collect the data it needs throughout the chain with sensors and display it in real time thanks to IoT technology, especially in sectors with a high number of suppliers and components, such as the automotive and spare parts sector (Sinha et al., 2020).

The control tower uses artificial intelligence to ensure appropriate decisions are made when deviations occur within the supply chain network (Abaku et al., 2024). Furthermore, predictive analytics, a key element of the application, is used to better respond to changes in supply and demand. Software companies offer cloud-based platforms for firms to establish control towers, often using the Software as a Service (SaaS) principle (Christopher, 2023). Companies such as GTNexus and Kinaxis are examples of companies that offer this service. Thanks to these companies, businesses get a customized system for themselves without the need to direct their internal resources to different areas. The following section of the chapter covers the role of the control tower in the supply chain in the automotive industry.

14.2.3 The Role of a Control Tower in Automotive Supply Chain Management

There are many different approaches to supply chain management, such as sourcing, quality leadership, chain management with single or multiple suppliers, core competence, horizontal and vertical integration, lean or agile manufacturing approaches, and resilient supply chains (Dametew et al., 2020). A pragmatic approach to supply chain management means that businesses choose the one(s) that suits them best (Sümen, 2020). What makes the automotive industry different from other industries is the considerable number of suppliers and business partners for both production and after-sales processes. Furthermore, the automotive industry is subject to intense competition in the global market. Vehicle production has become a multi-part production with electric and hybrid vehicle technologies, autonomous driving and safety equipment added to vehicles that cannot be compared to the practices of the past. This means that the number of suppliers is also increasing. All these actors working together must be managed correctly and an efficient production must be realized at a rate that is close to error-free with minimum stock to achieve optimum profitability in automotive production. Therefore, establishing and managing supplier relationships is becoming more important for automotive manufacturers. Supply chain control tower applications are a valuable tool for the smooth material flow and supplier management required by automotive manufacturers. In this context, industry-specific control tower applications are visualized in Fig. 14.2.

Supply chain software can be divided into two categories, namely planning and execution. Planning software supports the matching of demand and capacity (supply) forecasts, taking financial constraints into account. Sales and operations planning software (S&OP) is the most basic example of planning software. Execution software helps to realize plans and control deviations by monitoring real-time data for ordering, manufacturing, inventory and logistics operations (Blanchard, 2021). The features of control tower application not only enable the management of suppliers but also enables logistics operations in accordance with the just in time production philosophy.

Planning and execution were planned as sequential functions in traditional supply chain management approaches. However, nowadays high-performance computing software and cloud computing technologies have made planning and execution manageable together. In this case, the best supply chain plan for the automotive industry is an integrated supply chain plan. The following advantages are achieved with a single plan that covers the entire chain:

- Optimal operational efficiency is achieved across the chain.
- Data consistency is ensured; data does not need to be transferred between different business units, as each transfer can lead to data corruption.
- Communication delays are minimized; decisions made at the top are immediately visible at every point in the chain.

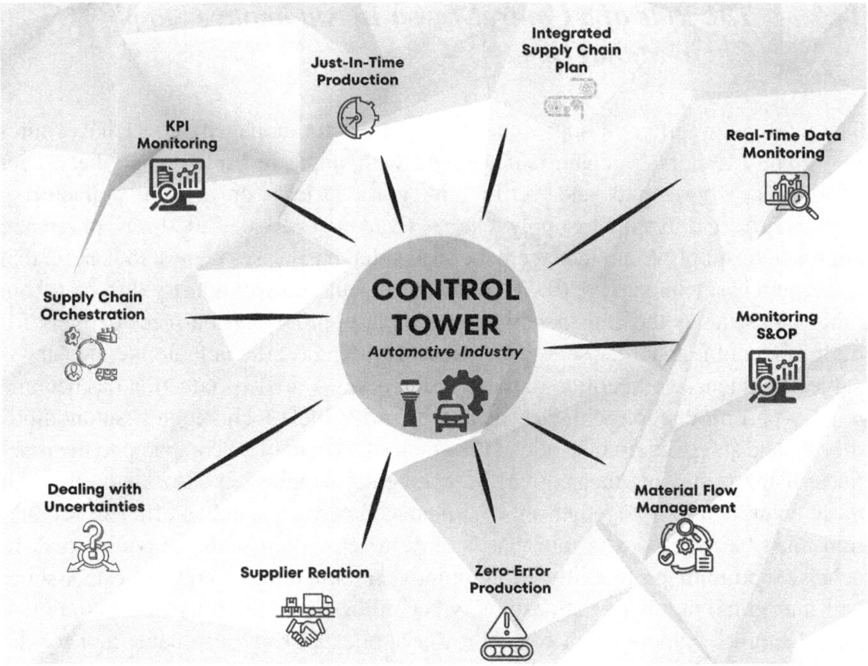

Fig. 14.2 Control towers in the automotive industry

An integrated design alone might not be enough to achieve all these benefits. The backbone of the integrated plan should be structured on one of the following objectives:

- Maximize inventory turnover.
- Maximize total profit for the planned period.
- Maximize on-time deliveries.

The main objective that can be chosen for automotive or automotive spare parts manufacturers is to maximize on-time deliveries. This is related to just-in-time production. Managing uncertainties in supply chain management for the automotive industry is also important (Dias et al., 2020). Although incomplete information can be managed for a reasonable period of time through statistical methods, control tower applications are prominent in avoiding possible supplier-induced delays. Until recently, the only way to deal with uncertainty was to reduce or eliminate it. However, the recently emerging concept of orchestration focuses on reducing the dependency on certainty. The full-time flow and management of information from suppliers required to reduce this dependency is possible with control tower applications (ASCM, 2024). The supply chain control tower enables orchestration through increased visibility, collaboration, and optimization across the entire chain. The control tower operates as a cross-functional tool (or digital platform) empowered to monitor, analyze and take appropriate action to improve predefined performance indicators (KPIs) across the extended value chain based on three key components—visibility, collaboration and optimization (Mrozek et al., 2020). For example, Renault Group, one of the world's largest automotive manufacturers, began implementing control tower applications into its processes in 2020. The system, which operates in conjunction with Google's artificial intelligence and machine learning technologies, can predict potential delays by accounting for variables such as weather conditions, strikes, border crossing delays, and traffic levels. As a result, the system enables early detection of potential disruptions and facilitates rerouting and expedited delivery for trucks. The company reports that, as of today, it has reduced line stoppages and unfinished vehicles by 50%, achieving €260 million in inventory savings and saving millions of euros in unforeseen costs (SupplyChainBrain, 2024). Another example of usage in the industry can be found in the Iveco Group. This automotive manufacturer partnered with Capgemini to implement a control tower system. Additionally, the company integrated Amazon Web Services (AWS) teams into the process to ensure the effective use of cloud technology. By leveraging AWS, the company achieved efficient cost management in data storage and made significant advancements in end-to-end visibility throughout the process (Capgemini, 2024). The following section explains the research methodology.

14.3 Methodology

The chapter uses a research design that integrates systematic analysis methods and multiple data sources to examine the implementation and strategic impact of digital control towers in automotive supply chain management. The methodology consists of three stages: (1) data collection, (2) SWOT analysis to assess key internal and external factors, and (3) TOWS analysis to develop strategic implications. The chapter examines the control tower in a more specific context, focusing on automotive supply chain management, which is characterized by a multi-layered and complex structure. First step, data was collected from multiple sources to ensure a comprehensive understanding of digital control. A systematic review of relevant literature, industry reports, and case studies was conducted to identify existing knowledge, theoretical foundations, and empirical findings related to digital control towers in supply chain management. However, it has been observed that studies on the use of control tower in the automotive sector are quite limited. It also found that the use of the control tower approach in supply chains is similarly limited. The control tower model is a preferred option for large businesses with supply, production and distribution networks that extend across different countries and even continents. This limits the scope of this approach and the number of academic studies on the control tower model. For all these reasons, the chapter also incorporates the professional expertise and field experience of the authors. The authors have extensive backgrounds in logistics, maritime administration, supply chain management and, in particular, control tower management. In summary, in developing Table 14.1, the control tower studies in the existing academic literature were consulted in detail, the

Table 14.1 SWOT analysis results and explanations

Strengths	Explanation	Reference
Ability to operate with optimum stock levels	A digital control tower enables real-time monitoring of stock levels, maintaining them at optimum levels. This reduces inventory costs and improves supply chain efficiency	Ivanov and Dolgui (2021) and Rejeb et al. (2019)
Easy communication with different suppliers	A digital control tower minimizes supply chain risks by identifying alternative suppliers	Vlachos (2023) and Ye et al. (2022)
Supports just-in-time production	A digital control tower supports JIT (just-in-time) production processes, optimizing production and delivery schedules	Monden (2012) and Pinto et al. (2018)
Fast data flow with actors	A digital control tower enables instant data sharing with all actors in the supply chain. This improves communication and collaboration	Authors of the chapter
Data accuracy checking	Advanced data analytics and validation systems acquired with a digital control tower improve decision-making by increasing the accuracy of data	Farahani et al. (2017)
Fast detection of problems	Supply chain issues can be quickly identified and resolved with real-time monitoring and analysis	Farahani et al. (2017)
Providing data for production planning	A digital control tower provides the necessary data for detailed and accurate production planning	Rustenburg (2016)
Process management with suppliers	A digital control tower enables the efficient management of processes with suppliers, thus increasing the reliability of the supply chain	Rustenburg (2016) and Vlachos (2023)
Entering data into the system	A digital control tower enables the data needed for different areas of the supply chain to be entered into the system	Kolehmainen (2013)
Document management	It ensures that documents related to shipments are obtained and available in the system	Authors of the chapter
Easy data sharing	The information and document flows provided by the control tower can also be used by different departments such as accounting and sales	Ojasalo et al. (2024)
Weaknesses	Explanation	
Difficulty controlling inaccurate information	Detecting and correcting inaccurate information in the digital control tower can be difficult, leading to incorrect decisions	Ojasalo et al. (2024)
Inaccurate information flow problems	Verification of inaccurate information from suppliers and delays in information flow from various locations can negatively impact the efficiency of the digital control tower	Bråtendal and Gustafsson (2020)
High start-up costs	Installing and implementing a digital control tower requires high initial costs	Ojasalo et al. (2024)

(continued)

Table 14.1 (continued)

Strengths	Explanation	Reference
Technological adaptation challenges	It may take time for suppliers to adapt to recent technologies and various challenges can be encountered along the way	Harmelink (2022)
Opportunities	Explanation	
Demand-driven production planning	A digital control tower enables rapid response to customer demands, enabling demand-driven production planning	Ma et al. (2024)
Increasing competitiveness	Responding quickly to sudden changes in the market is possible with a digital control tower	Patsavellas et al. (2021) and Zhang and Xiao (2023)
Adapting to changes in the market	It enables faster response to sudden changes in the market compared to competitors that do not use a digital control tower	Trzuskawska-Grzesińska (2017) and Vlachos (2022)
Management of special operations	Current digital control tower approaches enable the management of special operations required by differentiated products such as electric vehicles.	Authors of the chapter
Support for digital twin	The data provided by a digital control tower also helps manufacturing enterprises to establish "digital twins."	Maheshwari et al. (2023) and Vlachos (2022)
Threats	*Explanation*	
Confusion in stock tracking	Too many SKUs (stock keeping units) in the industry can sometimes cause confusion	Baluch et al. (2013) and Ojasalo et al. (2024)
Supplier resistance	Suppliers can resist being managed from a different center other than the customer	Ahmadi et al. (2023)
Bias of suppliers	Suppliers can be biased or uninterested in using a digital control tower	Authors of the chapter
Failure to track updates	Missed information update requests can put the risk supply chain management	Ojasalo et al. (2024) and Vlachos (2023)
Request for confidentiality of information	Some suppliers can be unwilling to transfer actual stock information to their customers through an intermediary	Authors of the chapter
Data security challenges	Protecting and storing data shared between supplier-customer-digital control tower in a way that is not accessible to third parties can be challenging	Richardson et al. (2024)

potential implications of these studies for the automotive sector were reviewed by the authors, and the results were structured within the framework of a SWOT analysis. Detailed explanation of the data and sources is presented in Table 14.1. Analyzing this data alone is insufficient to evaluate the current situation and develop strategic recommendations. For this reason, the second step was to evaluate the current situation through a SWOT analysis. Third step, TOWS analysis was employed to develop strategies determining the impact of a digital control tower in the automotive supply chain.

14.3.1 SWOT and TOWS Analyses

14.3.1.1 Strengths of the Control Tower in the Automotive Industry

It is clear from the previous sections that working in accordance with just-in-time production philosophy with minimum inventory is one of the primary success factors for the automotive industry. The two main functions of the control tower are the most suitable tools for achieving this success. Traditional tracking operations show the estimated arrival dates of ships at ports. Carriers often neglect to update possible changes in this information whereas businesses operating in the automotive industry expect proactive and timely information about possible disruptions and other events to ensure on-time deliveries (Mrozek et al., 2020). Thanks to the integrated information sharing infrastructure with carriers, the control tower can communicate changes in the estimated delivery times of shipments enroute to their customers instantly. This paves the way for dynamic changes in the production plan and makes businesses more agile.

Suppliers often make product shipments with goods or quantities that differ from customer demands. The control tower prevents bottlenecks by recognizing these differences in shipments long before they occur and presenting them to the customer for approval. Differences approved by manufacturers allow production plans to be updated, while alternative suppliers can be identified in case of delays or order cancellations.

A control tower is also integrated with systems such as ERP that businesses use within the supply chain network. Thanks to the visibility provided by the control tower throughout the entire supply chain, businesses do not need separate data entry for these modules. A control tower also ensures that the documents needed on the basis of each shipment (transportation documents, invoices, packing lists, customs declarations) are procured from the relevant actors (suppliers, logistics service providers, etc.) and archived. In addition, these documents and obtained information also help to provide the information needed by different departments of the enterprises.

14.3.1.2 Weaknesses of the Control Tower in the Automotive Industry

Information flows from suppliers must be accurate and timely to enable the control tower to operate without error. The control tower has to act on the assumption that the information coming from suppliers is correct. Therefore, the main challenge for businesses in the automotive industry is failure to transmit information about order changes to the control tower. The control tower is usually located in a single center or in several main centers. If they are not located in the same or close proximity to the manufacturers, the flow of information might not be instantaneous due to time differences, or emergency shipments from suppliers might not be responded to immediately.

14.3.1.3 Opportunities for Control Towers in the Automotive Industry

Large-scale automotive supply chain managements are usually planned as a push system. That is, manufacturers produce vehicles in versions that fit the dynamics of the market. However, this is not valid for every manufacturer. There are private brands that produce to order, and automotive manufacturers can also take orders for vehicles specified according to customer requirements. However, these orders require a long delivery time. Since automotive manufacturers using control towers have access to supplier stocks, they can provide a faster delivery by instantly transmitting spare parts required for production with the customer order to their suppliers. This is a competitive factor that will put companies ahead of their competitors.

The data flow that a control tower provides to businesses at every stage can also be used by different departments. Manufacturing enterprises use digital twins to predict potential problems that can occur in production. The data provided by the control tower is also used in the establishment of the digital twin.

14.3.1.4 Threats to the Control Tower in the Automotive Industry

Lots of parts are used in automotive manufacturing. Lots of parts means lots of product definitions (SKUs) in the system. This can lead to an increase in human error. Every piece of incorrect information that reaches the control tower can cause a chain of errors throughout the chain that are difficult to reverse if not corrected. However, technological tools and software that can be used in inventory movements will be enough to minimize these errors.

All suppliers involved in an enterprise's supply chain must use the programs and hardware required by the control tower. However, in some cases, suppliers can be reluctant to be managed by a company outside the business they serve or use programs they have not used before. In addition, learning to use these programs and entering data correctly can be a problem at the beginning for some suppliers. Therefore, it is often necessary for control tower service providers to train suppliers during the installation phase. The control tower should also always keep track of the fact that some suppliers do not want to transfer actual stock information to their customers through an intermediary company.

Another challenge in the use of control towers in the automotive industry is managing all suppliers collectively during the transition process (Bandur et al., 2021). Unlike traditional approaches, control towers require immediate and accurate access to data at the point of origin within the supply chain (Aljohani, 2023). In the past, data was transferred between the links of the chain, but in this new process, such adjustments weaken the chain's computational capabilities. Therefore, data must be forwarded seamlessly in its most accurate form. The precise management of data is too critical to be left to human effort alone. Cyber-physical systems are becoming increasingly important for collecting all types of data generated throughout the chain (Tonelli et al., 2021). Nevertheless, the incorporation of these physical systems into the supply chain leads to additional expenses, which poses a significant challenge. In the new system, the comprehensive collection of data is essential, and the ability to interpret this data is becoming increasingly important (Ojasalo et al., 2024).

The implementation of TOWS analysis in automotive supply chain management has the potential to enhance efficiency, adaptability, and competitiveness in the face of an increasingly complex and dynamic market. In essence, TOWS analysis underscores the leveraging of strengths and opportunities while addressing weaknesses and threats to optimize the implementation of digital control towers. The implementation of TOWS analysis in automotive supply chain management has the potential to enhance efficiency, adaptability, and competitiveness in the face of an increasingly complex and dynamic market.

14.4 Results

Figure 14.3 and Table 14.1 present the results of the SWOT analysis. To recommend strategies for the use of digital control towers in automotive supply chain managements, a TOWS analysis was conducted using the factors identified in the SWOT analysis. Detailed information on these strategies is presented in Table 14.2.

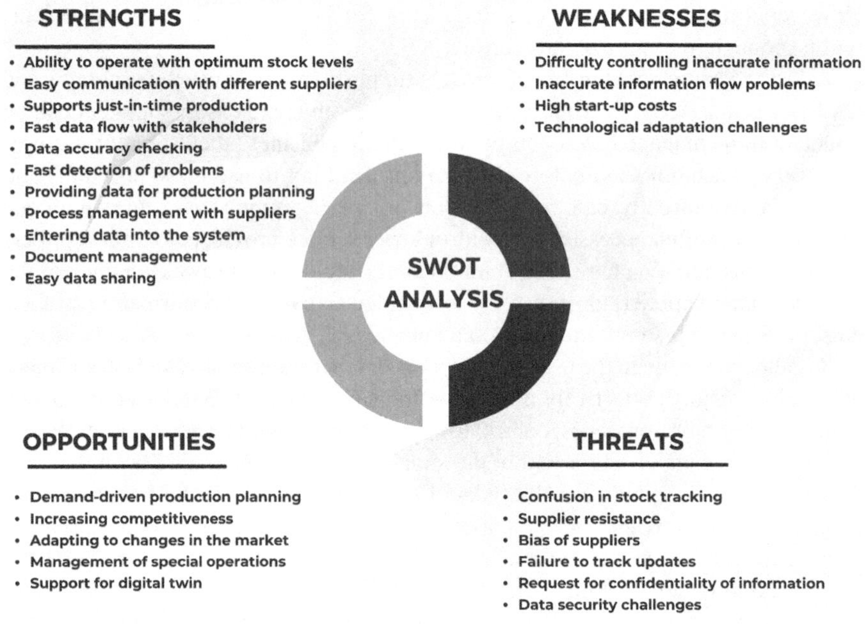

Fig. 14.3 SWOT analysis results

Table 14.2 TOWS analysis results

SO Strategies		ST Strategies	
SO1	Utilize real-time monitoring of stock levels to respond promptly to customer demands, facilitating demand-driven production planning	ST1	Implement stringent data accuracy checks and real-time monitoring to mitigate confusion in stock tracking
SO2	Exploit fast data flow with actors to swiftly adapt to market changes and enhance competitiveness	ST2	Use integrated systems such as ERP to streamline processes, reducing supplier resistance and bias
SO3	Implement digital twin by leveraging accurate data provided by control towers to improve production planning and manage special operations	ST3	Strengthen data security measures to address data security challenges and ensure confidentiality of information
WO strategies		WT strategies	
WO1	Enhance technological adaptation through comprehensive training programs to capitalize on demand-driven production planning and market adaptability	WT1	Develop robust training programs for suppliers to improve the accuracy and timeliness of information flow, addressing the risk of failing to track updates
WO2	Address high start-up costs by demonstrating long-term return on investment through optimized production and inventory management	WT2	Strengthen contractual agreements and data-sharing policies with suppliers to mitigate resistance and ensure data confidentiality
		WT3	Focus on reducing initial costs by demonstrating cost savings from optimized supply chain management

14.5 Conclusions

This chapter explores the impact and implementation of digital control towers in the management of automotive supply chains. It provides a comprehensive analysis of the challenges posed by the industry's intricate and global structure, emphasizing the significant role of digital control towers in modernizing automotive supply chain management. The chapter demonstrates that these digital control towers enhance visibility, adaptability, and operational resilience. While acknowledging the persistent challenges, it is emphasized that the strategic implementation of these systems, in collaboration with emerging technologies such as artificial intelligence and digital twins, will further unlock their full potential. In the context of the automotive industry, the chapter places particular emphasis on the necessity of adopting digital control towers to address the evolving demands of the industry and to achieve sustainable, competitive, and efficient supply chain operations. The results of the chapter also indicate that digital control towers facilitate real-time inventory monitoring, support JIT production principles, and reduce storage expenses. However, the chapter identified several critical weaknesses, including delays in data transmission, inconsistencies in supplier information, and a reluctance to adopt technological innovations. The limitations observed in this chapter—particularly the reliance on accurate supplier information, high implementation costs, and supplier reluctance—emphasize the necessity of targeted strategies for successful adoption. Addressing these limitations necessitates the implementation of advanced data validation mechanisms, supplier training programs, and the integration of digital control towers with ERP systems to enhance overall data consistency and system efficiency.

14.5.1 Recommendations and Future Studies

In consideration of the results obtained within the scope of this chapter, the recommendations are proposed for the automotive industry. First, to optimize JIT production processes and enhance real-time decision-making, it is crucial to integrate control towers with existing enterprise systems. This integration ensures data consistency, reduces manual interventions, and improves communication among supply chain actors. Second, practitioners should establish step-by-step digital adaptation guides for suppliers, supported by resolute supplier integration teams. These teams can oversee the onboarding process and resolve any discrepancies in data-sharing protocols. Third, a specific recommendation for addressing supplier resistance is the implementation of digital training programs. These programs should be customized to address the varying technological capacities of suppliers and provide direct workshops to minimize the learning curve. Fourth, by aligning production schedules with real-time demand patterns, companies can reduce overproduction and stockouts, improving operational efficiency. Control towers can monitor supplier inventories and dynamically adjust material flows to accommodate customized orders.

Recommendations for future studies are also recommended. First, the studies should focus on further integrating emerging technologies, such as blockchain and advanced artificial intelligence, into control tower systems. These technologies can enhance transparency and predictive capabilities while addressing persistent challenges in supplier coordination and data security. Second, comparative studies on the effectiveness of control towers in different supply chain structures will provide valuable insights for their optimization. Third, examining the adaptation processes of suppliers, particularly those involved in electric and hybrid vehicle production, will help refine implementation strategies.

References

Abaku, E. A., Edunjobi, T. E., & Odimarha, A. C. (2024). Theoretical approaches to AI in supply chain optimization: Pathways to efficiency and resilience. *International Journal of Science and Technology Research Archive, 6*, 092–107. https://doi.org/10.53771/ijstra.2024.6.1.0033

Ahmadi, M., Pahlavani, M., Karimi, A., Moradi, M., & Lawrence, J. (2023). The impact of the fourth industrial revolution on the transitory stage of the automotive industry. In H. Gholami,

G. Abdul-Nour, S. Sharif, & D. Streimikiene (Eds.), *Sustainable manufacturing in industry 4.0: Pathways and practices* (pp. 79–96). Springer. https://doi.org/10.1007/978-981-19-7218-8_5

Aljohani, A. (2023). Predictive analytics and machine learning for real-time supply chain risk mitigation and agility. *Sustainability, 15*, 15088. https://doi.org/10.3390/su152015088

ASCM. (2024). *SCOR Model [WWW Document]*. URL https://scor.ascm.org/processes/orchestratesupplychain/OE9. Accessed 6.5.24.

Baluch, N. H., Abdullah, C. S., & Mohtar, S. (2013). Evaluating effective spare-parts inventory management for equipment reliability in manufacturing industries. *European Journal of Management and Business Economics, 5*, 69–76.

Bandur, V., Selim, G., Pantelic, V., & Lawford, M. (2021). Making the case for centralized automotive E/E architectures. *IEEE Transactions on Vehicular Technology, 70*, 1230–1245. https://doi.org/10.1109/TVT.2021.3054934

Blanchard, D. (2021). *Supply chain management best practices*. John Wiley & Sons.

Bråtendal, J., & Gustafsson, F. (2020). *Development of a tactical strategy for a supply chain control tower*. Chalmers University of Technology.

Capgemini. (2024). *Iveco Group's intelligent supply chain ensures efficiency and sustainability [WWW Document]*. URL https://www.capgemini.com/news/client-stories/iveco-groups-intelligent-supply-chain-ensures-efficiency-and-sustainability/. Accessed 12.15.24.

Christopher, M. (2023). *Logistics and supply chain management* (6th ed.). FT Publishing.

Dametew, A. W. W., Kitaw, D., & Ebinger, F. (2020). Enhancing basic metal industry global competitiveness through total quality management, supply chain management & just-in-time. *Journal of Optimization in Industrial Engineering, 13*, 27–46. https://doi.org/10.22094/joie.2019.741.1472

Dias, G. C., Hernandez, C. T., & Oliveira, U. R. D. (2020). Supply chain risk management and risk ranking in the automotive industryGestão & Produção. *Gestão & Produção, 27*, e3800. https://doi.org/10.1590/0104-530X3800-20

Dolgui, A., & Ivanov, D. (2021). Ripple effect in the supply chain: Definitions, frameworks and future research perspectives. *Ann. Oper. Res., 292*, 313–332. https://doi.org/10.1007/978-3-030-14302-2_1

Farahani, P., Meier, C., & Wilke, J. (2017). Digital supply chain management agenda for the automotive supplier industry. In G. Oswald & M. Kleinemeier (Eds.), *Shaping the digital Enterprise* (pp. 157–172). Springer. https://doi.org/10.1007/978-3-319-40967-2_8

Guan, D., Wang, D., Hallegatte, S., Davis, S. J., Huo, J., Li, S., & Gong, P. (2020). Global supply-chain effects of COVID-19 control measures. *Nature Human Behaviour, 4*, 577–587. https://doi.org/10.1038/s41562-020-0896-8

Handfield, R. B., Graham, G., & Burns, L. (2020). Corona virus, tariffs, trade wars and supply chain evolutionary design. *International Journal of Operations & Production Management, 40*, 1649–1660. https://doi.org/10.1108/IJOPM-03-2020-0171

Harmelink, R. (2022). *Constructing the service control tower*. University of Twente.

Ivanov, D. (2020). Viable supply chain model: Integrating agility, resilience and sustainability perspectives—Lessons from and thinking beyond the COVID-19 pandemic. *Ann. Oper. Res., 293*, 1–21. https://doi.org/10.1007/s10479-020-03640-6

Ivanov, D., & Dolgui, A. (2020). Ripple effect and supply chain disruption management: New trends and research directions. *International Journal of Production Research, 59*, 102–109. https://doi.org/10.1080/00207543.2021.1840148

Ivanov, D., & Dolgui, A. (2021). A digital supply chain twin for managing the disruption risks and resilience in the era of industry 4.0. *Production Planning and Control, 32*, 775–788. https://doi.org/10.1080/09537287.2020.1768450

Kepczynski, R., Dimofte, A., Jandhyala, R., Sankaran, G., & Boyle, A. (2018). *Implementing integrated business planning: A guide exemplified with process context and SAP IBP use cases*. Springer.

Khajavi, S. H., Partanen, J., & Holmström, J. (2014). Additive manufacturing in the spare parts supply chain. *Computers in Industry, 65*, 50–63. https://doi.org/10.1016/j.compind.2013.07.008

Kolehmainen, K. (2013*). Automotive supply chain through the" control tower"-model: case Valmet Automotive.*

Ma, G., da Silva, E. R., & Møller, C. (2024). Digital twin-based demand-driven production planning and scheduling system: A new conceptual framework. *Procedia CIRP, 130*, 705–710. https://doi.org/10.1016/j.procir.2024.10.152

Maheshwari, P., Kamble, S., Kumar, S., Belhadi, A., & Gupta, S. (2023). Digital twin-based warehouse management system: A theoretical toolbox for future research and applications. *International Journal of Logistics Management, 35*, 1073–1106. https://doi.org/10.1108/IJLM-01-2023-0030

Monden, Y. (2012). *Toyota production system: An integrated approach to just-in-time* (4th ed.). CRC Press.

Mrozek, T., Seitz, D., Gundermann, K. U., & Dicke, M. (2020). *Digital supply chains: A practitioner's guide to successful digitalization*. Campus Verlag.

Ojasalo, H., Varis, K., & Ekblom, P. (2024). *Control tower business in logistic services: Simulated strategies and business plans*. Cambridge Scholars Publishing.

Panwar, R., Pinkse, J., & De Marchi, V. (2022). The future of global supply chains in a post-COVID-19 world. *California Management Review, 64*, 5–23. https://doi.org/10.1177/00081256211073355

Patsavellas, J., Kaur, R., & Salonitis, K. (2021). Supply chain control towers: Technology push or market pull—An assessment tool. *IET Collaborative Intelligent Manufacturing, 3*, 290–302. https://doi.org/10.1049/cim2.12040

Pinto, J. L. Q., Matias, J. C. O., Pimentel, C., Azevedo, S. G., Govindan, K., Pinto, J. L. Q., & Govindan, K. (2018). Lean Manufacturing Tools. In *Just Time Fact. Implement. Through Lean Manuf. Tools* (pp. 39–112). https://doi.org/10.1007/978-3-319-77016-1_4

Rejeb, A., Keogh, J. G., & Treiblmaier, H. (2019). Leveraging the internet of things and blockchain technology in supply chain management. *Future Internet, 11*, 161. https://doi.org/10.3390/fi11070161

Richardson, R., Soluade, O., & Shin, H. (2024). 11 Ensuring transportation cybersecurity: Air, auto, and rail, In *Volume V Cybersecurity Risk Management: Enhancing Leadership and Expertise* (pp. 213–230). https://doi.org/10.1515/9783111289069-011.

Rustenburg, J. W. (2016). Planning services: A control tower solution for managing spare parts. In *Logistics and supply chain innovation. Lecture notes in logistics* (pp. 239–259). Springer. https://doi.org/10.1007/978-3-319-22288-2_14

Sarkis, J., Cohen, M. J., Dewick, P., & Schröder, P. (2020). A brave new world: Lessons from the COVID-19 pandemic for transitioning to sustainable supply and production. *Resources, Conservation and Recycling, 159*. https://doi.org/10.1016/j.resconrec.2020.104894

Seyedan, M., & Mafakheri, F. (2020). Predictive big data analytics for supply chain demand forecasting: Methods, applications, and research opportunities. *Journal of Big Data, 7*, 53. https://doi.org/10.1186/s40537-020-00329-2

Sinha, A., Bernardes, E., Calderon, R., & Wuest, T. (2020). *Digital supply networks: Transform your supply chain and gain competitive advantage with disruptive technology and reimagined processes* (1st ed.). McGraw Hill.

Sümen, H. H. (2020). Geçmişten Günümüze Dijital Dönüşüm. In G. Telli & S. Aydın (Eds.), *Dijital Dönüşüm* (pp. 185–212). Kapital Medya.

SupplyChainBrain. (2024). *Renault Group Acquires a Supply Chain Control Tower to Manage Transportation in Europe and Beyond [WWW Document]*. URL https://www.supplychainbrain.com/articles/40609-renault-group-acquires-a-supply-chain-control-tower-to-manage-transportation-in-europe-and-beyond. Accessed 12.14.24.

Taschner, A., & Charifzadeh, M. (2020). *Management accounting in supply chains*. Springer Fachmedien Wiesbaden.

Tonelli, F., Demartini, M., Pacella, M., & Lala, R. (2021). Cyber-physical systems (CPS) in supply chain management: From foundations to practical implementation. *Procedia CIRP, 99*, 598–603. https://doi.org/10.1016/j.procir.2021.03.080

Toygar, A., & Yıldırım, U. (2022). Rusya Ukrayna çatışmasının deniz lojistiğine etkisi. *Mersin Üniversitesi Denizcilik ve Lojistik Araştırmaları Derg, 4*, 163–180. https://doi.org/10.54410/denlojad.1170425

Toygar, A., & Yildirim, U. (2023). Examining the effects of the Russia-Ukraine conflict on global supply chains. In *Handbook of research on war policies, strategies, and cyber wars* (pp. 184–199). IGI Global. https://doi.org/10.4018/978-1-6684-6741-1.ch010

Trzuskawska-Grzesińska, A. (2017). Control towers in supply chain management–past and future. *Journal of Economics & Management Strategy, 27*, 114–133.

Vlachos, I. (2022). Supply Chain Control Towers: Integrative Human, Operational, and Technological Capabilities. In *Proceedings of the International Conference on Industrial Engineering and Operations Management. İstanbul* (pp. 1563–1570).

Vlachos, I. (2023). Implementation of an intelligent supply chain control tower: A socio-technical systems case study. *Production Planning and Control, 34*, 1415–1431. https://doi.org/10.1080/09537287.2021.2015805

Yamamoto, K., Milstead, M., & Lioyd, R. (2019). A review of the development of lean manufacturing and related lean practices: The case of Toyota production system and managerial thinking. *International Journal of Management Reviews, 15*, 21–90.

Ye, C., Zaraté, P., & Kamissoko, D. (2022). A DSS based on a control tower for supply chain risks management. In A. P. Cabral Seixas Costa, J. Papathanasiou, U. Jayawickrama, & D. Kamissoko (Eds.), *Decision support systems XII: Decision support addressing modern industry, business, and societal needs. ICDSST 2022. Lecture notes in business information processing* (pp. 124–136). Springer. https://doi.org/10.1007/978-3-031-06530-9_10

Zhang, X., & Xiao, G. (2023). Digital control tower empowers supply chain resilience improvement from a dynamic capability perspective. In *Digitalization and management innovation II* (pp. 186–194). IOS Press. https://doi.org/10.3233/FAIA230730

GPSR Compliance

The European Union's (EU) General Product Safety Regulation (GPSR) is a set of rules that requires consumer products to be safe and our obligations to ensure this.

If you have any concerns about our products, you can contact us on ProductSafety@springernature.com

In case Publisher is established outside the EU, the EU authorized representative is:

Springer Nature Customer Service Center GmbH
Europaplatz 3
69115 Heidelberg, Germany

Batch number: 10607204

Printed by Printforce, the Netherlands